Antiperspirants and Deodorants

COSMETIC SCIENCE AND TECHNOLOGY SERIES

Series Editor

ERIC JUNGERMANN

Jungermann Associates, Inc.
Phoenix, Arizona

Volume 1: Cosmetic and Drug Preservation: Principles and Practice, *edited by Jon J. Kabara*

Volume 2: The Cosmetic Industry: Scientific and Regulatory Foundations, *edited by Norman F. Estrin*

Volume 3: Cosmetic Product Testing: A Modern Psychophysical Approach, *by Howard R. Moskowitz*

Volume 4: Cosmetic Analysis: Selective Methods and Techniques, *edited by P. Boré*

Volume 5: Cosmetic Safety: A Primer for Cosmetic Scientists, *edited by James H. Whittam*

Volume 6: Oral Hygiene Products and Practice, *by Morton Pader*

Volume 7: Antiperspirants and Deodorants, *edited by Karl Laden and Carl B. Felger*

Other Volumes in Preparation

Antiperspirants and Deodorants

edited by

KARL LADEN

Consultant in Technology Transfer
and New Product Development
Haifa, Israel

CARL B. FELGER

Manager Scientific Affairs
Gillette Research Institute
Gaithersburg, Maryland

MARCEL DEKKER, INC. NEW YORK AND BASEL

Library of Congress Cataloging-in-Publication Data

Antiperspirants and deodorants / edited by Karl Laden, Carl B. Felger.
p. cm. -- (Cosmetic science and technology series ; v. 7)
Includes bibliographies and index.
ISBN 0-8247-7893-6
1. Antiperspirants. 2. Deodorants. 3. Sweat glands. I. Laden, Karl. II. Felger, Carl B. III. Title: Antiperspirants and deodorants. IV. Series.
TP986.A58 1988
668'.55--dc19 88-14678
CIP

MARCEL DEKKER, INC.
270 Madison Avenue, New York, New York 10016

Current printing (last digit):
10 9 8 7 6 5 4 3 2 1

PRINTED IN THE UNITED STATES OF AMERICA

About the Series

The Cosmetic Science and Technology series was conceived to permit discussion of a broad spectrum of current knowledge and theories of cosmetic science and technology. The Series consists of books by a single author as well as edited volumes. Well-known authorities from industry, academia, and the government participate in writing these books.

The aim of this Series is to cover the many facets of cosmetic science and technology. Topics are drawn from a wide range of disciplines ranging from chemical, physical, analytical, and consumer evaluations to safety, efficacy, and regulatory questions. Organic, inorganic, physical and polymer chemistry, emulsion technology, microbiology, dermatology, and related fields all play a role in cosmetic science. There is little commonality in the scientific methods, processes, or formulations required for the wide variety of cosmetics and toiletries manufactured. Products range from preparations for hair, mouth, and skin care to lipsticks, nail polishes and extenders, deodorants, body powders, and aerosols to over-the counter products such as antiperspirants, dandruff and acne treatments, antimicrobial soaps, and sunscreens.

Cosmetics and toiletries represent a highly diversified field with many subsections of science and "art"; indeed, even in these days of high technology, much "art" and instinct are used and needed in the formulation, evaluation, and selection of cosmetic products although there is a strong move toward the "scientific method," particularly in such areas as claim substantiation and product evaluation and analysis.

Emphasis is placed on reporting the current status of cosmetic technology and science in addition to historical reviews, The series includes or will include books on safety, product testing, oral hygiene products, sunscreens, deodorants and antiperspirants, hair care, and claim substantiation. Contributions range from highly sophisticated and scientific treatises to primers, practical applications, and pragmatic presentations. Authors are encouraged to present their own concepts as well as established theories. They have been asked not to shy away from fields that are still in a state of development or transition, nor to hesitate to present detailed discussions of their own work. Altogether, we intend to develop in this series a collection of critical surveys by noted experts covering most phases of the cosmetic business.

The seventh book in the series, *Antiperspirants and Deodorants*, is edited by Dr. Karl Laden, consultant in technology transfer and new product development, Haifa, Israel; and by Dr. Carl B. Felger, Manager Scientific Affairs, Gillette Research Institute, Gaithersburg, Maryland. The book covers all aspects of deodorant and antiperspirant technology, including a historical review of these products, their formulations, safety and efficacy testing, and regulatory considerations, as well as discussions of theories on the causes of odor formation and perspiration. Special care has been taken to provide insights into the correlation of subtle differences in chemical structures and efficacy. This volume's intention is to provide a better understanding to chemists and researchers working in this field and to serve as a useful reference book for medical school libraries and for practicing dermatologists.

Finally, I want to thank all the authors and editors who are participating in this series, the editorial staff at Marcel Dekker, Inc., and above all, my wife, Eva, without whose editorial help and constant support I would never have undertaken this project.

ERIC JUNGERMANN, Ph.D.

Preface

Antiperspirants and Deodorants is the first comprehensive review of this complex and multidisciplinary subject. Each chapter author has been carefully selected to provide the expertise necessary to deal with the issues of the chemistry and physiology of antiperspirants, the microbiology of deodorants, and the federal regulations that impact on the marketing and safety of such products, to mention but a few. All the contributors to *Antiperspirants and Deodorants* are recognized authorities in their fields and have been encouraged to review and critique the current state of the art as well as present their own concepts and theories based on personal research.

This book is designed to be an authoritative text that will be a valuable addition to the libraries of all—scientists, regulators, or managers—concerned with the development or marketing of antiperspirants or deodorants. The nature of the subject matter is such that it is not possible to review comprehensively all the information available on antiperspirants and deodorants. To assist the reader in obtaining the maximum benefit, each chapter concludes with an extensive list of references that lead to additional information in specific subject areas.

The editors would like to acknowledge the efforts of each author. Without the dedication of the many individuals who contributed to *Antiperspirants and Deodorants*, it would not have been possible to compile such a complete and authoritative review of the many facets of this subject.

KARL LADEN, Ph.D.

CARL B. FELGER, Ph.D.

Contents

Contributors

Carl B. Felger Department of Scientific Affairs, Gillette Research Institute, Gaithersburg, Maryland

John J. Fitzgerald Department of Chemistry and Chemical Engineering, South Dakota School of Mines and Technology, Rapid City, South Dakota

Herman E. Jass Technical Management Consultant, Skillman, New Jersey

John N. Labows, Jr. Colgate-Palmolive Company, Piscataway, New Jersey

Karl Laden* Carter Products Division, Carter-Wallace, Inc., Cranbury, New Jersey

James J. Leyden Department of Dermatology, University of Pennsylvania, Philadelphia, Pennsylvania

---

**Current Affiliation:* Consultant in Technology Transfer and New Product Development, Haifa, Israel

Richard P. Quatrale* Shulton Skin Care Center, American Cyanamid Company, Clifton, New Jersey

Paul M. Quinton Division of Biomedical Sciences, University of California at Riverside, Riverside, California, and Department of Physiology, University of California at Los Angeles, Los Angeles, California

David I. Richardson† Dial Technical Center, The Dial Corporation, Scottsdale, Arizona

Janice G. Rogers Department of Clinical Sciences, Gillette Medical Evaluation Laboratories, Gaithersburg, Maryland

Earl Philip Seitz, Jr. Personal Care Technology Development Department, Dial Technical Center, The Dial Corporation, Scottsdale, Arizona

John E. Wild Dermal Clinical Testing Division, Hill Top Research, Inc., Cincinnati, Ohio

Ronald J. Wulf Carter Products Research, Carter-Wallace, Inc., Cranbury, New Jersey

---

Current Affiliations:
*Corporate Scientific, Cosmair, Inc., Clark, New Jersey
†Vipont Pharmaceuticals, Inc., Fort Collins, Colorado

Antiperspirants and Deodorants

1

Introduction and History of Antiperspirants and Deodorants

Karl Laden* *Carter-Wallace, Inc., Cranbury, New Jersey*

> *"Quam paene admonui ne trux caper iret in alas"*
>
> "How nearly did I warn you that no rude goat find its way beneath your arms"
>
> Ovid, *Artis Amatorial* (1) (The Art of Love)

I. INTRODUCTION

The above admonition, given almost 2000 years ago in a chapter dealing with personal hygiene and good grooming, suggests that the problem of offensive underarm odor is not the invention of 20th century advertising.

However, while the preparation and use of cosmetics is an art that even predates recorded history, the formulation of specific products to control underarm wetness and odor began only about 100 years ago. Although it is more than likely that fragrance oils and other aromatics have been used through the centuries to mask body odors as well as impart pleasant scents, the specific use of products to control and regulate underarm odor and wetness had its beginnings in the latter part of the 19th century.

Today, antiperspirants and deodorants represent one of the largest cosmetic categories with a volume well over $1 billion in

Current affiliation: Consultant in Technology Transfer and New Product Development, Haifa, Israel the United States. Few classes of cosmetic or toiletry products have grown so quickly and with such diversity. Over 90% of the US population currently uses an antiperspirant or deodorant every day, and the number of forms and methods of application is without parallel in the cosmetic industry.

Although the terms antiperspirant and deodorant are frequently used interchangeably by the public, they represent two different and distinct actions. *Antiperspirant* specifically refers to materials or formulations which reduce the amount of underarm perspiration. *Deodorant* refers to materials or formulations which mask or inhibit the formation of unpleasant body odors. Today, virtually all antiperspirant products also have deodorant qualities. However, many deodorant products only prevent or mask malodors and have no effect on perspiration.

It is the purpose of this book to give the reader a detailed overview of this dynamic field of cosmetic science.

II. EARLY PRODUCTS

The first trademarked brand of underarm deodorant was introduced in the United States in 1888 under the brand name of Mum (2). It was a waxy cream containing zinc oxide and probably controlled odor by virtue of the weak antimicrobial properties of the zinc oxide. The waxy nature of the base was also supposed to plug the sweat pores and thereby control sweating. It is interesting to note that 100 years later there continues to be marketed a cream deodorant under the Mum brand name (Bristol-Myers Co.). The product now uses aluminum chlorohydrate in a cream base to achieve its antiperspirancy and deodorancy.

Less than 15 years later the first brand name antiperspirant appeared on the market as Everdry (3). It was a simple aqueous solution of aluminum chloride and was applied to the underarm by dabbing with cotton. From these modest beginnings of a waxy cream and a simple liquid have evolved an array of antiperspirant and deodorant products representing one of the greatest assortment of formulations and packaging forms in the cosmetic industry. These include

Stick—solids
Stick—creams
Pads
Dabbers
Aerosols

Roll-ons
Pump Sprays
Creams
Squeeze bottles
Powders

Early antiperspirants based on aluminum chloride solutions were not very aesthetic products. They needed to be dabbed on with cotton and were slow-drying, sticky, and cold. Even worse, they were highly acidic. This led to problems of burning, stinging, and irritation as well as damage to clothing caused by the acidity. It is no wonder that it has been speculated that early use tended to be confined to people in the entertainment field.

In spite of the numerous deficiencies in product attributes, the use of these underarm products increased to the point where in 1914 the Odo-Ro-No brand of antiperspirant achieved enough volume to support national advertising (4).

The first scientific evidence for the antiperspirant effectiveness of aluminum salts was reported by Stillians in 1916 in the *Journal of the American Medical Association* (5). Stillians observed that a 25% solution of aluminum chloride in distilled water, dabbed gently on the armpit every second or third day, reduced excessive sweating. Stillians was well aware of the highly corrosive nature of aluminum chloride solutions and cautioned against the itching, stinging, and burning that it might cause. He even warned against excessive or careless use.

With time, less corrosive aluminum salts were used, and as early as 1921 attempts were made to buffer the highly acidic solutions with borax and alum (6).

III. ANTIPERSPIRANT CREAMS

In the early 1930s, the first of what was to become a long line of major changes in product form was introduced. The product was called Arrid Cream and it contained aluminum sulfate (7). Compared to the simple solutions of aluminum salts that existed until that time, Arrid Cream represented a major step forward in terms of cosmetic elegance. It was easier to apply, dried faster, and more closely resembled the appearance and aesthetics of conventional cosmetic products. Until then many aesthetically pleasing deodorant creams were available; however, these products only controlled odor and did nothing to prevent perspiration.

The cream form of antiperspirant offered consumers a product with odor and wetness control in a form that was easy to use. Cream antiperspirants rapidly gained in popularity and remained the dominant form for almost 20 years. Introduced in 1934, Arrid Cream represented over 34% of the antiperspirant-deodorant market in 1945 (8). During that same period the use of deodorant creams (as opposed to antiperspirant creams) began to decline. The total dominance of this new cream form is best seen in a 1946 survey, which showed cream antiperspirants and deodorants to represent 88% of the total market (9).

Since the 1940s, a variety of new formulations and product forms has largely displaced cream from its dominant position. However, in 1984, 50 years after its introduction, Arrid Cream remained the largest selling brand in the cream category.

Other examples of this phenomenon are available in the antiperspirant-deodorant industry. That is, the first brand name introducing a new form tends to maintain a long dominance of that form. Other examples of this will be seen later in this chapter.

The introduction of the cream form also focused on another trend that was to continue in the category. Starting with creams, new product forms have almost all focused on improved aesthetics and convenience rather than improved efficiency.

IV. SQUEEZE SPRAYS

By 1947, antiperspirants and deodorants represented about a $30 million business (10) and new forms and methods of application began to appear. Pads saturated with liquid and packaged in jars (the best known brand was 5-Day Deodorant Pads) somewhat simplified the application of liquid deodorants, but still were slow-drying, sticky, and required consumers to apply the product with their fingers. The plastics developed during World War II began to find their way into packaging systems for cosmetic products. One of the earliest applications of these new packaging materials was a squeeze spray bottle for application of deodorants. An adaptation of the old rubber bulb atomizer, the squeeze spray bottle allowed the consumer to dispense a low-viscosity liquid as a somewhat coarse spray. The major brand was Stoppette Spray Deodorant (11). The main advantage of the squeeze bottle over all previous forms was that it allowed users to apply the antiperspirant without getting it on their fingers.

This single advantage represented a major step forward in the aesthetics of application for this category. The squeeze spray dispenser, however, had a number of significant drawbacks. The coarse spray felt cold and wet. The squeeze dispensing required using low-viscosity liquid formulations. These often were slow-drying and allowed product to drip down the sides of the user's body.

The considerable, albeit short-lived, popularity of squeeze sprays demonstrated the significant desire of consumers to apply underarm products without getting them on their fingers. This is best reflected in the fact that since the introduction of the squeeze spray in the late 1940s, every newer product form (regardless of method of application) has had the common feature that it is not necessary to apply the product directly with your hands.

During the same period of time (early 1940s) formulators continued to wrestle with the problem of low pH (high acidity) of antiperspirant formulations. Most approaches involved using buffers to try to ameliorate the fabric damage and skin irritation frequently resulting from the application of antiperspirants. Urea was widely used for this purpose; however, none of the approaches was entirely successful until a buffered, less acidic form of aluminum chloride was discovered and found to have antiperspirant efficacy. Starting in 1947, aluminum chlorohydrate began to replace the more acidic aluminum salts and rapidly became the antiperspirant ingredient of choice (10).

V. ROLL-ONS

In the late 1940s a product appeared that offered a new way to apply a viscous liquid onto a surface. It was called the ballpoint pen, and Reynolds was the most common name associated with it. Bristol-Myers fashioned a new antiperspirant applicator using the same principle and introduced the first antiperspirant roll-on into a test market in 1952 as Mum Roullette (2). Apparently, the gamble failed, and the product never went national. Technical problems with the ball and fitment mechanism needed to be solved, and several years later a product based on the same principle was introduced as Ban Roll-On (1955). The product was an almost immediate success; and the roll-on form soon dominated the category. By 1963 roll-ons represented 37% of all deodorants and antiperspirants sold, outpacing their nearest rival, the cream form, by almost 10 percentage points.

The advantages of the roll-on form over other product forms were obvious. It permitted application without directly using the fingers. It was quick and easy to apply. Higher-viscosity and more elegant emulsions could be used, which not only dispensed more evenly but also solved the dripping and running-down-the-sides problem. The emulsion-type formulations could be made to dry reasonably fast and with relatively low levels of stickiness. As was the case with the first cream antiperspirant, the first brand of roll-on, Ban, has remained the leading roll-on some 30 years later.

The introduction of roll-ons continued to expand the use of antiperspirant and deodorant products. In the early 1960s, the underarm control business reached $100 million.

VI. AEROSOLS

The next major new product form had its origins in 1942, when Sullivan and Goodhue developed the first practical insecticide aerosol for use in World War II. (The idea that a spray form as a method of application for antiperspirants would have great convenience had been put forth as early as 1934 by Auch [12].) However, the development of mass-produced valves and pressurized cans for consumer use required many years of work after World War II ended. By the mid-1950s the aerosol can had begun to find application in the cosmetic industry for shave creams and perfume sprays. However, scientists in the industry were not able to solve the problems associated with packaging high concentrations of aluminum salts in aerosol cans without corrosion and clogging.

In the early 1960s, the Gillette Company introduced a new aerosol product called Right Guard. It attempted to solve the earlier corrosion and clogging problem by using a lower concentration of a less corrosive salt. The salt chosen was zinc phenolsulfonate, which, by virtue of its solubility in alcohol, simplified the formulation problems. It was combined with hexachlorophene, a relatively new and powerful deodorant. The product was initially claimed to be an antiperspirant on its label. Testing indicated, however, that it had little or no antiperspirant efficacy and soon the antiperspirancy claims were dropped. However, the product rapidly achieved success. The aerosol form offered a fine, light spray (unlike the earlier squeeze bottles). The high concentration of alcohol in the formula gave quick drying with virtually no stickiness.

The product was quick and easy to apply and, lastly, it capitalized on another feature lacking in roll-ons. Because the applicator ball in a roll-on is placed directly against the armpit to apply the product, the hygienic aspect of multiple use within a family could be questioned. Aerosol sprays lent themselves to family use, and as pointed out in a popular claim, "nothing touches you but the spray itself."

Even though the product did not control wetness, large numbers of consumers deserted the use of roll-on antiperspirants to take up a deodorant spray. They apparently either did not perceive or did not care about the loss of antiperspirant activity.

In a few short years (1962–1967) the aerosol form represented over 50% of the category. In addition, the particular appeal of the aerosol form to men broadened the whole antiperspirant-deodorant category so that between 1962 and 1967 the volume grew from just over $100 million to just over $200 million.

The huge success of deodorant sprays in the mid-1960s raised an interesting question for antiperspirant product formulators. It suggested that application aesthetics and form could be more important to the consumer than antiperspirant efficacy.

In terms of active ingredients for efficacy, new compounds were also appearing. As early as 1950, Van Mater (13) described the use of zirconium salts as antiperspirants. Early testing indicated that zirconium salts tended to be more powerful antiperspirants than aluminum salts. In the mid-1950s, sodium zirconyl lactate was used in alcoholic soap gel deodorant sticks. Granulomatous eruptions in the axillae began to be reported from the use of these products. Since then most zirconium antiperspirants have been based on aluminum-zirconium mixed salts or buffered systems. The use of zirconium-aluminum complexes in antiperspirant systems continued to increase. In 1978, three of the top four roll-ons contained zirconium salts.

Some five years after the introduction of Right Guard, the first true antiperspirant aerosol was introduced. It solved the corrosion and clogging problem in a novel way. Aluminum chlorohydrate at a relatively low concentration was dispersed into an anhydrous vehicle containing propellant and an emollient oil. Because no water was present, the aluminum salt did not cause corrosion problems and did not suffer from crystal buildup in the valve orifices. The high spray rate, in addition to helping keep the valve orifices clear, delivered considerably more product to the skin surface, thereby compensating for the relatively low aluminum concentration. The product was called Arrid Extra Dry and it was an immediate success. In addition to delivering antiperspirant efficacy, this new product had another advantage over aerosol deodorants. Because it contained no alcohol, the spray which reached the skin was considerably warmer. It also had a much drier feel since the propellant rapidly volatilized, leaving only a dry-feeling emollient and aluminum salt on the skin. In contrast, deodorant sprays with their high alcohol content not only went on colder but took longer to dry and felt cold during the drying process. In addition, alcohol was more prone to cause stinging problems on freshly shaved axillae.

Within three years after the introduction of the first aerosol antiperspirant, the market for aerosol sprays represented 75% of the total category with the antiperspirant sprays being 45% and deodorant sprays 30%. The total market for antiperspirants and deodorants also grew rapidly during this period, well exceeding the $300 million level in 1970.

A number of factors remain unanswered concerning the rapid increase in popularity of antiperspirant aerosols as well as the rapid growth in the total antiperspirant-deodorant market. The popularity of Right Guard deodorant spray at the expense of roll-ons clearly demonstrated that form and aesthetics were more important than antiperspirant efficacy to many consumers. The introduction of an antiperspirant spray, while now offering antiperspirant efficacy to the same spray users, also offered them improved aesthetics via a warmer and drier spray. It must be remembered that if antiperspirant efficacy were the prime concern of the consumer, the roll-on form would have remained dominant. Even antiperspirant aerosols were considerably less effective than roll-ons.

It is still unclear whether it was the aesthetic or antiperspirant benefits which resulted in such a quick and widespread acceptance of aerosol antiperspirant sprays over alcoholic deodorant sprays. Since many of these alcoholic-spray users had originally abandoned the use of roll-on antiperspirants, it can be argued that they were, in fact, more responsive to aesthetics than to antiperspirant efficacy. Therefore, it may be more likely that they were attracted by the aesthetics of the new antiperspirant aerosol as opposed to its antiperspirant efficacy. The aerosol antiperspirant also now offered the roll-on user a more convenient, faster drying form without the need to give up wetness control.

It is not surprising that by the early 1970s, aerosol sprays accounted for over 82% of the total antiperspirant-deodorant market. Of the total aerosol market, 75% was in antiperspirant sprays and 25% in deodorant sprays.

VII. EROSION OF AEROSOL ANTIPERSPIRANT MARKET

In the mid- to late 1970s, a major disruption occurred in the antiperspirant product field which, for the first time in the history of this category, was not initiated by the introduction of a new product form or formulation. Rather, the disruption was caused 1. by questions concerning the safety of aluminum-zirconium salts in aerosols, 2. by concern that the fluorinated hydrocarbon propellants could cause a depletion of the earth's protective ozone layer, and 3. by concerns raised by a Food and Drug Administration (FDA) advisory review panel concerning the safety of long-term exposure to inhalation of aerosols containing aluminum salts.

By the early 1970s, several antiperspirant manufacturers had brought out aerosol sprays in which aluminum-zirconium salt complexes replaced the previously used aluminum chlorohydrate. This step was taken to bolster the antiperspirant activity of aerosols to a range approximating that achieved by aluminum chlorohydrate roll-on products. As a result of longer-term inhalation testing of an aluminum-zirconium aerosol by Gillette, some questions were raised concerning safety. Based on this information, the FDA began examining the safety of aerosolized zirconium salts. Motivated by reports of a possible FDA ban, manufacturers quickly started to revert to straight aluminum chlorohydrate in their aerosol products. In 1977, after all manufacturers had converted back to aluminum chlorohydrate, the FDA finally banned the use of zirconium-based aerosol antiperspirants (14). Since much of the deliberations of the FDA and its advisory panels was reported in the lay press, public concern about the overall safety of antiperspirant aerosols began to build.

This public concern was further compounded when two scientists, Molina and Rowland (15), based on limited experimental data, formulated a theory which predicted that the continued release of chlorofluorinated hydrocarbons (the propellants used in antiperspirant sprays) could lead to a breakdown of the earth's protective ozone layer. This disruption of the ozone layer could allow more ultraviolet radiation to reach the earth's surface, causing any number of harmful effects including increases in skin cancer rates.

The merits of this theory and speculations based on it were widely debated in the scientific community. Finally, in the late 1970s, federal regulatory agenices banned the use of chlorofluorinated hydrocarbon propellants in aerosol products (16).

In anticipation of this ban, most manufacturers shifted to using hydrocarbon propellants in place of the chloroflurocarbons. This change in propellant resulted in an even warmer, drier spray than had previously been offered; however, the product also became more dusty. In spite of the smooth conversion to hydrocarbon propellants, consumers continued to hear concerns about aerosol safety. It is likely that many did not distinguish between safety for the atmosphere and direct inhalation safety.

This eroding confidence in the aerosol form was manifested in a rapid decline in aerosol antiperspirant use. In 1974 aerosols represented about 82% of the market; by 1977 they had dropped to under 50% of the market.

Further contributing to this concern for safety was the publication by the FDA in 1978 of the recommendations of the Advisory Review Panel on Over-the-Counter Antiperspirant Drug Products (17). The expert panel expressed concern about the possible consequences of long-term inhalation of aerosolized aluminum antiperspirants. Based on this concern, they placed aluminum-containing aerosols in Category III (insufficient safety data available to permit them to be classified as safe and effective). Although a subsequent review of the safety data, as well as additional data given to the FDA, eventually resulted in the FDA classifying aerosol aluminum antiperspirants in Category I (safe and effective) in 1982 (18), by that time aerosol antiperspirants had declined to 32% of the market.

VIII. NEW CATEGORIES

In the mid-1970s, pump sprays tried to fill in the void left by the declining aerosol market. By then, vastly improved sprays could be obtained compared to the original squeeze sprays. The mechanical pumps produced excellent fine sprays which could come close to mimicking aerosols. However, the formulations that could be dispensed by these pumps had to be thin aqueous/alcoholic solutions containing either aluminum chlorohydrate or some newer alcohol-soluble aluminum salts. Aesthetically, these systems resembled the original deodorant sprays. That is, they were cold, wet, and slow-drying. The popularity of the pumps was therefore short-lived.

Consumer disenchantment with the aerosol spray (without a new form ready to fill in the gap) proved to be a boon for the roll-on market. In 1974, before much of the aerosol controversy influenced consumers, roll-ons represented less than 10% of the antiperspirant-deodorant market. Four years later, in 1978, roll-ons represented over 30% of the market. Another form that benefited from the decline in aerosol popularity was the deodorant stick. Soap-based deodorant sticks had been on the market for many years. They were based on alcohol or propylene glycol soap gels containing an antimicrobial deodorant material. They did not possess antiperspirant activity and were generally considered products for men (Mennen Deodorant Stick being the best-known brand). Deodorant sticks had many application advantages. They were easy to apply, not cold or wet, and fast-drying. During the four years alluded to above (1974–1978), the stick form of deodorant increased from 4% to over 10% of the market, again signaling that form and aesthetics were as important as (if not more important than) antiperspirant efficacy.

In the 1970s a new class of cosmetic raw materials came into use that shaped the way for the next major changes in antiperspirant formulations and forms. The materials were organic silicone compounds that were liquid at room temperature but volatilized readily on skin without producing a cooling sensation. They were called volatile silicones. By the mid-1970s they had made their way into aerosol antiperspirants, where they partially replaced some of the emollient oils that had been used.

In 1978 a new type of roll-on was introduced based on this material. The idea was that if consumers liked the warm, dry feeling of an aerosol antiperspirant, but were disenchanted with aerosols, why not apply an antiperspirant aerosol concentrate from a roll-on container? The product was called Dry Idea. Unlike conventional roll-ons, the formulation contained almost no water and was essentially a suspension of an antiperspirant salt in volatile silicone. It had several advantages over aqueous roll-ons. Because the volatile silicone evaporated from the skin without feeling cool, the product did not feel as wet. It also did not go through a sticky phase as regular roll-ons do during drying. And, since it evaporated, the volatile silicone did not leave an oily residue on the skin as other potential emollients might have done. This had the added benefit of allowing the antiperspirant salt to exhibit high levels of antiperspirant efficacy by not interfering with its dissolution in sweat, something which nonvolatile emollient oils frequently did. Dry Idea quickly established itself with a significant share of the roll-on market and eventually led to other manufacturers introducing nonaqueous suspension roll-ons based on volatile silicone.

The most recent product form to take up the slack in aerosol use, as Dry Idea did, is not a new form but an older product form improved by the use of volatile silicones. Antiperspirant sticks had been on the market for many years. The first commercial sticks date back to the 1930s. They were usually based on suspending an antiperspirant salt in a waxy matrix containing emollient oils. Because the emollient oils did not evaporate, the products tended to be somewhat greasy and oily. The emollient oils also tended to interfere with the antiperspirant salts exhibiting their full efficacy.

In the late 1970s antiperspirant sticks, now called solids, began to reappear with a matrix of cetyl alcohol and a castor wax and using volatile silicone as the emollient oil. They had better application aesthetics than the older stick antiperspirants in that they were less greasy, faster drying, and more efficacious. They also had distinct advantages over the conventional deodorant sticks.

The total popularity of both stick deodorants and solid antiperspirants grew very rapidly. In 1974 sticks and solids comprised less than 5% of the total market. Less than 10 years later, the sticks and solids became the most popular form, representing over 35% of the deodorant-antiperspirant category.

The growth in popularity of the roll-on, stick, and solid forms over the past 10 years represents a unique phenomenon in the history of antiperspirant and deodorant products. Until the mid-1970s new product forms based on improved packaging, aesthetics, and efficacy had led the way in the expansion and growth of the antiperspirant and deodorant market. It was the introduction of these new forms (creams, roll-ons, aerosols) that naturally displaced their predecessors in popularity. In the mid-1970s the predominant form (aerosol) lost its popularity, not to a newer product form but to consumer concern for safety. With no new form ready to take its place, consumers and manufacturers searched for new directions. In the case of both suspension dry roll-ons and sticks and solids, the answer was not a totally new form but an aesthetic improvement on an already existing form. Fortunately, all of this did not interfere with the overall expansion of the antiperspirant and deodorant category, which went from $30 million in 1947 to over $1 billion in 1983.

After almost 50 years of product innovation in the antiperspirant field being spearheaded by new packaging forms, we are now in a period where formulation aesthetics has replaced packaging innovation in setting trends. With this discontinuity in product trend, it is difficult to predict the direction of future new products in the field.

REFERENCES

1. Ovid, *Artis Amatorial,* Liber III, v. 193.
2. Carson, H. C. *Household Pers. Prod. Ind., 18:* 33 (1981).
3. Jass, H. E. *Cosmet. Toiletries, 95:* 25 (1980).
4. Schubert, W. R. *Cosmet. Perfum, 88:* 69 (1973).
5. Stillians, A. W. *J. Am. Med. Assoc., 67:* 2015 (1916).
6. Tate, L. U.S. Patent 1,371,822 (1921).
7. *Redbook Magazine,* August (1962).
8. *Dell Modern Magazine,* 14th Survey (1945).
9. *Fawcett Beauty Reader Survey* (1946).
10. Govett, T., and deNavarre, M. G. *Am. Perfum. Essent. Oil Rev., 49:* 365 (1947).
11. *Mod. Packag., 26:* 136 (1953).
12. Auch, R. *Am. Perfum. Essent. Oil Rev. 29:* 187 (1934).
13. Van Mater, H. L. U.S. Patent 2,498,514 (1950).
14. Food and Drug Administration, Aerosol Drug and Cosmetic Products Containing Zirconium, Final Rule, *Fed. Regist., 42:* 41375 (1977).
15. Molina, M. J., and Rowland, F. S. *Nature (London), 249:* 810 (1974).
16. *Fed. Regist., 43:* 11301–11326 (March 17, 1978).
17. *Fed. Regist., 43:* 46694–46732 (October 10, 1978).
18. *Fed. Regist., 43:* 36492–36505 (August 20, 1982).

2

Safety Testing of Antiperspirants and Deodorants

Ronald J. Wulf *Carter-Wallace, Inc., Cranbury, New Jersey*

I. INTRODUCTION

In 1972 the Food and Drug Administration published in the *Federal Register* a procedure for rule making for the classification of over-the-counter (OTC) drugs as generally recognized as safe (GRAS) and/or effective (GRAE) and not misbranded under prescribed, recommended, or suggested conditions of use. The regulation established procedures for categorization of OTC drugs and appointment of advisory review panels of "qualified experts" to evaluate safety and efficacy and spell out regulatory definitions of safety, efficacy, benefit-to-risk ratio, combinations, and labeling (1).

The Antiperspirant Review Panel first convened on March 15, 1974 and had many additional meetings, the last of which was held nearly four years later on January 26, 1978. The antiperspirant industry responded dutifully to a notice published in the *Federal Register* on September 7, 1973 requesting the submission of data and information on OTC antiperspirant drugs. The review panel performed a comprehensive, thorough review with industry cooperation and participation, which culminated in the publication in the *Federal Register* of proposed (2) and tentative final (3) monographs. The final monograph has not yet been published (4).

A brief overview of the antiperspirant review is being presented to emphasize the very recent regulatory atmosphere which must be considered when deciding on a course of action or strategy for safety testing of an antiperspirant and/or deodorant product. Although the review panels are advisory to the FDA, there have been instances of the FDA not adopting their recommendations.

Antiperspirants are drugs which have a deodorant component in their activity. Deodorants are cosmetics. A drug is defined in section 201(g)(1) of the FDC Act, 21 U.S. Code (USC) 321(g)(1) as follows:

> The term "drug" means...articles intended for use in the diagnosis, cure, mitigation, treatment, or prevention of disease in man...and...articles (other than food) intended to affect the structure of any function of the body of man... and...articles intended for use as a component of any (such) articles.
>
> A cosmetic is defined in section 201(i) of the Federal Food, Drug and Cosmetic Act (FDC Act), 21 USC 321(i) as follows:
>
> The term "cosmetic" means (a) articles intended to be rubbed, poured, sprinkled, or sprayed on, introduced into, or otherwise applied to the human body or any part thereof for cleansing, beautifying, promoting attractiveness, or altering the appearance, and (b) articles for use as a component of any such articles; except that such term shall not include soap.

II. "COSMETIC DRUGS"

Antiperspirants are drug products that are also cosmetic products, depending on the claims made. These products, often called "cosmetic drugs," are regulated by the FDA both as drugs and as cosmetics. This is particularly important in complying with labeling requirements (5).

The terminology "cosmetic drug" is also important in considering the safety testing of this product category. These products, often thought of as cosmetics (e.g., antiperspirants as well as sunscreens), are nonprescription drugs and are therefore subject to the drug monographs. They frequently may be cosmetics as well because both cosmetic and drug claims are made in labeling and advertising. If no cosmetic claims are made, these products are regulated only as drugs. It is the intended use or claim which determines whether an article is a drug; thus, foods and cosmetics may also be subject to the drug requirements of the law if health claims are made for them.

When the drug monographs are made final, they will be published in 21 CFR, Part 330 of the code of Federal Regulations (CFR). A drug product covered by the final monograph system must conform to the conditions of the monograph, or it may not be marketed without an approved New Drug Application (NDA). Manufacturers of antiperspirant drug products should therefore follow developments in the OTC Drug Review very carefully, as they will be required to comply with these monographs as they become final.

III. INGREDIENT SAFETY BY OTC ADVISORY PANEL REVIEW

Table 1 lists the OTC Antiperspirant Review Panel recommended Category I antiperspirant actives (3). *Aluminum chlorohydrate* has been designated as a *generic* term for the various aluminum chlorohydrate compounds reviewed by the FDA-OTC Review Panel. Since their review indicated that the various aluminum chlorohydrates were similar in properties and that the available submitted toxicity data reviewed by the panel suggested they have identical or very similar toxicities, the panel concluded that their risk potentials were the same. This conclusion was also reached for the *aluminum zirconium chlorohydrates* (2). Aluminum chloride and buffered aluminum sulfate currently have more historical interest than wide usage interest.

The FDA's ultimate purpose was to assure consumers that all over-the-counter drugs on the market are safe, effective, and properly labeled. The review was of *active* ingredients. Formulated products containing the active ingredients must be tested for safety and efficacy as formulated. It is the responsibility of the manufacturer to ensure safety of the formulated product when a Category I safe and effective active ingredient is used in a formulation.

Table 1 Category I Antiperspirant Actives[a]

Active ingredient	Concentration	Ratio range	
		Al:Cl	Al:Zr
Aluminum cnlorohydrate[b]	25% or less concentration (calculated on an anhydrous basis, omitting from the calculation any buffer component present in the compound) of an aerosol and nonaerosol dosage form	2.1 down to but not including 1.9:1	
Aluminum dichlorohydrate[b]	Same as above	1.25 down to and including 0.9:1	
Aluminum sesquichlorohydrate[b]	Same as above	1.9 down to but not including 1.25:1	
Aluminum zirconium trichlorohydrate[c]	20% or less concentration (calculated on an anhydrous basis, omitting from the calculation any buffer component present in the compound) of a nonaerosol dosage form	2.1 down to but not including 1.5:1	2.0 up to but not including 6.0:1

Aluminum zirconium tetrachlorohydrate[c]	Same as above	1.5 down to and including 0.9:1	2.0 up to but not including 6.0:1
Aluminum zirconium pentachlorohydrate[c]	Same as above	2.1 down to but not including 1.5:1	6.0 up to and including 10.0:1
Aluminum zirconium octachlorohydrate[c]	Same as above	1.5 down to and including 0.1:1	6.0 up to and including 10.0:1
Aluminum chloride	15% or less concentration (calculated on the hexahydrate form) of an aqueous solution nonaerosol dosage form		
Aluminum sulfate buffered	8% concentration of aluminum sulfate buffered with 8% concentration of sodium aluminum lactate in a non-aerosol form		

[a]Category I, generally recognized as safe and effective; Category II, not generally recognized as safe and effective; Category III, available data are insufficient to classify as either Category I or Category II.
[b]Complexes with propylene glycol and polyethylene glycol are also Category I.
[c]Complex with glycine is also Category I

The vast majority of antiperspirant active salts contain aluminum, atomic number 13, the third most abundant element in the earth's crust. Numerous reviews on the biological effects of aluminum have appeared that contain information on general uses, hazard (or better, safety) data, and regulatory information pertinent to its use in manufacturing and its unique role in the environment.

In 1957 Campbell et al. (6) published an exhaustive review entitled "Aluminum in the environment of man" in which 1500 books, reports, and articles by investigators in more than 20 countries were assembled and abstracted. These authors concluded that after consideration of the wide distribution of aluminum in the normal environment—soil, atmosphere, vegetation, and water as well as in human food, food processing, food containers, and medicines—there were "no problems associated with aluminum in the environment of man and that none appeared on the horizon." Estimates of the quantity of aluminum in the diet of the ordinary adult, including amounts derived from aluminum utensils, range from 10 to 100 milligrams per day.

The acute oral and intraperitoneal 50% lethal dose (LD_{50}) of aluminum chlorohydrate in the mouse is between 3000 and 10,000 mg/kg. Similar magnitudes of safety have been found for other aluminum salts: aluminum acetate, 5000 to 15,000 mg/kg and aluminum chloride, 1000 to 3500 mg/kg. Acute toxicity determinations have, in fact, been hampered by the relative insolubility of the compound being tested and, in addition, by the complicating effects of the acid radical. No delayed or chronic form of aluminum poisoning has been observed or described in connection with the many and extensive investigations of the effects of feeding aluminum compounds to animals or humans.

The two most likely routes by which aluminum enters the human body are ingestion and inhalation. No evidence for topical absorption has been reported. Aluminum compounds are poorly absorbed from the gastrointestinal tract, producing insignificant increases in blood and urine levels. Therefore, it is to be expected that most or all of an orally administered dose would be elminated in the feces. Also, aluminum inhaled and then ingested due to pulmonary mucociliary clearance and swallowing of the escalated material would be expected to be eliminated in this manner. Experiments in humans indicate little danger from ingestions of quantities of aluminum 25 to 30 times those commonly present in the normal diet and environment, which are estimated to be 10 to 100 mg daily. No interference with assimilation of such food components as proteins, carbohydrates, or fats has been reported.

Little more than trace quantities of aluminum has been reported to be absorbed from the respiratory tract. The quantities of aluminum found in the lung usually are attributed to dust in the environment due to the large quantities of aluminum in the soil. Although analytical methods used for detecting trace quantities of aluminum have been criticized as being inadequate for determining the minute quantities of aluminum present in body fluids and tissues, they tend to err on the high side.

Tipton and co-workers (7), in a series of papers about trace elements in human tissues, reported on the use of emission spectroscopy to measure the concentration and distribution of trace elements, including aluminum. Data were reported for a population of about 250 normal adult cadavers, carefully collected and sampled. A normal cadaver was defined as an individual who had died suddenly and who had no apparent diseased condition at the time of death. Aluminum was observed *above* the limit of detection of 0.10 micrograms of aluminum per gram of dry tissue in every sample of tissue. In any one tissue the variation in the concentration of aluminum from sample to sample was wide, and tissues which are exposed to the outside environment showed the greatest concentration. For example, lung tissues contained 8.3 to >196 micrograms of aluminum per gram of dry tissue, and skin contained 3.8–83.3 μg Al/g. The median value for lung of 98 μg Al/g dry tissue was two orders of magnitude higher than for all other tissues except skin, with a value of 12.7 μg Al/g. Tipton et al. used the median in most comparisons because it is a better estimate of the central tendency for the data, which is skewed on the high side. We too have observed this skewness in our measurement of the aluminum content of tissues and other materials. Since aluminum is the third most abundant element on the earth's surface and is widely distributed in nature as soft, low-density materials, we would *expect* to find the highest concentrations in the lungs and on the skin.

The available raw data on aluminum from the published accounts of the studies by Tipton et al. from six cities in the United States (Richmond, Virginia; Seattle-Tacoma, Washington; Baltimore, Maryland; Dallas, Texas; Miami, Florida; and Denver, Colorado) have been subjected to statistical analysis (8). This analysis found no detectable difference between cities for aluminum in the lung. There also did not appear to be any relation between age of the subject at autopsy and concentration of aluminum in lung tissue.

The normal blood level of aluminum is approximately 17 μg Al/100 ml. The human body burden of aluminum is 50 to 150 mg and appears to be *unaffected* by normal daily intake levels, estimated to be approximately 10 to 100 mg, or considerably higher doses (9). The absorption of ingested aluminum and its salts is minimal. Shaver's disease (fibrosis related to bauxite fume inhalation) is the only aluminum-induced industrial disease. It has been recommended that the maximum atmospheric concentration over an 8-hour period of time be 50 million particles per cubic foot (10).

Occasionally, controversies have arisen in the literature as to a possible adverse effect of aluminum. For example, Drew et al. (11) described the effects in hamsters and rabbits of a respirable aqueous solution of a propylene glycol complex of aluminum chloride hydroxide (Alchlor) and, more recently, an aqueous solution of aluminum chlorohydrate alone. Generation of a true aerosol dose with the Laskin generator they used ensures maximum access to the lung. If the concentration of aluminum is high enough, there is a tissue response to the aluminum chlorohydrate. Obviously, *any* substance in sufficient quantity can be expected to give an adverse toxicological response, including many normally innocuous materials, e.g., water and sodium chloride.

The most relevant data for the estimation of the aluminum chlorohydrate lung burdens in animals were developed by the Lovelace Inhalation Toxicology Research Institute for the Food and Drug Administration and the Consumer Product Safety Commission (12). Additional data were developed in the chronic inhalation toxicity study of particulate aluminum chlorohydrate conducted by Becton-Dickinson for the National Institute of Environmental Health Services (NIEHS) (13).

The Lovelace Institute studies determined the deposition characteristics and clearance rates for insoluble particulate materials in several mammalian species. These data were incorporated into a computer simulation model which linked the exposure environment, dose to target organ, and biological parameters for a given species. Compartments in the simulation model were designed to correspond to anatomical units, particularly those of the lung, so that deposition characteristics of particles and clearance kinetics following deposition could be determined. The expected consumer exposure to various aerosolized products, including antiperspirants, under simulated human use conditions was also determined (14–16). The estimated human lung burden resulting from exposure to aluminum chlorohydrate was 0.6 μg Al/g lung. The simulation model was capable of relating the lung burden of aluminum chlorohydrate achieved in an animal to that estimated to be reached in humans. Furthermore, the model could calculate conditions for experimental exposure which would result in lung burdens in animals 100 times or more than the predicted human lung burdens resulting from typical antiperspirant use.

In addition to the Lovelace Institute study on aluminum chlorohydrate, a chronic toxicity study on aluminum chlorohydrate was conducted by Becton-Dickinson for NIEHS (17). The study employed a more conventional inhalation exposure regimen based on daily exposures of 6 hours per day, 5 days per week, for 2 years at three concentration levels, 0.25, 2.5, and 25 mg/m^3 (18).

In summary, it may be concluded that the lesions examined in aluminum chlorohydrate-exposed animals are similar to those of foreign body reactions. These are considered to be the response of the normal lung defense mechanisms to inert particulate inhalation. The FDA-OTC Review Panel and the FDA have agreed with this conclusion and classified aluminum chlorohydrate as safe for use in aerosol formulated products.

Much attention has been focused on the *supposed* link between elevated amounts of aluminum in the body with age and Alzheimer's disease. It is true that some people with Alzheimer's disease are found to have high levels of aluminum in many tissues in the body. However, when the aluminum content of drinking water and the use of aluminum cooking utensils by individuals suffering from Alzheimer's disease was examined, it was concluded that these factors were not related (19). Aluminum workers, for example, do not have a higher rate of Alzheimer's disease than the general population (20). Aluminum levels are increased in patients with Alzheimer's disease for unknown reasons and may represent an effect of the disorder rather than a cause of the disease. The general conclusion is that aluminum does not cause Alzheimer's disease and that the use of aluminum cooking utensils is safe. The contribution of the *use* of the aluminum salts used as antiperspirants/deodorants, which are highly insoluble salts, is infinitesimally small and insignificant (21).

The second most widely used antiperspirant salts are those containing zirconium. Zirconium, atomic number 40, ranks twentieth in abundance of all elements and ninth of all trace elements. Its mean concentration in the earth's crustal rocks is 170 parts per million (ppm) and in soils 300 ppm. Its concentration in seawater is 4 parts per billion (ppb) and in marine sediments 132 ppm. Concentration in organs of animals and humans has been determined. Zirconium is present in the human brain. There is a high content in human liver, fat, and red blood cells. Zirconium is absent in the urine, since it is excreted in the feces. Zirconium is widely distributed in foods; as a result, the daily intake by humans ranges from 1 to 6 mg/day. An adult man's body probably contains 250 mg of zirconium. Zirconium and hafnium occur together in nature in a ratio of 100 to 0.5–1.5. They have almost complete chemical similarity, and most chemical methods do not distinguish between them. Zirconium is not usually available for incorporation into plant and animal tissues, because it occurs mostly in insoluble forms.

The average body burden of zirconium is 250 mg. Concentrations of zirconium in the body vary between 19 and 2 μg/g of tissue (wet weight), fat and testes containing the greatest amounts of zirconium. A homeostatic mechanism has been suggested in humans (9).

Granulomatous lesions have been observed following use of the salt sodium zirconium lactate. Granulomatous lesions have been observed in rabbits with this salt. Inhalation exposure in animals has been studied extensively and reviewed by several authors (22, 23). The zirconium-containing deodorant sticks recognized as inducing a granulomatous response in the axillae of humans contained the salt sodium zirconium lactate. The zirconium-containing salts which were declared "new drugs" or "adulterated cosmetic" products by the FDA in 1977 were aluminum-zirconium complexes and not sodium zirconium lactate, as used in the earlier sticks which produced the axillary granulomas (24, 25). The aluminum-zirconium complexes have not been shown to induce this response in the axillae.

Over-the-counter aerosol drug products containing zirconium salts were not considered to be generally recognized as safe (GRAS) antiperspirant drug products because of lack of toxicological data adequate to establish a safe level for human use. It was recognized that certain zirconium compounds have caused human skin granulomas and toxic effects in the lungs and other organs of experimental animals, which led to concern that some zirconium salts, when used as an aerosol, would reach the deep portions of the lungs. Unlike the skin, the lung will not reveal the presence of granulomatous changes until they have become advanced and possibly permanent. Adequate evidence could not be presented that zirconium-containing aerosol antiperspirants would not produce hidden lung disease in some users. Evidence to refute this conclusion would be difficult to obtain and, if obtainable, would require a lengthy period of time and perhaps unreasonable cost. The FDA issued regulations declaring that aerosol drugs or cosmetic products containing zirconium are new drugs requiring an NDA (22, 23). No approved NDA has been forthcoming.

The various categories of salts—those containing only aluminum salts, those containing only zirconium salts, and those containing both aluminum and zirconium—vary considerably in their chemistry and physical properties. However, the review panel concluded that the skin is relatively impermeable to metallic salts and complexes and that there was no evidence to suggest that the direct application of antiperspirants to intact skin has been associated with systemic toxic effects (3). Although not in the published literature, the percutaneous dermal toxicity test in animals has been performed for many antiperspirant formulations without reported ill effects (26).

Although antiperspirant ingredients have been associated with or blamed by some users for induction of papular or follicular eruptions in the axillae, the OTC review panel observed that the antiperspirant ingredients are not implicated in standard dermatologic texts (3). In current medical thought, these papular or follicular eruptions indigenous to the axillae, e.g., hidradenitis suppurativa, Fox-Fordyce disease, and seborrheic dermatitis, are not believed to be caused by antiperspirants.

Giovacchini (27) recently reviewed the literature on abuse problems associated with the use of volatile organic chemicals as propellants. The paper reviews the history of the perfected use of commercial products such as aerosolized antiperspirants and deodorants, the inhalant abuser, the signs and symptoms of abuse, the pharmacotoxicology, and the various approaches to solutions. It concludes that inhalation abuse is a social psychological problem with educational programs as the best solution. Even a ban on aerosol product forms would not stop the "sniffer."

IV. DEODORANT "COSMETICS"

Presently, the most extensively used deodorant actives are the generic aluminum chlorohydrates and aluminum zirconium chlorohydrates. These are followed by Triclosan (Ciba Geigy), zinc phenolsulfonate, and the so-called deo-colognes (fragrance mixtures specifically formulated to modify or mask axillary odor).

The generic aluminum chlorohydrates and aluminum zirconium chlorohydrates were reviewed for safety as deodorants as part of the FDA-OTC review of antiperspirants. It was later concluded that deodorancy claims are cosmetic claims. Triclosan was reviewed for safety by the OTC Topical Antimicrobial Products and Drug and Cosmetic Products Panel and given a safety classification of Category III (28). Additional studies to answer questions about percutaneous absorption are in progress.

Zinc phenolsulfonate is being reviewed for safety by the Cosmetic, Toiletry and Fragrance Association (CTFA) in the Cosmetic Ingredient Review. Results of the safety review (unpublished) to date indicate no toxicity-related problems with use topically at 5% maximum concentration (29).

Deo-colognes are mixtures of fragrance components and must be reviewed on obtaining a confidential disclosure of their identities in fragrances on an individual basis.

This recent history of antiperspirant and/or deodorant review brings us to the testing required when developing for market a cosmetic drug (antiperspirant/deodorant) or cosmetic (deodorant) product. If the antiperspirant contains a Category I active and all additional components of the formulation are well-known recognized as safe ingredients, a very different set of testing procedures will be recommended than for a new active ingredient, e.g., some hitherto untested and little known or studied metal salt. The latter case requires an NDA application, while the former requires a lesser degree of testing.

V. NEW DRUGS

Active antiperspirant ingredients not covered by the FDA-OTC review must be considered as "new drugs." Approval by the FDA is based on a New Drug Application (21 CFR 314) submitted by the sponsor, containing acceptable scientific data including the results of tests to evaluate its safety, and substantial evidence of effectiveness for the conditions for which the drug is to be offered. "Substantial evidence" is defined by the law (30) as

> evidence consisting of adequate and well-controlled investigations, including clinical investigations, by experts quali-

fied by scientific training and experience to evaluate the effectiveness of the drug involved, on the basis of which it could be fairly and responsibly concluded by such experts that the drug will have the effect it purports or is represented to have under the conditions of use prescribed, recommended, or suggested in the labeling or proposed labeling thereof.

VI. SAFETY TESTING

The CTFA Safety Testing Guidelines are the best compiled set of guidelines for antiperspirant/deodorant safety testing (3). They are available from the CTFA, Suite 800, 1110 Vermont Avenue, N.W., Washington, DC 20005. An updated revised edition is about to be published (31).

These guidelines were prepared by the CTFA Pharmacology/Toxicology Committee in response to industry requests to provide manufacturers of cosmetic, toiletry, and fragrance products with guidance regarding the use of animal and human testing as a means of substantiating the safety of finished products. Safety substantiation is a complex process requiring sound professional judgment in the evaluation of the relevant safety data and information.

No guidelines should be looked at as providing protocols for the conduct of a study, nor should they be looked at as the last definitive word in providing the data and information required for safety substantiation. They are also not intended to be construed as an established minimum of testing for the final marketed product. Rather, the guidelines are intended to identify representative animal and human tests which are useful for substantiating the safety of products, and to guide in the selection and design of appropriate safety testing protocols.

A protocol is prepared in response to a specific safety problem or need to accomplish a specific objective. A guideline serves only as an outline or framework from which a protocol can be derived. It is the proper conduct of the protocol's objective that establishes a safety document on which acceptable risk evaluation can be accomplished.

The pharmacological principle of dose should be kept in mind when designing a study and preparing a protocol. A demonstration of a dose-response relationship for a toxic effect establishes a better rationale for modification or, for that matter, documentation of an effect.

Some of the tests described in the CTFA guidelines would be inappropriate for particular product categories, and the applicability of a particular test would depend on such factors as the nature and frequency of use of the product, site of application, and nature of ingredients. There may also exist variations of the tests or other relevant and meaningful tests which would be acceptable for judging particular aspects of product safety.

The monographs (proposed, tentative final, and final) address only the safety and efficacy aspects of the *active* ingredients. Since it is the formulated product that will be placed in the marketplace, additional safety studies will be required before a formulated product can be assumed to have sufficient safety and acceptable risk to expose a group of consumers. It is therefore recommended that safety evaluations be conducted to ensure safety with regard to the potential for irritation and possible contact allergic reaction.

In the final analysis, it is the marketplace that determines the ultimate safety. Limited safety-in-use exposure therefore precedes test marketing.

Interpretation of guidelines for the safety testing of an antiperspirant/deodorant cosmetic drug is a matter of some variability, with some requiring the bare minimum of safety testing and others requiring extensive data in excess of the studies necessary for an assurance of safety. While this is partially a reflection of preferential priorities and is not directly related to meeting guideline requirements, it does appear that more consideration could be given to more practical use of limited toxicological resources. This could lead to more intelligent use of fewer studies rather than a shotgun approach to safety testing. Subacute studies conducted at the same or similar doses in the same species for the same duration, for no obvious practical reason, provide little knowledge about the safety of the drug.

The ultimate goal of the safety evaluation is an assurance that a formulation (product) is safe for the expected, recommended, and/or customary conditions of use, as well as under reasonably expected conditions of misuse.

An updated version of the CTFA Safety Testing Guidelines is in preparation. The CTFA Safety Testing Guidelines consist of the following specific guidelines:

Guidelines for Evaluating Primary Skin Irritation Potential
Guidelines for Evaluating Eye Irritation Potential
Guidelines for Evaluating Contact Sensitization Potential

Guidelines for Evaluating Photodermatitis
Guidelines for Evaluating Mucous Membrane Toxicity
Guidelines for Evaluation of Percutaneous Toxicity
Guidelines for Evaluating Oral Toxicity
Guidelines for Evaluating Inhalation Toxicity
Guidelines for Evaluating Controlled Use Studies

Note that not all of these guidelines are applicable to all formulated antiperspirant/deodorant products.

Updated medical literature searches and developments in safety and toxicological evaluation techniques should be followed in order to keep a product in the marketplace under the best of safety assessment know-how. The complaint file can be an excellent and reliable indicator of problems in the marketplace. Additional safety testing and follow-up are ongoing necessities.

Appropriate care and protection against inadvertent or deliberate misuse of animals used in safety testing is partially ensured by the Good Laboratory Practice (GLP) regulations. A program for alternative methods to the use of animals in product development and safety testing is supported by a majority of the industry through the CTFA's program at the Johns Hopkins Center for Alternatives to Animal Testing.

As an indication of industry support for this research program, E. E. Kavanaugh, the CTFA president, recently stated in a CTFA New Release (32):

> Although the cosmetic industry is only responsible for less than *one* percent of all animal testing, it has been in the forefront of the search for alternatives to any use of animals in product testing. The industry has established and given continual support to the Johns Hopkins Center for Alternatives to Animal Testing. Through CTFA, the industry has given the Center almost $2,000,000 and other Association members have given individual grants. These have enabled the Center to fund 30 research projects on non-whole animal test methods.
>
> In addition to the Johns Hopkins Center, cosmetic industry members support eight other universities and groups that are searching for alternatives to animal tests. Also, a number of companies are conducting in-house research to develop alternatives to the use of animals in safety testing procedures.

A large amount of the safety testing performed by the industry is carried out in test laboratories throughout the United States and Europe. The recently published *Directory of Toxicology Testing Institutions* (33) is very useful. The first edition, published in 1983, is about to be revised.

VII. SUMMARY

The OTC drug review brought out the best of the industry, demonstrating that responsible scientists have dominated an industry that has matured to uphold the ethical prescription in its adherence to good science with responsible scientists. In the book *Of Acceptable Risk—Science and the Determination of Safety* prepared during his association with the Committee on Science and Public Policy, National Academy of Sciences, as an Alfred P. Sloan Foundation Resident Fellow in 1973, Lawrence (34) summarizes well the appropriate and necessary nature of safety evaluation.

Lawrence states that "peeled back to their essential nature, all of the many questions about hazards...divide into four lines of investigation." In assessing risk, measurements are made in order to

1. Identify conditions of exposure
2. Identify the adverse effects
3. Relate exposure with effect
4. Estimate overall risk.

Answers to these key lines of investigation will ensure marketing of safe products or better products with an *acceptable risk*.

REFERENCES

1. *The OTC Review*, The Proprietary Association, Washington, D.C. (July 1973).
2. Food and Drug Administration. Antiperspirant drug products for over-the-counter human use, establishment of a monograph, proposed rule, *Fed. Regist. 43:* 46694 (1978).
3. Food and Drug Administration. Antiperspirant drug products for over-the-counter human use, tentative final monograph, *Fed. Regist. 47*: 36592 (1982).

4. Cosmetic, Toiletries and Fragrance Association, Washington, D.C., personal communication.
5. McNamara, S. H., When is a cosmetic also a drug—what you need to know. *CTFA, Cosmet. Toiletr. Fragrance Assoc. Cosmet. J. 12*: 6–14 (1980).
6. Campbell, I. R., Cass, J. S., Cholak, J., and Kehoe, R. A., Aluminum in the environment of man, *Ind. Health, 15:* 359 (1957).
7. Tipton, I. H., Cook, M. J., Steiner, R. L., Boye, C. A., Perry, H. M., Jr., and Schroeder, H. A., Trace elements in human tissue—Part I. Methods, Part II. Adult subjects from the United States, Part III. Subjects from Africa, the Near and Far East and Europe, *Health Phys. 9:* 89 (1963); *9:* 103 (1963); *11:* 403 (1965).
8. Unpublished data, on file at Carter Products Research, Carter-Wallace, Inc., Cranbury, New Jersey.
9. Hammond, P. B., and Beliles, R. P., Metals, *Toxicology—The Basic Science of Poisons,* 2nd ed. (J. Doull, C. D. Klaassen, and M. D. Amdur, eds.), Macmillan, New York, (1980).
10. *TLVS^R: Threshold Limit Values for Chemical Substances in Workroom Air Adopted by ACGIH for 1981,* American Conference of Governmental Industrial Hygienists, Cincinnati, Ohio (1981).
11. Drew, R. T., Gupta, B. N., Bend, J. R., and Hook, G. E. R., Inhalation studies with a glycol complex of aluminum chloride hydroxide, *Arch. Environ. Health, 28:* 321 (1974).
12. Inhalation Toxicology Research Institute, Lovelace Biomedical and Environmental Research Institute, Inc. "Inhalation Toxicology Studies of Aerosolized Products, Final Report, prepared for FDA and CPSC via contract EY-76-C-04-1013 (July 1979).
13. National Institute of Environment Health Services, "Final Report on Aluminum Chlorohydrate Study," prepared for NIEHS via contract NIH-NIEHS-N01-ES-4-2164 (November 21, 1978).
14. Mokler, B. V., Wong, B. A., and Snow, M. J., Respirable particulates generated by pressurized consumer products. I. Experimental method and general characteristics, *Am. Ind. Hyg. Assoc. J., 40:* 330–338 (1979).
15. Mokler, B. V., Wong, B. A., and Snow, M. J., Respirable particulates generated by pressurized consumer products. II. Influence of experimental conditions, 339–347 (1979).

16. Troy, W. R., and Mokler, B. V., Comments—"In regard to generation of respirable particulates..., *Am. Ind. Hyg. Assoc. J., 40:* 748–750 (1979).
17. Steinhagen, W. H., Cavender, F. L., and Cockrell, B. Y., Six month inhalation exposure of rats and guinea pigs to aluminum chlorohydrate, *J. Environ. Pathol. Toxicol., 1(3):* 267 (1978).
18. Stone, C. J., McLaurin, D. A., Steinhagen, W. H., Cavender, F. L., and Haseman, J. K., Tissue deposition patterns after chronic inhalation exposures of rats and guinea pigs to aluminum chlorohydrate, *Toxicol. Appl. Pharmacol., 49:* 71 (1979).
19. Hecht, A., Searching for clues to Alzheimer's disease, *FDA Consumer,* November: 23–26 (1983).
20. Liss, L., ed., *Aluminum Neurotoxicity,* Pathotox Publishers, Park Forrest South, Illinois (1980).
21. Kolate, G., Clues to Alzheimer's disease emerge, *Science, 219:* 941–942 (1983).
22. Food and Drug Administration. Aerosol drug and cosmetic products containing zirconium, proposed determination, *Fed. Regist. 40:* 21328 (1975).
23. Food and Drug Administration. Aerosol drug and cosmetic products containing zirconium, final rule, *Fed. Regist. 42:* 41375 (1977).
24. Prior, J. T., Rustad, H., and Cronk, G. A., Pathological changes associated with deodorant preparations containing sodium zirconium lactate, an experimental study, *J. Invest. Dermatol., 29:* 449 (1957).
25. Shelley, W. B. and Hurley, J. H., Experimental evidence for an allergic basis for granuloma formation in man, *Nature (London) 160:* 1060 (1957).
26. In-house reports, Carter-Wallace, Inc., Cranbury, New Jersey.
27. Giovacchini, R. P., Abusing the volatile organic chemicals, *Regul. Toxicol. Pharmacol. 5:* 18–37 (1985).
28. Food and Drug Administration, O-T-C topical antimicrobial products and drug and cosmetic products, *Fed. Regist. 39:* 33102 (1974).
29. McEwen, G. N., Jr. *CIR and Industry—A Sourcebook,* Cosmetic, Toiletry and Fragrance Association, Washington, D.C. (1982).
30. 21 CFR, p. 314.

31. Hagnes, C. R., and Estrin, N. F., eds., *CTFA Technical Guidelines*, Cosmetic, Toiletry and Fragrance Association, Washington, D.C. (1983).
32. Personal communications.
33. *Directory of Toxicology Testing Institutions*, Texas Research Institute, P.O. box 20165, Houston, Texas, 1983.
34. Lawrence, W. W., *Of Acceptable Risk—Science and Determination of Safety*, Karfmann, Los Altos, California (1976).

3

Regulatory Considerations for Antiperspirants and Deodorants

Herman E. Jass *Technical Management Consultant, Skillman, New Jersey*

I. HISTORY

A. Food, Drug and Cosmetic Act of 1938

Regulation of antiperspirants and deodorants by the federal government began with the passage of the Federal Food, Drug and Cosmetic Act of 1938 (1). Although a Food and Drugs Act had been enacted in 1906, it had defined a drug as any substance intended to be used for the cure, mitigation, or prevention of disease. An antiperspirant, intended to modify a normal physiological function in normal human subjects, could not easily be viewed as a drug under this definition. Cosmetics were not included in the 1906 act.

However, the 1938 act added to the definition of a drug the following description: "articles intended...to affect the structure or any function of the body of man...." This was taken to mean that antiperspirants, which apparently reduce the amount of sweat secretion by an effect on the sweat glands of the skin, were drugs. Cosmetics were also regulated for the first time by the 1938 act and were defined as "articles intended to be rubbed, poured, sprinkled, or sprayed on, introduced into, or otherwise applied to the human body thereof for cleansing, beautifying, promoting attractiveness, or altering the appearance" and thus could be interpreted as including antiperspirants also, as well as deodorants.

The official interpretation of the 1938 act with respect to antiperspirants and deodorants was clarified in an item of Food and Drug Administration (FDA) trade correspondence (2). In this item, FDA stated: "If the action of the preparation is to stop perspiration, such a product would in our opinion be a drug; if, however, the only action of the deodorant powder is to absorb perspiration or to mask its odor, it would probably be a cosmetic." This interpretation has been maintained essentially intact by FDA to the present time.

B. The Drug Amendments of 1962

The 1938 act had required that a new drug application (NDA) be approved by the Food and Drug Administration for safety prior to marketing of a new drug. Approval was given after review of full reports of investigations made to show whether or not the drug was safe to use. In addition, a description of the composition and manufacture of the drug and its labeling were to be included in the new drug application. Many new antiperspirant-deodorant composition NDAs were submitted to FDA after passage of the 1938 Act.

In 1962 Congress passed an act, the Drug Amendments of 1962 (3), which required that new drugs be found effective for their intended purpose, as well as safe. FDA then instituted a program to evaluate the effectiveness of all drugs approved as new drugs after 1938. This evaluation was performed by a number of expert committees appointed by the National Academy of Sciences-National Research Council (NAS-NRC) and was called the Drug Efficacy Study Group. The program under which FDA evaluated the NAS-NRC reports and determined the status of the affected drugs was named the Drug Efficacy Study Implementation (DESI).

Under the DESI program, a number of marketed antiperspirant-deodorant products which had been subject to approved NDAs were evaluated. Among the compositions studied and found effective as well as safe were zirconium oxychloride, aluminum chlorhydroxide and hexachlorophene, aluminum chlorhydroxide and dibromsalan, and aluminum sulfamate (4). An antiperspirant-deodorant composition not found to be sufficiently safe as a result of this review was that containing the antibiotic neomycin sulfate(5). It is interesting that none of these compositions is on the market today, mostly as a result of subsequent regulatory actions.

C. Deodorant Products

1. *Deodorant Claims for Antiperspirant Products*

As noted above, some new drug applications had been submitted to FDA and approved which contained antimicrobial agents as auxiliary deodorant active ingredients. These agents were included in the label listing of active ingredients despite earlier FDA statements that deodorancy is a cosmetic claim. The rationale for this apparent contradiction is that the deodorant ingredient was included as an active ingredient for the suppression of microorganisms which can cause malodor. This antimicrobial claim, therefore, was the basis for regarding the deodorant ingredient as a drug. Many antimicrobial substances were utilized as deodorant components together with antiperspirant materials. In addition to hexachlorophene and dibromosalan and neomycin, quarternary ammonium salts and other antimicrobials were used in conjunction with aluminum salts. Although most of these ingredients have vanished because of potential toxicity problems, other antimicrobial agents are used in cosmetic deodorant products today. However, the labels of these products do not list an active ingredient, nor do they carry antimicrobial or other drug claims.

Not only dibromosalan but also a variety of halogenated salicylanilides were banned by FDA in 1975 for being photosensitizers (6). Earlier, bithionol suffered the same fate (7). Also in 1975, hexachlorophene was limited to preservative use only at a maximum level of 0.1% because of skin penetration and neutrotoxicity (8). Because of numerous reports of topical sensitization, the use of neomycin sulfate as a component of deodorant-antiperspirant drugs ended with the termination of its antibiotic certification for this indication (9).

As a result of these negative findings, the use of antimicrobials as auxiliary deodorant active ingredients was largely discontinued. This has not hurt the ability of antiperspirant products to continue to make deodorant claims, because the aluminum and zirconium active antiperspirant ingredients are sufficiently effective in suppressing axillary bacterial populations to be effective deodorants when used on a regular daily basis.

2. *Cosmetic Deodorant Products*

As noted in Sec. I.C, cosmetic deodorant products still contain antimicrobial ingredients to control odor-causing bacteria. However, they maintain their cosmetic status by limiting their claims to deodorizing and do not list an active ingredient. The deodorant ingredient is treated in the same way as any other cosmetic ingredient with respect to cosmetic labeling regulations. It is included in the list of ingredients according to its predominance in the formula.

Some cosmetic products make deodorant claims based on physical odor-controlling phenomena. Thus, a deodorant cologne may claim to control odor by virtue of the odor-masking effect of high levels of fragrance. Body powders may control odor by absorbing odorous materials and keeping the skin dry. Soaps and other cleansers may control odor by removal of odorous accumulations and bacteria if used often enough. In all of these cases, it is necessary to establish the details of the specific deodorant claim by appropriate clinical testing in order to avoid challenges by other parties or regulatory agencies.

3. *Feminine Deodorant Sprays*

Another form of deodorant product which suffered marketing setbacks because of regulatory action is the feminine deodorant spray. Meant to reduce odor in the anogenital region, these products usually contained hexachlorophene until an FDA regulation restricted their use (10). At the same time, FDA issued additional restrictions on the labeling of such products. Use of the word "hygiene" was prohibited and a lengthy warning statement was required for the labeling of feminine deodorant sprays.

4. *Deodorant Soaps*

Soap is excluded from the definition of "cosmetic" in the FDC act, which specifically states in a clause at the end of the definition: "except that such term shall not include soap" (1). Therefore, soap per se is not subject to FDC act cosmetic regulation, including the requirement for ingredient listing. However, FDA has published regulations in which soap is defined very precisely (11). First, the bulk of the nonvolatile matter in the product consists of an alkali salt of fatty acids and the detergent properties of the article are due to the alkali-fatty acid compounds; second, the product is labeled, sold, and represented only as soap. Thus, compositions which contain other surfactant materials besides alkali-fatty acid compounds which contribute to detergency are cosmetics, not soaps.

With respect to deodorant soaps, the label claims determine whether the item is a soap or cosmetic. If a soap label lists an active ingredient, i.e., an antimicrobial agent, which is the basis for deodorant claims, the product would probably be regarded as a drug by FDA. If no ingredient is listed but the soap's label defines the article as a "deodorant soap," FDA would regard it as a cosmetic because in their interpretation the conferring of deodorant protection is "promoting attractiveness," which is part of the definition of cosmetic in the FDC act.

5. *Aerosol Products*

During the 1960s and 1970s aerosol deodorants emerged as the most popular form on the market. However, research predicted that release of fully halogenated chlorofluorocarbons, the primary propellants in self-pressurized containers, into the atmosphere could cause depletion of stratospheric ozone with serious climatic and public health sequelae (12). As a result, the various federal regulatory agencies having jurisdiction over commercial aerosol products issued regulations banning the use of such propellants (13).

II. THE OVER-THE-COUNTER DRUG REVIEW

A. Background

The DESI review described above included only 512 over-the-counter (OTC) drugs, of which only 25% were found to be effective for their intended uses. Since it was estimated at the time that more than 300,000 OTC drug products were being marketed with about 700 active ingredients claimed on their labels, FDA decided to institute a new review of safety and efficacy, focusing on the active ingredients used in OTC products (14). Seventeen advisory panels of outside experts were established, of which one was an antiperspirant panel. Although these panels have long since completed their work and disbanded, the OTC review is far from finished.

B. Monographs

The objective of the OTC drug review panels was to review the safety and effectiveness of marketed drugs and to make recommendations to FDA regarding the active ingredients, packaging, and labeling which would then enable FDA to establish a monograph including all of the acceptable criteria for the continued marketing of the reviewed drugs, including antiperspirant drugs. The monographs for each drug category are meant to provide standards for OTC drug products not covered by new drug applications. Even new drug products which are in conformance with a monograph may be marketed without preclearance by FDA because a monograph establishes conditions for an OTC drug class which are considered to be generally recognized as safe and effective (GRASE).

The final report of an advisory panel is published by FDA without changes and usually without comment as an Advance Notice of Proposed Rulemaking, which contains a proposed monograph dealing with ingredients and labeling based on panel recommendations. After reviewing the panel recommendations and public comments on them, FDA will publish a Proposed Rule which includes a preamble containing FDA's comments and a tentative final monograph reflecting FDA's views. Eventually, a final monograph will be published. In the case of the antiperspirant panel, only an Advance Notice (15) and a Proposed Rule have been published (16).

C. OTC Review Rules

1. *Effective Date*

A final monograph is effective 12 months after the date of its publication in the *Federal Register*. Therefore, all drug products on the market at the time of publication of the monograph must be brought into conformance with the provisions of the monograph on or before the effective date. All new products introduced into the market after the effective date must be in conformance with the monograph or else be the subject of an approved new drug application. Until 12 months after the publication date, however, drugs may continue to be marketed even with apparent nonmonograph conditions until the effective date of the monograph, unless FDA has issued a specific regulation which sets an earlier deadline.

2. *Inactive Ingredients*

Although the OTC review process has as its objective the evaluation of active drug ingredients only and not inactive ingredients, FDA has published a proposal to regulate the use of inactive ingredients (17). According to 21 CFR 330.1(e) (18), OTC human drug products are considered safe and effective if "the product contains only suitable inactive ingredients which are safe in the amounts administered and do not interfere with the effectiveness of the preparation...." Under the new proposal, however, "suitable" also means that each inactive ingredient should perform a specific function. The proposal defines a safe and suitable inactive OTC ingredient as one that is listed in an official compendium as a pharmaceutical aid or performs one or more of 24 physical or technical functions listed in the proposal.

At this time, this proposed regulation is dormant and is not expected to be issued as a final regulation.

3. *Amending a Monograph*

After a final monograph is published, any of its conditions may be amended or repealed or a new condition may be introduced, either on the FDA Commissioner's own initiative or as a result of a petition from an interested party. A new drug application may be submitted in lieu of a petition if the drug product containing the new condition has not been marketed prior to publication of the monograph. FDA may handle an NDA as a petition to amend a monograph if provisions of 21 CFR 330.1 preclude an NDA approval but permit the granting of a petition.

4. *Labeling*

Labeling of OTC drugs, including antiperspirant drugs, must confirm to 21 CFR 201, particularly subparts A, C, and D. The specific nature of labeling for antiperspirant OTC drugs is described in Sec. II,C,1 below.

5. *Cosmetic Versus Drug*

According to the definitions of the FDC act in sections 291(g) and (i), drugs and cosmetics are two distinct categories. However, as pointed out in Sec. I,A above, the drug and cosmetic definitions hinge on the intended use of the article in question. If the intended use is to "promote attractiveness" by virtue of an ingredient "intended to affect...any function of the body of man," then the article may be regulated as both a cosmetic and a drug. This is the case for OTC antiperspirant products.

Thus, antiperspirants are subject not only to all the regulations pertaining to OTC drugs, including expiration dating and drug good manufacturing practices (GMP), but also to cosmetic regulations, including cosmetic ingredient labeling.

A deodorant product that makes no antiperspirant or other drug claims is a cosmetic. Likewise, a deodorant claim on the label of an antiperspirant product is a cosmetic claim. Confirmation of this statement is found in all of the publications dealing with the OTC review, including the antiperspirant tentative final monograph (19).

6. *Prescription-to-OTC Switches*

The OTC review provides a mechanism by which prescription drugs may be reviewed and found suitable for OTC marketing. Although this mechanism has been applied a number of times in other drug categories, no such prescription-to-OTC switches have taken place in the antiperspirant drug group, nor are there any prospects for such switches.

D. The Antiperspirant Tentative Final Monograph

As pointed out earlier, no final monograph has been published for antiperspirant drugs. However, the tentative final monograph (TFM) (16) provides a preview of what we may expect in a final monograph and is reviewed below.

1. *Active Ingredients*

FDA has accepted the antiperspirant panel's recommendations for Category I active ingredients (generally recognized as safe and effective and suitable for monograph inclusion), which it classified into two groups for the purpose of evaluating safety and effectiveness: the aluminum chlorohydrates and the aluminum zirconium chlorohydrates. Table 1 lists these ingredients. The individual ingredients are listed in the first column according to the nomenclature system of the *Cosmetic Ingredient Dictionary* (20), which was the recommendation of the antiperspirant panel. However, the TFM utilizes a different system of

Table 1 Antiperspirant Active Ingredients

Active ingredient	Ratio range Al:Cl	Al:Zr
Aluminum chlorohydrate	2.1 down to but not including 1.9:1	—
Aluminum dichlorohydrate	1.25 down to and including 0.9:1	—
Aluminum sesquichlorohydrate	1.9 down to but not including 1.25:1	—
Aluminum chlorohydrex PG[a]	2.1 down to but not including 1.9:1	—
Aluminum dichlorohydrex PG	1.25 down to and including 0.9:1	—
Aluminum sesquichlorohydrex PG	1.9 down to but not including 1.25:1	—
Aluminum chlorohydrex PEG[b]	2.1 down to but not including 1.9:1	—
Aluminum dichlorohydrex PEG	1.25 down to and including 0.2:1	—
Aluminum sesquichlorohydrex PEG	1.9 down to but not including 1.25:1	—
Aluminum zirconium trichlorohydrate	2.1 down to but not including 1.5:1	2.0 up to but not including 6.0:1
Aluminum zirconium tetrachlorohydrate	1.5 down to and including 0.9:1	2.0 up to but not including 6.0:1
Aluminum zirconium pentachlorohydrate	2.1 down to but not including 1.5:1	6.0 up to and including 10.0:1
Aluminum zirconium octachlorohydrate	1.5 down to and including 0.9:1	6.0 up to and including 10.0:1
Aluminum zirconium trichlorohydrex GLY[c]	2.1 down to but not including 1.5:1	2.0 up to but not including 6.0:1

Table 1 (continued)

Active ingredient	Ratio range Al:Cl	Al:Zr
Aluminum zirconium tetrachlorohydrex GLY	1.5 down to and including 0.9:1	2.0 up to but not including 6.0:1
Aluminum zirconium pentachlorohydrex GLY	2.1 down to but not including 1.5:1	6.0 up to and including 10.0:1
Aluminum zirconium octachlorohydrex GLY	1.5 down to and including 0.9:1	6.0 up to and including 10.0:1
Aluminum chloride	—	—
Aluminum sulfate buffered	—	—

Source: Ref. 22.
[a]PG = Propylene glycol.
[b]PEG = Polyethylene glycol.
[c]GLY = Glycine.

names. It is not clear at this time what the final monograph will use, but it is expected that the nomenclature of Table 1 will prevail.

Most of the antiperspirant active ingredients are complex chemical compounds which can have variable molecular proportions, depending on their manner of preparation and purification. Therefore it is necessary to define their composition both by metal:halide ratio range (Table 1, column 2) and by aluminum:zirconium ratio range (column 3) for the compounds containing both of these elements.

Concentrations of Active Ingredients

As proposed in the tentative final monograph, the accepted concentration of active ingredients in OTC antiperspirant drug products is defined as follows:

1. Aluminum Chlorohydrates: 25% or less, calculated on an anhydrous basis, omitting from the calculations any buffer component of the compound. By buffer component, the TFM means the molecular groups propylene glycol and polyethylene glycol.

2. Aluminum zirconium chlorohydrates: 20% or less, calculated on an anhydrous basis, omitting from the calculation any buffer component present in the compound, the buffer components including propylene glycol, polyethylene glycol, and glycine.
3. Aluminum chloride: 15% or less calculated as the hexahydrate form.
4. Aluminum sulfate, buffered: 8% of aluminum sulfate buffered with 8% of sodium aluminum lactate.

Restrictions on Use of Active Ingredients

Aluminum zirconium chlorohydrates and aluminum sulfate buffered are presently restricted to use in nonaerosol dosage forms (21). In this context, aerosol includes all types of packages which deliver a spray on use: self-pressurized containers, pump sprays, and squeeze bottle sprays. Aluminum chloride is restricted to aqueous solution nonaerosol dosage forms. All antiperspirant products are restricted to underarm application only. Only one antiperspirant active ingredient may be used in an antiperspirant product.

2. *Labeling*

Since antiperspirant-deodorant products are considered to be cosmetic drugs, the labeling must contain the information required for both an OTC antiperspirant drug and a cosmetic, whereas a product that is labeled as a deodorant only would bear label information pertaining just to a cosmetic. For comparative purposes, labeling for products that are cosmetic deodorants (nonantiperspirants) are included in this section.

Statement of Identity

1. Antiperspirants. The label contains the established name of the drug and identifies the product as an "antiperspirant" (22) on the principal display panel of the outer package or wrapper (23).
2. Deodorants. A statement of identity is required for a cosmetic also, in terms of its common or usual name or appropriately descriptive name (24). For a deodorant, the usual name is, appropriately, "deodorant." For a product for which a manufacturer wishes to claim both deodorant and antiperspirant action, the usual name can be "antiperspirant deodorant."

a. Vaginal douches and suppositories. The Advisory Review Panel on OTC Contraceptives and Other Vaginal Drug Products, in its final report and recommended monograph (25), concluded that if a vaginal douche label claimed only cleansing, soothing and refreshing, and deodorizing effects, it was a cosmetic. Similarly, a vaginal suppository which claimed soothing and refreshing effects and deodorizing was also a cosmetic. However, claims such as relief of serious irritations of the vagina, lowering of surface tension, mucolytic effect, pH change, or astringency are drug claims. Douches or suppositories which make both cosmetic and drug claims must be labeled as cosmetic drugs. It should be noted that these recommendations are not official until accepted by FDA in a tentative final monograph.
b. Feminine deodorant sprays. Regulations (10) state that any spray deodorant product whose labeling represents or suggests that the product is for use in the female genital area or for use all over the body shall be termed a "feminine deodorant spray." Such products are cosmetics but may not use the word "hygiene" or "hygienic" or similar words. Use of such terms renders such products illegal new drugs.

Indications for Use

Indications for use are required for drugs only. Therefore, they are applicable to an antiperspirant product (25, 26), but not to a product that is solely a deodorant. Under the heading "Indications" list one or more of the following phrases:

(Any one of the following terms may be used: "Reduces," "Decreases," "Diminishes," or "Lessens,") "underarm wetness," or "underarm dampness," or "underarm perspiration."

Exclusivity Policy: Note that the TFM calls for drug label wording to be exactly as set forth in a monograph. This rule, particularly as it refers to indications wording, is called the "exclusivity policy" unofficially within FDA and in trade circles.

Recently, however, FDA has amended this policy to allow other wording or phrases for indications for use as long as the alternative wording is truthful and nonmisleading (27). Thus, drug labeling may be presented in three alternative ways, according to this regulation, as follows:

1. The first alternative permits the use of the exact indications wording contained in a monograph. Such exact wording must appear in a boxed area contained in a prominent and conspicuous label location and is designated on the label as "APPROVED USES." At the manufacturer's discretion, other labeling requirements (warnings, etc.) could be included in the boxed area, in which case the label designation of the boxed area would be "FDA-APPROVED INFORMATION" or "USES (or "INFORMATION") APPROVED BY THE FOOD AND DRUG ADMINISTRATION." It is important to note that because the language included in the box is designated as FDA-approved, this alternative may not be used until a final monograph for antiperspirants is published, because the final monograph constitutes the establishment of FDA approval.
2. A second alternative permits other truthful and nonmisleading statements describing the indications for use approved in a monograph. That is, alternative wording can be used, provided that it does not go beyond the indications established in a monograph. Such additional alternative statements must not be boxed nor contain the designations pertaining to approved uses or FDA-approved uses or information.
3. As a third alternative, a manufacturer can combine the two alternatives described above, i.e., include the exact monograph wording in a boxed area, and other statements somewhere on the label.

Note that this policy change refers only to the indications-for-use labeling. All other label statements—identity, warnings, and directions for use—must be included in the exact language of an applicable monograph. However, any or all of the additional required label information, such as the warnings legend, may appear within the boxed area described in item 1 above, provided it is worded exactly as specified by the monograph. If such information is included within the box, the designation must be "FDA-APPROVED INFORMATION" rather than "FDA-APPROVED USES."

Use of Antiperspirants on Other Parts of the Body: The OTC antiperspirant panel placed the use of antiperspirant products on the hands and feet in Category III (insufficient data on effectiveness to make a recommendation). The antifungal advisory panel recommended monograph status for combinations of an antifungal active ingredient with an antiperspirant active ingredient for the treatment of athlete's foot, jock itch, and ringworm (28). However, in the comments in the preamble to this panel's report (29), FDA advised that data are needed to establish the safety and effectiveness of this type of combination. However, it also stated that such combinations already on the market would be permitted to remain there "at this time." The duration of "at this time" was not defined.

Warnings

1. Antiperspirants. Under the heading "Warnings" include the following:
   a. "Do not apply to broken skin. If rash or irritation develops, discontinue use."
   b. For products in an aerosolized form: "Avoid excessive inhalation."
   c. "Keep this and all drugs out of the reach of children."
2. Deodorants
   a. Safety substantiation warning: FDA has issued a regulation (30) which states that any cosmetic ingredient or product whose safety is not adequately substantiated prior to marketing is misbranded unless it contains the following conspicuous statement on the principal display panel:

      Warning—The safety of this product has not been determined.

   b. Feminine deodorant sprays: A spray deodorant whose label represents or suggests that the product is for use in the female genital area or for use all over the body shall bear the following statement (10):

Caution—For external use only. Spray at least 8 inches from skin. Do not apply to broken, irritated or itching skin. Persistent, unusual odor or discharge may indicate conditions for which a physician should be consulted. Discontinue use immediately if rash, irritation, or discomfort develops.

The sentence "Spray at least 8 inches from skin" need not be included if the spray does not contain a liquefied gas propellant such as a halocarbon or hydrocarbon.

3. Aerosol products. Self-pressurized products, whether cosmetic or drug, must bear a number of warning statements on their labels (31), as follows:

   Warning—Avoid spraying in eyes. Contents under pressure. Do not puncture or incinerate. Do not store at temperature above 120°F. Keep out of reach of children.

   In the case of products packaged in glass containers, the word "break" may be substituted for the word "puncture."

   If the propellant in the container consists in whole or in part of a halocarbon or a hydrocarbon, the following additional warning must be added:

   Use only as directed. Intentional misuse by deliberately concentrating and inhaling the contents can be harmful or fatal.

   This warning is not required for containers of less than 2 ounces with metered valves or for barrier packages from which propellant does not escape with use.

Directions for Use

1. Antiperspirants. The FDA-proposed monograph directions for use are:

   Directions: Apply to underarms only.

   Additional directions to ensure proper use of a specific package must also be included (26).
2. Deodorants. There are no specific requirements for directions for cosmetic products, which include deodorants (32).

Ingredient Listing

1. Antiperspirants. The active ingredient of the product shall be listed, followed by the listing of all the inactive ingredients under the heading "Other ingredients," in descending order of predominance in the formula. Ingredients of not more than 1% and color additives may be grouped at the end of the list without regard to their order of predominance in the formula (33). In addition, flavor or fragrance may be listed as flavor or fragrance. The cosmetic ingredients should be named according to the nomenclature in the *Cosmetic Ingredient Dictionary*, 2nd ed. (34).
2. Deodorants. The ingredients are listed in descending order of predominance in the formula, as described in the preceding paragraph.

Other Required Labeling

1. Net contents. A statement of the net quantity of contents is required for both drugs (35) and cosmetics (36).
2. Manufacturer. The name and place of business of the manufacturer, packer, or distributor are required for both drugs (37) and cosmetics (38).
3. Expiration date. Antiperspirants, as OTC drugs, must bear an expiration date unless appropriate stability data support a finding that the product is stable for at least 3 years (39). The aluminum and zirconium salts currently proposed for monograph listing are inherently quite stable, which is reflected in little use of expiration dating on marketed antiperspirant products.

National Drug Code

FDA requests but does not require that the NDC number appear on all drug labels (40).

3. *Effectiveness Testing*

Antiperspirants in finished dosage form may be tested for effectiveness according to guidelines provided by FDA and available on request from that agency (41). Such testing will not be required by the monograph, according to the proposal in the TFM. However, FDA agreed with the advisory panel's recommendation that, to be considered effective, an antiperspirant product must be shown to produce a minimum of 20% reduction in perspiration in at least half of the users on gravimetric tests.

Special claims for effectiveness to be included on the label, such as "effective against emotional sweating," lack adequate data for substantiation, in the judgment of FDA. Therefore, to be approved for monograph inclusion, they must be submitted to FDA as a petition to amend the monograph, together with clinical substantiation according to FDA guidelines, including user perception test data.

III. OTHER FDA REGULATORY CONSIDERATIONS

A. Good Manufacturing Practices

Regulations (42) set forth the minimum good manufacturing practices for methods to be used in, and the facilities and controls to be used for, the manufacturing, processing, packing, or holding of a drug to ensure that such drug meets the requirements of the FDC act as to safety and has the identity and strength and meets the quality characteristics that it purports or is represented to possess. As drugs, antiperspirants are subject to these regulations.

There are no GMPs for cosmetics, including deodorants. However, FDA has a program and budget for factory inspections of cosmetic manufacturing establishments (43). For a detailed explanation of how to cope with an FDA inspection, see McNamara (44).

B. Plant Registration

Owners and operators of all drug establishments that engage in the manufacture, preparation, propagation, compounding, or processing of a drug, including antiperspirants, are required to register with FDA and to submit a list of every drug in commercial distribution (45). Foreign drug establishments whose drugs are imported into the United States must also comply with the registration regulations (46).

C. Voluntary Cosmetic Regulations

FDA has established a set of three voluntary reporting regulations for cosmetic products (47). These include voluntary registration of cosmetic product establishments, voluntary filing of cosmetic product ingredient and cosmetic raw material composition statements, and voluntary filing of cosmetic product experiences. The term "experiences" refers to reports to the marketer of consumer experiences with the marketer's product which may have had an adverse user effect. Among the cosmetic product filing categories listed are *deodorants (underarm)* and *feminine hygiene deodorants.* Antiperspirants as drugs, are not included in the reportable category list, but may be filed as *deodorants (underarm)* at the manufacturer's discretion. A number of such filings have been made.

IV. OTHER FEDERAL REGULATORY CONSIDERATIONS

A number of federal agencies other than the FDA can affect the drug and cosmetic industry. Since a detailed discussion of these agencies lies outside the scope of this chapter, a brief listing of those which most directly affect the manufacturing and marketing of antiperspirants and deodorants is given here.

A. Federal Trade Commission

FTC is charged, under the FTC act, with enforcing the act's prohibition of unfair methods of competition (48) and false and misleading advertising (49). Claims made in product labeling are regulated by FDA, not FTC. For a thorough discussion of the FTC as it has affected drug and cosmetic advertising, including advertising of both deodorants and antiperspirants, see Donegan (50).

B. Occupational Safety and Health Administration

The Occupational Safety and Health Act issues regulations to ensure that employers furnish their employees a safe and healthful workplace (51). OSHA operates as part of the Department of Labor. It is authorized to enter and inspect a workplace at any reasonable time without prior notice to determine whether the requirements of its regulations and standards are being met and to assess penalties for violations (52).

C. Environmental Protection Agency

EPA administers a number of regulations under many federal statutes which can affect manufacturing plants, both drug and cosmetic. These include regulations under the Clean Water Act which define limitations on pollutant discharge and effluents into surface waters and public sewage treatment works (53), the Superfund Act, which deals with the release of toxic substances into the environment (54), and the Solid Waste Disposal Act, which regulates the generation, transportation, storage, treatment, and siposal of hazardous wastes.

D. Bureau of Alcohol, Tobacco and Firearms

This bureau, part of the Department of the Treasury, regulates the use of alcohol in drugs and cosmetics. The regulations covering the use of specially denatured alcohol are particularly important to manufacturers of cosmetics, such as deodorant colognes (56).

E. Department of Transportation

DOT administers the Hazardous Materials Transportation Act, which regulates transportation of hazardous materials by defining types and sizes of permissible packing and required warning labels. Flammable materials used in drugs and cosmetics, such as alcohol and hydrocarbon aerosol propellant gases, are subject to specific regulations, as are aerosolized products (57).

V. NON-GOVERNMENTAL REGULATION

A. National Advertising Divison

The National Advertising Divison (NAD) of the Council of Better Business Bureaus is an organization supported by industry to self-regulate the national advertising of consumer products and services. It will investigate advertising to determine whether claims are accurate and adequately substantiated. Such investigations may be initiated by the NAD or result from challenges by consumers or competitors. The NAD has been an important force in the maintenance of substantiated and accurate claims in advertising, possibly even more influential than the FTC. It has challenged deodorant and antiperspirant advertising on more than one occasion (58).

B. Cosmetic Ingredient Review

The Cosmetic Ingredient Review (CIR) of the Cosmetic, Toiletry and Fragrance Association is an industry-sponsored system for reviewing and evaluating the safety of cosmetic ingredients. It exercises an important influence on formulators of cosmetics and antiperspirants in their selection of ingredients. However, CIR does not review ingredients regulated or evaluated for safety by government agencies. Therefore, antiperspirant active ingredients are not reviewed by CIR (59).

REFERENCES

1. Federal Food, Drug and Cosmetic Act, 21 U.S. Code (USC) 301 et seq.
2. Trade Correspondence 26, February 9, 1940, *The Federal Food, Drug and Cosmetic Act, 1938–1964* (S. A. Weizman, E. G. Murphy, and N. Mark, eds.), Vol. A, Food and Drug Law Institute, Washington, D.C. (1978), p. 581.
3. 21 USC 352.355.
4. *Fed. Regist., 36:* 18022 (September 8, 1971).
5. *Fed. Regist., 37:* 25820 (December 5, 1972).
6. *Fed. Regist., 40:* 50531 (October 30, 1975).
7. *Fed. Regist., 39:* 10054 (March 15, 1974).
8. *Fed. Regist., 40:* 14033 (March 27, 1975); *Fed. Regist., 42:* 63773 (December 20, 1972).
9. *Fed. Regist., 40:* 8929 (March 3, 1975).
10. 21 Code of Federal Regulations (CFR) 740.12.
11. 21 CFR 701.20.
12. Molina, M. J., and Rowland, F. S, *Nature (London), 249:* 810 (1974); National Research Council. *Halocarbons: Effects on Stratospheric Ozone*, National Academy of Sciences, Washington, D.C. (1976).
13. *Fed. Regist. 43:* 11301–11326 (March 17, 1978).
14. *Fed. Regist. 37:* 9464 (March 11, 1972).
15. *Fed. Regist., 43:* 46694–46732 (October 10, 1978).
16. *Fed. Regist., 47:* 36492–36505 (August 20, 1982).
17. *Fed. Regist., 42:* 19156–19160 (April 12, 1977).
18. 21 CFR 330.1(e).
19. *Fed. Regist., 47:* 36494 (August 20, 1982).
20. *Cosmetic Ingredient Dictionary*, 3rd ed., Cosmetic Toiletry and Fragrance Association, Washington, D.C. (1982).
21. CFR 700.16; *Fed Regist. 40:* 24328 (June 5, 1975); *Fed. Regist. 42:* 41376 (August 16, 1977).

22. *Fed. Regist.*, *47:* 36504 (August 20, 1982).
23. 21 CFR 201.60.
24. 21 CFR 701.11.
25. *Fed. Regist.*, *48:* 46701 (October 13, 1983).
26. 21 CFR 201.5
27. *Fed. Regist.*, *51:* 16258 (May 1, 1986); Jass, H. E., *Cosmet. Toiletries*, *101*: 21–22 (July 1986).
28. *Fed. Regist.*, *47:* 12565 (March 23, 1982).
29. *Fed. Regist.*, *47:* 12480–12481 (March 23, 1982).
30. 21 CFR 710.10.
31. 21 CFR 740.11.
32. 21 CFR 701 et seq.
33. 21 CFR 701.3.
34. *Cosmetic Ingredient Dictionary*, 2nd ed., Cosmetic, Toiletry and Fragrance Association, Washington, D.C. (1977).
35. 21 CFR 201.62.
36. 21 CFR 701.13.
37. 21 CFR 201.1.
38. 21 CFR 701.12.
39. 21 CFR 211.134(g).
40. 21 CFR 201.2; 207.35 (b)(3).
41. *Fed. Regist.*, *47:* 36494–36495, 36502, 36504 (August 20, 1982).
42. 21 CFR 210–211.
43. *Cosmetic Handbook*, Food and Drug Administration, Bureau of Foods (HFF-720), Washington, D.C. (1984).
44. McNamara, S. H., When an FDA inspector calles, *The Cosmetic Industry* (N. F. Estrin, ed.), Marcel Dekker, New York (1984), pp. 465–478.
45. 21 CFR 207.
46. 21 CFR 207.40.
47. 21 CFR 710, 720, 730.
48. 15 USC§45.
49. 15 USC §52.
50. Donegan, T. J., Jr. Regulation of advertising by the Federal Trade Commission, *The Cosmetic Industry* (N. F. Estrin, Ed.), Marcel Dekker, New York (1984), pp. 91–107.
51. 29 USC 651 et seq.
52. 29 USC 657(a), 666.
53. 33 USC 1251 et seq.
54. 42 USC 9601 et seq.
55. 42 USC 6901 et seq.
56. 27 CFR Parts 11 and 12.

57. 49 CFR 172.101, 173.306, 173.1200.
58. National Advertising Division, Council of Better Business Bureaus. *NAD Case Report, 13:* 24 (August 15, 1983).
59. Elder, R. L., and Busch, J. T., The cosmetic ingredient review, *The Cosmetic Industry* (N. F. Estrin, Ed.), Marcel Dekker, New York (1984), pp. 203–213.

4

Structure and Function of Eccrine Sweat Glands in Humans

Paul M. Quinton *University of California at Riverside, Riverside California, and University of California at Los Angeles, Los Angeles, California*

I. INTRODUCTION

It is intriguing to note that in evolutionary development humans are superior to all other animals in two capabilities, sweating and thinking—possibly in that order. When we speak of sweating, we generally refer to the elaboration of a fluid secretion on the general body surface. The eccrine sweat glands are responsible for most of this capacity. Such organs are developed only in mammals, and while most mammals possess at least rudimentary sweat glands, it is only in humans and horses (and perhaps a few cows) that the organ plays a truly major role in the critical physiological function of thermoregulation. In other mammals the function of the gland seems to be primarily limited to lubricating contact surfaces, e.g., footpads, eyelids, or palms, and the gland is much less involved in temperature control (1).

Sweating is clearly of selective advantage since it permits work to be done under conditions otherwise adverse to heat dissipation, that is, above the thermoneutral zone. Mathematically, and perhaps more dramatically, only about 1.0% of the body weight need be evaporated as sweat in order to prevent a 10°C rise in body temperature. No doubt, the ability of humans to occupy virtually every climate on earth, while other animals are generally limited to relatively specific zones, is due to their ability to tolerate extremes in temperature. It is clearly the sweat gland that has given humans the ability to excel in the torrid climates.

The ability to elaborate secretions carries with it a potentially heavy expense. That is, it is biologically impossible to move water without coupling it to the movement of salt (2). The principal salts in body fluids are potassium and sodium chloride. As we shall see, the initial steps in the formation of sweat require elaboration of a fluid containing relatively high concentrations of sodium chloride. It is at this point that humans become markedly superior in sweating to lower animals and most primates since the sweat gland has uniquely developed the ability to conserve most of the salt required to form these primary secretions before the fluid is lost on the surface of the skin. Contrary to most common thought, salt is not an abundant substance biologically. For nonmarine animals it must be sought and conserved. Without conservatory mechanisms, large quantities of NaCl would be obligatorily lost during the sweating process in humans. Since salt is the principal osmotic ingredient of the blood and extracellular fluid, proper circulation crucially depends on an adequate salt content in the body. Uncompensated losses invariably lead to circulatory collapse. The ability to conserve salt avoids circulatory compromise and allows humans to cope with comparatively enormous heat loads. For example, to our knowledge humans are the only animals capable of running a marathon in temperatures above the thermal neutral zone. Thus, they are superior to other animals in two regards. First, general surface sweating allows precise thermoregulation in the presence of unfavorable thermal gradients. Second, conservation of salt during sweating minimizes the potentially dangerous effects of salt loss on systemic circulation. These properties of sweating endow humans with mobility, flexibility, and endurance in warm temperatures that is unsurpassed among other animals.

Lest we become too enthralled with our excellence in sweating, we should point out that it has recently been found that another primate, the *Patas* monkey, has developed sweating almost on a par with humans (3). It is striking to note that this monkey lives in a torrid zone and hunts by giving sustained running chase to its prey—conditions which require a highly efficient system for heat management.

Since this volume is devoted to cosmetology, it is understandable that readers may be less interested in why we sweat than in how to prevent sweating. In this chapter, we will try to deal with what is known about how we sweat in terms of morphological, physiological, and pharmacological components in order to assist the readers in this endeavor to stop sweating. Several previous reviews have been published on physiology (4–6), pharmacology (4,7), innervation (8), morphology (9–11), and pathology (12).

II. GENERAL BACKGROUND

It may have been Malpighi in 1687 who first reported that the skin contains a large number of pores. But it was almost 150 years later before sweat glands were discovered by Purkinje in 1833 (5). Depending on body type, it is estimated that between 2 million and 4 million sweat glands are distributed over almost the entire body surface of humans. Sweat glands are not found on the surfaces of highly vascularized (red) skin. The density of sweat glands over the body surface varies between about 100 and 200 per cm^2. Highest densities of glands are found on the back, chest, and forehead and lowest densities occur on the extremities (5). The total number of active sweat glands is fixed and does not increase with age. Thus, the gland density in children is significantly higher than in adults. The glands in children are smaller and secrete at lower maximal sweat rates than those in adults. Maximal sweat rates seem to be relatively constant from maturation throughout midlife, but rates fall in the elderly (13).

III. EMBRYOLOGY

The eccrine gland begins differentiation in about the third to fourth month of embryonic life (10,14,15). The anlagen of the eccrine sweat gland begins as proliferative cell buds, which first appear on the lower surface of the epidermis of the hands and feet. About a month later, similar formations begin in the region of the axilla and, slightly later, over the general body surface. The buds form columns of cells which invade the hypodermis. However, it is not until the seventh or eighth month that the continuous lumen of the gland tubule is formed and the glands begin to assume the characteristic structure of mature glands (16). Glands are structurally and physiologically intact at birth, so sweating can usually be detected within the first 3 days of life (17).

IV. MORPHOLOGY

The simple structure of the eccrine sweat gland makes it the least complicated of all exocrine glands. It consists of a single tubule ranging from about 4 to 8 mm in total length with an outside diameter ranging between 30 and 60 μm. The length and diameter vary not only among individuals but also among glands (18). The bulk of the tubule is collected into a closely associated random coil, which forms the fundus of the gland and is embedded in the connective tissue of the dermis.

A. The Secretory Coil

1. *Tissue Organization*

The fact that the sweat gland is a single tubule in some ways simplifies the microanatomy. The closed, or blind-ended, proximal terminus of the tubule is the *secretory coil* and is continuous with the reabsorptive duct (Fig. 1). The coil and the duct each constitute about 50% of the gland mass. As will be discussed later, these two distinct structures correspond directly with two distinct physiological functions. That is, the secretory coil elaborates a fluid which is somewhat similar to an ultrafiltrate of plasma, while the reabsorptive duct, the distal portion of the tubule, modifies that primary secretion by reabsorbing the principal solute, sodium chloride, in excess water before the sweat reaches the epidermis. It is not surprising, then, that the proximal and distal portions of the sweat gland tubule exhibit distinctive cell and tissue structures. The tubule of the secretory coil varies considerably among individuals and is reported to be between 2 and 4 mm in length and between 30 and 60 μm in diameter in men (18). The epithelium of the secretory coil consists of three predominant cell types (19). About half consists of the so-called *dark* cells and about half is composed of *clear* cells, but a small percentage of cells are *myoepithelial* cells. The terminology "dark" versus "clear" or "light" cells emerges from the reaction of this tissue with stains such as eosin, methylene, or toluidine blue when examined by light microscopy (10,11,20). With such stains, about half of the cells stain intensely while the remaining cells stain relatively poorly. The dark cells contain numerous diastase-resistant, PAS-positive granules (21) which are also highly osmophilic at the ultrastructural level (10). Such cells are also occasionally referred to as *mucus* cells (10,22). In contrast, the remaining cells of the epithelium contain very few granules, appear relatively clear or unstained, and are sometimes referred to as *serous* cells (22). Since dark and clear cells can easily be identified by the presence or absence of granules and dark cells in the light microscope may appear as lighter cells in the electron microscope, it seems more appropriate to refer to these two cell types as *granular* and *agranular* (23). The third class, the myoepithelial cells, comprises only a very small fraction of the total number of cells found in the secretory coil and are not epithelial cells at all, but rather a special class of smooth muscle.

2. *Luminal Surface*

The major portion of the luminal surface of the secretory epithelium consists of granular cells, with only a small fraction of the total surface being composed of nongranular cells (10,23). Most of the cell-to-cell contacts made at the luminal surface involve either granular-to-granular cells or granular-to-agranular cells. It is seldom that two agranular cells are found in contact at the luminal surface (24). Nonetheless, extensions or continuities of the lumen termed *intracellular canaliculi* created by pockets which invaginate from the lumen between adjacent agranular cells (Figs. 1 and 2). These pockets are formed by application of a small portion of the lateral surface of two or more agranular cells in a junctional complex that is continuous with the tight junction complex of the apical border (22,24). The cell surfaces of the intracellular canaliculi are characterized by a moderate population of microvilli that is significantly greater than that observed in the main lumen. An anionic carbohydrate surface coat associated with the intracellular canaliculi is sensitive to neuraminidase digestion (25). Intracellular canaliculi do not occur between graular cells. Thus, while it is held that the epithelium of the secretory coil is a pseudostratified one in which all cells presumably extend from the luminal surface of the epithelium to the contraluminal surface, the arrangement of cells is such that the luminal surface is composed predominantly of granular cells and and the contraluminal surface is predominantly agranular cells (9).

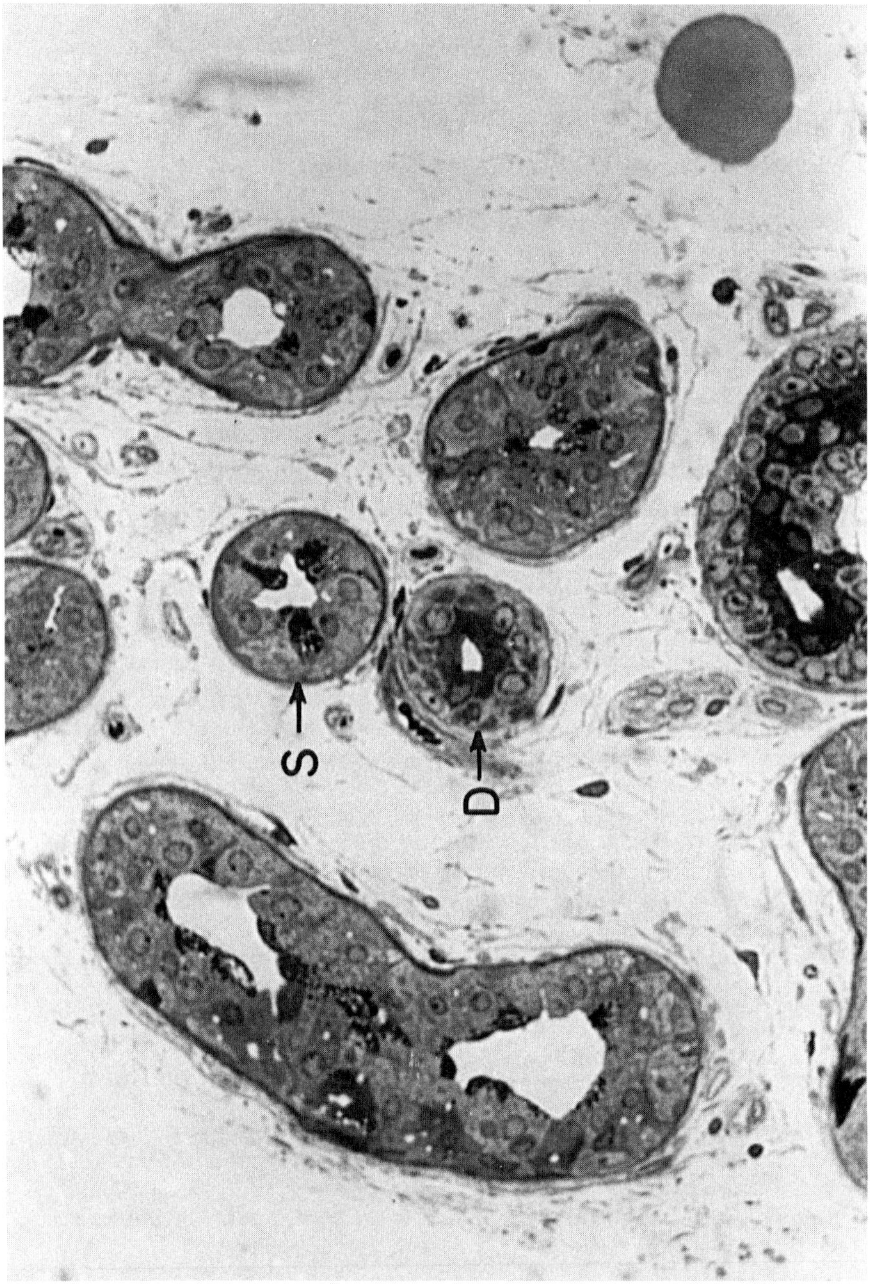
S
D

Figure 1 Light micrograph showing cross-sectional profiles of a reabsorptive duct (D) and a secretory coil (S). The remaining secretory and ductal tissue can be identified by comparison. Of note in the ductal tissue is the concentric layer of epithelial cells, which appear as a bilayer with the inner layer of cells staining more intensely. In the secretory cells it may be seen that the epithelium is a single layer characterized by two predominant cell types, granular and agranular cells. The granular cells are clearly identified by the small intensely staining granules appearing in their cytoplasm. Small unstained regions that might ordinarily be considered as vacuoles are actually portions of intracellular canaliculi between cells that are continuous with the secretory lumen. The tissue is contained in a loose ground substance presumably laid down by fibroblasts, some of which are visible. Arterioles and venules, some of which contain blood cells, are also visible. The micrograph was supplied by the kind courtesy of Dr. James Tidball. Toluidine blue, ×200.

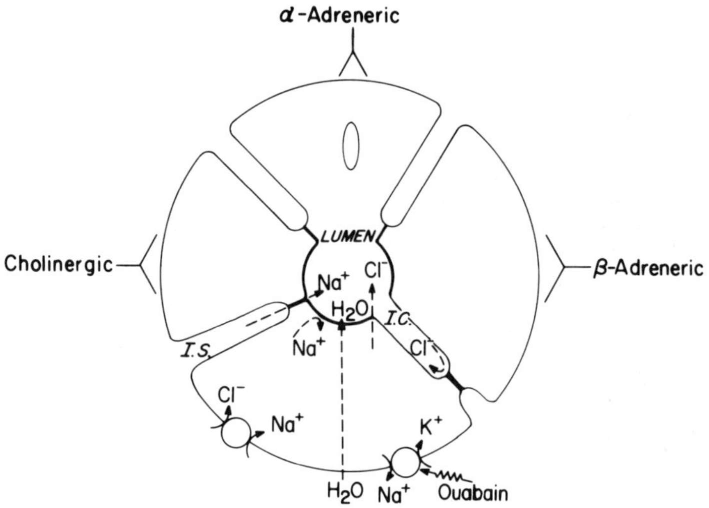

Figure 2 Diagram illustrating the essential components of the secretory coil of the eccrine sweat gland and the process of sweat formation. The tubule of the coil consists of a single layer of epithelial cells joined at the luminal surface and the borders of the intracellular canaliculi (I.C.) to form the luminal compartment. The driving force for the secretory process is supplied by Na^+/K^+-ATPase located in the basolateral membrane of the secretory cells. This Na/K pump is inhibitable by ouabain and serves to keep the intracellular concentration of sodium low. The sodium gradient into the cell thus established is used by a coupled Na/Cl carrier also located in the basal membrane to transport chloride against its electrochemical gradient into the cell. Since the apical membrane is permeable to chloride, chloride ions will then diffuse out of the cell into the lumen. This movement renders the lumen electrically negative with respect to the interstitial fluid. Since the apical membrane should be relatively impermeable to Na, Na can reach the lumen only via the permeable tight junctions as indicated. The accumulation of sodium and chloride in the lumen establishes an osmotic gradient for the influx of water primarily across the cell. These concerted movements produce the volume of primary sweat. As indicated, sweating can be stimulated by cholinergic, α-adrenergic, or β-adrenergic agonists. See text for further details.

3. *Tight Junctions*

The space in the main lumen and intercellular canaliculi is separated from the extracellular fluid compartment by these two cells and by the relatively complex, specialized region of cell-to-cell contacts between them. Using freeze fracture techniques, it may be seen that these contacts (tight junctions) are themselves composed of a continuous network of strands or filaments formed by the true points of contact between the plasma membranes of two adjacent cells. These strands form the principal extracellular barrier to the extracellular movement of water and solutes to and from the lumen. When whole sweat glands are incubated for a sustained period in a lanthanum-containing solution, lanthanum migrates freely into the intercellular spaces up to, but not through, the terminal tight junction at the luminal border between granular and agranular cells (10,26,27). Despite cell death and sloughing into the lumen, the integrity of the epithelium is apparently maintained since lanthanum injected into the serosal space does not gain access to the gland lumen even after intense sweating (27). Molecules moving from the main lumen to the serosal compartment must cross about six of these strands, while molecules moving across the junctions bordering the intracellular canaliculi must cross about nine strands (28,29).

4. *Granular Cells*

At the apical surface, granular cells are characterized by a membrane with stubby microvilli. In the resting state these cells contain granules distributed more or less throughout the apical half of the cytoplasm. A prominent Golgi apparatus is present with a few mitochondria scattered throughout the cytoplasm. The nucleus tends to be situated in the basal half of the two cells (10,17,22). On stimulation, granules accumulate in the apical zone. After prolonged stimulation, the cells do not degranulate but the Golgi apparatus becomes less prominent (23). The basolateral border of granular cells is amplified by a surface that interdigitates with the surface of adjacent cells. On intense prolonged stimulation, it appears that cytoplasmic vacuoles form within the dark cells (21) and that certain cells may be sloughed off, leading to an accumulation of debris in the lumen (23).

5. *Agranular Cells*

The agranular cells are distinguished by the presence of relatively few granules, but they have a high glycogen content (21). Characteristically, these cells have a higher number of mitochondria than the granular cells. The mitochondria seem to be mostly associated with the basolateral membrane and its amplified, interdigitating surfaces. Golgi apparatuses are not usually seen in these cells and the nucleus is more or less centrally positioned. Recently, a form of smooth endoplasmic reticulum termed the tubulocisternal endoplasmic reticulum (TER) has been described in several fluid-transporting epithelia. Agranular, but not granular, cells contain TER in close association with the intercellular canaliculi (30). The function of the organelle is unknown.

After prolonged stimulation, the nongranular cells become depleted of glycogen and their mitochondria appear less electron-opaque. It was not possible to determine whether the cellular debris accumulating in the lumen resulting from the sloughing of granular, agranular, or both cell types during prolonged secretion (23). Agranular cells appear to accumulate iron deposits in patients with iron overload (31). A high concentration of low-activity carbonic anhydrase II is found in the cytoplasm of agranular cells but very little of this enzyme is identified in the granular cells (32).

6. *Myoepithelial Cells*

The myoepithelial cells lie as flattened cells with processes extended between the layer of secretory epithelium and the surrounding layer of basement membrane. Since this cell does not form a continuous sheet, regions of direct contact between the basal surface of granular and agranular cells and the basal membrane occur. Myoepithelial cells are found only in the secretory coil and are not associated with the reabsorptive portion of the sweat gland (9). With the onset of stimulation of sweating, the myepithelial cells appear to shorten and grow thicker. The shortening appears to reverse during the course of prolonged sweating so that the cells become more elongated and thinner as in the resting state (23). Sato et al. (33) have shown that on stimulation, the tissue develops tension. Myoepithelial cells may also make connections with adjacent epithelial cells by desmosome contacts (23). Myoepithelial cells show microscopic properties of smooth muscle, having staining properties consistent with the presence of myosins rather than keratins

(34). A high concentration of high-activity carbonic anhydrase I is found in the cytoplasm of myoepithelial cells (32). It seems likely that the function of this cell is to maintain the structure of the epithelium in the presence of the relatively large hydrostatic gradients that are generated during secretion.

The border between the cells of the gland and the serosal space is marked by a continuous basement membrane which is birefringent, with an inner layer that is reactive and an outer layer that is unreactive to PAS staining (11). Very fine elaunin fibers surround the basement membrane (11,35).

B. Coiled Duct

The entire duct of the sweat gland extends continuously from the transition zone to the orifice or pore on the epidermis. The transition zone is usually a relatively short region of columnar cells which connects the secretory tubule to the duct (36). The epithelium of this region contains no granular or myoepithelial cells or intercullar canaliculi. On the other hand, unlike the duct, it consists of only a single layer of cells even though small flat cells are occasionally seen between the basal portions of the epithelial cells and the basement membrane. Like the duct cells, however, the luminal membrane retains numerous stubby microvilli and the basal lateral membrane contains numerous gap junctions and desmosomes (11,19). These cells may be progenitors of mature duct cells (36).

The coiled duct is the first portion of the duct system (Fig. 1) and, like the secretory coil, is completely contained within the fundus of the gland. The coiled duct is generally described as an epithelium formed of two layers of cuboidal epithelial cells (9,19,37), although Montgomery et al. (38) report that the inner luminal cells are more columnar and the outer surrounding cells may appear as two layers of flat basal cells. In any case, the cells throughout the duct epithelium are connected at numerous sites by desmosomes and gap junctions such that the cells probably function physiologically as a syncytium (12), as had been described for other multilayered epithelia (39,40). Such intercellular coupling has not been conclusively demonstrated, but recent studies of the iontophoretically injected fluorescent dye lucifer yellow have shown dye movement from a single cell to surrounding cells without movement to the extracellular space (41). These studies argue that the intercellular junctions provide cell-to-cell communication for small solutes at least. Solutes in the extracellular space may partially penetrate the junctions but do not pass into the cell, as shown by the restriction of lanthanum added to the external solution (42). As in the secretory coil, lanthanum is prevented from moving into the lumen by the tight junction or *zonula occludens*, which is formed at the apical pole of the luminal cells and thereby creates a continuous barrier between the luminal and serosal extracellular compartments (42). Using freeze fracture techniques, the tight junctions of the coiled duct appear similar to those of the secretory coil. The depth of the junctions was about the same (0.25 μm) and the average number of strands in the region was about six (28–30).

The cells of the inner layer, the luminal cells, are distinguished by the presence of a relatively thick terminal web (10) or cortical zone (22,37). This zone, which occupies about one-fourth of the apical cytoplasm, is relatively devoid of organelles but is rich in tonofilaments (11). Most of this zone is filled with tubulocisternal endoplasmic reticulum, but a clear region containing almost no structures stretches continuously between the apical membrane and the TER across the entire apical pole (30). Numerous mitochondria, smooth and rough endoplasmic reticulum, and glycogen granules are dispersed in the remaining cytoplasm around a centrally located nucleus. The basal region of these cells shows intense immunocytochemical reaction for carbonic anhydrase I (32).

The apical membrane is moderately populated with stubby, irregular microvilli. The surface of basal lateral membranes is significantly amplified by much folding and interdigitation with surfaces of adjacent cells. The lateral surfaces are heavily populated with desmosomes, particularly toward the apical end. Gap junctions are more common at the basal pole between adjacent luminal cells and underlying basal cells (30,38).

The surrounding layer of basal cells is very similar in morphology to the luminal layer of cells with the following notable distinctions. There is no apical membrane, and hence there is no distinct cell polarity or tight junction structure. The cell membrane appears uniformly the same on all sides of the cell and appears identical to the basal lateral membrane of the luminal cells. There is no cortical zone or TER (30). The basal cells contain almost no detectable carbonic anhydrase (32). There are more mitochondria and glycogen than in any other cell type in the gland (20,22,37).

The basal cells rest on a continuous basement membrane which surrounds the duct. The basement membrane in turn is surrounded by a collagenous capsule (10). There are no myoepithelial cells associated with any of the duct segments (37,38).

During prolonged stimulation, the apical membrane of the luminal cells tends to lose microvillar fomrs and occasionally to develop amorphous blebs. The cortical zone appears to collapse and becomes more electron-opaque. Small clusters of up to four electron-opaque granules appear in the apical cytoplasm. The basal cells appear to remain relatively unchanged except, perhaps, for an increase of small cytoplasmic vesicles (38).

C. Straight Duct

There is much similarity between the structure of the straight duct and that of the coiled duct, but there is a gradual transition in both the microanatomy of the cells and the architecture of the duct tissue as it courses from the fundus through the dermis toward the epidermis. The number of layers of basal cells increases from two to four near the fundus to six to eight near the epidermis. The basal cells composing these layers become less cuboidal, thinner, and flatter. Tonofilaments become more prevalent in both luminal and basal cells, while the number of mitochondria in the cells decreases with distance from the fundus. Likewise, there is a corresponding increase in the number of desmosomes and a decrease in the number of gap junctions between both luminal and basal cells (37,38). The changes suggest a transition from a tissue that is less metabolically active to one that is structurally more stable.

D. Acrosyringium (Intradermal Duct)

As the terminal segment of the duct enters the intraepidermal region of the skin it begins to assume a spiraling course which presents something of a corkscrew appearance. The length and degree of spiraling depend on the location. It is most prominent in glands of the palm, sole, and scalp, where the epidermis tends to be thickest (20,38). The cells of the duct in this region continue to resemble the preceding cells in the straight duct. However, the cortical zone of the luminal cells is less prevalent and evidence of keratinization in these cells increases.

V. INNERVATION AND PHARMACOLOGY

A. Central Control

Although the earlier literature contains numerous reports of lesions in the brain which suggest that sweating is controlled by centers distributed relatively widely from the cortex through the medulla (43), it is now well established that the principal center for thermoregulation is in the hypothalamus. Lesions of the preoptic and anterior regions of the hypothalamus generally result in anhidrosis. There is some suggestion of inhibitory centers in the cortex, since mental stress reduces sweat rate during exercise or heat while lobotomy is often associated with hyperhidrosis (5). Nerves from the hypothalamus apparently pass through the pons and medulla into the cervical cord, since injury to certain areas of these structures results in anhidrosis of specific regions of the body surface (43,44).

Nerves leaving the ventral ramus of the spinal nerve cord pass through the chain of sympathetic ganglia so that from thoracic roots T2–T4 the head and neck are innervated and from T2–T8 the upper limbs are supplied. The trunk receives nerves from T6–T10 and the lower extremities are controlled via segments T11–L2 (14,45). There is evidence of some innervation of the face and upper extremities from T1, even though autonomic function is presumed to arise only below the first thoracic root (46). For example, destruction of stellate ganglia (C8–T1 or T2) produces anhidrosis of the upper body and its extremities (47, 48). Despite these generalizations, the supply of nerves to small areas such as a finger may originate from as many as seven spinal segments (5). It may also be very important to recognize that the anatomy of the sympathetic chain is highly varied and that many nerves may bypass the ganglia entirely, thus accounting for numerous discrepancies in the literature concerning pathways and control.

List and Peet (43,44) concluded from lesions at various levels that section of the spinal cord and specific lesions within the cord such as deterioration of the gray matter (intermediolateral columns) result in loss of sweating in respose to heat, but not to exogenous drugs. On the other hand, destruction of peripheral nerves by interruption of the nerve trunk results in loss of sweating in response to both heat and drugs within 2 weeks. The structure of the sweat gland may undergo degenerative changes after such denervation (49).

B. Peripheral Innervation

In 1878, Ott and Wood clearly demonstrated the neural control of sweating by stimulating the exposed sciatic nerve of severed limbs (7). Ranvier (50) presented the initial description of nerves associated with myoepithelial cells. But these observations were followed by almost 100 years of contradictory reports on the innervation of the sweat glands. The discrepancies arose primarily from the well-accepted presumption that sympathetic innervation was mediated by adrenergic agonists (catecholamines) and parasympathetic innervation was mediated by cholinergic agonists (acetylcholine). It was established that the sweat glands were under sympathetic control, but it was also known that acetylcholine, not adrenaline, was most effective in stimulating sweating (51,52). Some clarification of the problem came with the demonstration in cats that stimulation of the sympathetic trunk caused release of acetylcholine to the venous blood, which was prevented by ligation of the footpad of the cat, where sweat glands are localized (53). These results gave good evidence that the sweat glands are innervated by cholinergic fibers from the sympathetic chain, a clear exception to the accepted rule.

In addition to control from higher centers, there is good evidence that local regional sweating may be stimulated by axon reflex. If acetylcholine is injected into the skin in a small wheal, sweating not only occrs in the area of the wheal but also extends radially beyond the wheal for several centimeters. The sweating response outside the wheal, but not in it, is abolished by previous anesthetization with procaine (54,55).

C. Pharmacology

While the stimulatory effect of acetylcholine was demonstrated consistently, the effects of adrenergic stimulation were considerably less uniform. Several investigators (56–59) found no response to adrenergic stimulation. Others, even after confirming reports that adrenergic stimulation was effective (60), were reluctant to accept such evidence as proof that adrenaline (or more accurately noradrenaline) was a natural agonist (61). In recent years, however, it has become clear, largely as a result of the work of Uno and Montagna and of Sato, that the eccrine sweat glad receives separate, dual innervation. Using fluorescent histochemistry to reveal plexuses of adrenergic nerve terminals, Uno and Montagna (62) demonstrated the presence of adrenergic terminals in close association with the secretory coil of the macaque monkey sweat gland. The nerves could be chemically resected with isomers of hydroxydopamine, leaving cholinergic terminals and gland structures intact. With the electron microscope, these investigators were able to identify both adrenergic and cholinergic nerves with the same Schwann cell. In humans the glands are surrounded by a dense network containing acetylcholinesterase (11). Monoamine oxidase, an enzyme associated with the breakdown of catecholamines, is also found throughout the gland (63).

It is well established that systemic administration of cholinomimetics rapidly evokes generalized sweating while systemic administration of atropine completely blocks generalized thermal sweating (5,64,65). Catecholamines given at the maximal concentrations that can be tolerated do not induce systemic sweating in humans (5). The β-adrenergic blocker propranolol has no inhibitory effect (65) and may be slightly stimulatory when administered systemically (66). Sato and Sato (67) have pharmacologically dissected the secretory response of single glands in vitro to cholinergic and adrenergic agonists. By selectively blocking the cholinergic components with atropine and the adrenergic components with phentolamine (α-blocker) and propranolol (β-blocker), these workers showed that mecholyl, a cholinomimetic, stimulates a maximal sweat rate about 2.5 times greater than that stimulated by isoproterenol (β-argonist) and 5 times greater than that stimulated by phenylephrine (α-argonist).

Some of the discrepancies in previous studies attempting to dissect pharmacological responses in vivo are more understandable now in view of Sato and Sato's discovery that subjects with cystic fibrosis (CF) do not respond to β-adrenergic stimulation (68,69). In a group of 28 CF subjects, a few subjects (number unspecified) responded to intradermal injections of isoproterenol (β-adrenergic agonist) plus theophylline (an inhibitor of phosphodiesterase), but no CF subject responded to the injection if atropine was injected simultaneously. This result strongly suggests that either the adrenergic drugs or the procedure excites cholinergic nerves which stimulate sweating. In contrast, about half of 21 control subjects stimulated by intradermal injection of isoproterenol alone failed to respond with sweating; however, when theophylline was added to the injection solution, all control patients responded to the β-adrenergic stimulation. Since β-adrenergic stimulation is usually accompanied by an increase in intracellular cyclic AMP (cAMP), theophylline inhibits degradation cAMP, and cAMP is likely to be an intracellular mediator of the sweat response, it seems likely that much of the previous variability in response to adrenergic drugs (70) was due to the inability of such drugs when applied alone to elevate intracellular cAMP levels above a threshold for secretory response. Thus, not only does there appear to be interaction between adrenergic and cholinergic stimulation that must be eliminated if unambiguous responses are to be obtained, but also pharmacological application of certain adrenergic agonists may not attain sufficient concentrations to result in a physiological response.

VI. STIMULUS-SECRETION COUPLING

As noted above, there is now good evidence that α- and β-adrenergic and cholinergic agonists are effective in stimulating sweat secretion in vitro as well as in vivo. This evidence, in addition to the fact that the response to each agonist can be blocked by specific antagonists, implies the existence of at least three different receptors in the cell membranes of the secretory coil. Unfortunately, the biochemical linkage between receptor binding and activation and sweat secretion is far from clear. Largely by analogy with other systems and from the work of Sato and Sato on simian sweat glands from the palm, we may surmise that in a general sense the cholinergic and α-adrenergic responses seem to be mediated by increases in intracellular Ca while β-adrenergic responses involve elevated levels of intracellular cAMP.

In view of the recent hypothesis advanced by Berridge (71), it seems likely that agonists binding to the cholinergic and α-adrenergic receptors activate a membrane-bound phosphodiesterase, phospholipase C, which converts phosphatidylinositol-4,5-biphosphate (PIP_2) to 1,2-diacylglycerol (DG) and inositol triphosphate (IP_3). The IP_3 acts to release Ca from intracellular stores, possibly the endoplasmic reticulum (72,73). Free cytoplasmic calcium may activate a phosphorylase and/or stimulate other effects such as membrane depolarization by increasing K permeability during stimulation (74). The DG acts on membrane-bound protein kinase C to enhance the phosphorylation of, and thereby activation of, cytoplasmic proteins. Exactly how these changes evoke events leading to secretion, or even whether they are directly involved, is not known. However, extracellular Ca is required for cholinergic and α-adrenergic stimulation (67,75). Elevated external bath Ca (75), admission of external Ca to the cytoplasm by application of the Ca ionophore A23187 (67), and substitution of Sr for Ca (76) in the bath all stimulate sweating directly.

While there is direct evidence that cytoplasmic cAMP increases with β-adrenergic stimulation in the sweat gland, the exact action of cAMP in inducing secretion is not known. Addition of membrane-permeable analogs of cAMP stimulates glucose utilization (77) as well as low levels of sweat secretion which are enhanced slightly by theophylline (68). Although calcium appears not to be required to stimulate β-adrenergic sweating, increased cytoplasmic levels of Ca or simultaneous cholinergic stimulation leads to a severalfold increase in the cytoplasmic cAMP concentrations. In separate studies, Sato and Sato (78) found that cholinergic stimulation produced almost no cytoplasmic cAMP accumulation, but Kealey (79) found that cholinergic stimulation produced about half as much cAMP as β-adrenergic stimulation, and Khullar et al. (80) found that mecholyl (cholinergenic) produced slightly more cAMP than β-stimulation. All three groups found that β-adrenergic stimulation was effective in increasing cAMP accumulation. Analogs of cGMP were ineffective in stimulating sweating (81) even though stimulation with mecholyl increased cell levels of the guanidine nucleotide. Adrenergic stimulation had no effect on cytoplasmic cGMP levels (79,81). Again, it seems clear that levels of cyclic nucleotides change in the course of stimulation, but the role they play in coupling the event to secretion seems far from defined at this point.

VII. MECHANISMS OF SWEAT SECRETION

It is only within the past decade that the ionic basis for secretion has begun to be understood. The question of whether sweat was formed principally by a process of ultrafiltration, as was known to occur with urine in the kidneys, was seriously entertained much earlier (5). However, since the process of filtration requires a hydrostatic pressure derived from the arterial pressure, the notion of ultrafilatration can be put aside on the basis of a few simple observations. First, sweating can be stimulated, although at reduced maximal rates, after blocking blood flow and hydrostatic pressure by arterial occlusion (82, 83). Second, isolated portions of skin (75,84) and isolated single sweat glands can be stimulated to secrete in vitro. Thus, a hydrostatic pressure is not required for secretion and sweat formation must occur at the level of the cell via active processes of fluid movement.

A fundamental in biological systems is that water per se is never transported. That is, there are no biological water pumps (2). Rather, water is moved subsequent to the transport of solutes and the consequent buildup of an osmotic gradient. The question of the mechanism of fluid secretion then becomes one of how, and which, solutes are moved by the secretory cell to support the movement of water.

It has been known for many years that ouabain, a cardiac glycoside, blocks active solute processes. Ouabain is a highly specific inhibitor of Na^+/K^+-ATPase (85). Thus, it seems that virtually all, if not all, transport processes are ultimately dependent on this enzyme, which is integrally associated with the cell membrane. Its function is to transport two Na ions out of and three K ions into the cell at the expense of the hydrolysis of one ATP molecule. In most cells the action of this enzyme is believed to be primarily responsible for cell volume regulation. However, in fluid-transporting tissues, i.e., epithelia, its distribution around the cell is asymmetric and, in addition to volume regulation, it thereby serves as the driving force to move solutes vectorially across the cell. On the basis of this reasoning, one might anticipate that in cells of secretory epithelia, Na^+/K^+-ATPase should be located on the membrane surface of the cell facing the lumen so that Na could be pumped out of the cell and accumulated in the lumen with water following toward osmotic equilibrium. Unfortunately, even though Na^+/K^+-ATPase is found in the secretory coil of the sweat gland in high activity (86) and ouabain completely blocks sweat secretion in vivo (87) and in vitro (84), the enzyme is not found on the luminal membrane but only on the opposite basal membrane (88), where its orientation should be exactly opposite that required for the secretion of Na into the lumen. In almost all other secretory systems, the enzyme is similarly located on the "wrong" membrane, that is, the basal membrane (89).

Our current view of fluid secretion in the sweat must be taken largely by analogy from other systems where data are more adequate. The basis for most of this model is taken from work on the rectal gland, the intestine, and the trachea (90,91). In essence, as shown in Fig. 2, an Na gradient into the secretory cell is set up and maintained by the action of the Na^+/K^+-ATPase in the basal membrane. The gradient for passive Na movement into the cell is used to move Cl into the cell also, but against an unfavorable electrochemical gradient. The Cl movement is accomplished by coupling its transport to Na in an electrically neutral carrier, which presumably moves Na and Cl in stoichiometrically equivalent amounts. Consequently, energy for the uphill electromechanical movement of Cl is derived from the downhill chemical gradient for Na. Once Cl accumulates inside the cell at an electrochemical activity greater than that of Cl outside the cell, its movement is directed into the lumen by the asymmetric Cl permeabilities of the luminal and basal membranes (Fig. 2). That is, if the luminal membrane is Cl-permeable and the basal membrane is Cl-impermeable, Cl exit down its electrochemical gradient can be directed only across the apical membrane into the lumen. Since the principal osmotic component of sweat (and other fluid secretions) is NaCl, Na must also enter the lumen. The basal membrane must be relatively impermeable to Na (otherwise no Na gradient could be maintained) and the apical membrane must also be relatively impermeable to Na (since the electrochemical gradient across the apical membrane is into the cell for Na). Thus, Na must enter the lumen via the tight junction region. The luminal potential in the secretory coil is reported to be about -6 to -10 mV and the intracellular potential is reported to be -60 mV below the extracellular fluid (92). These values, taken with estimates of intracellular Na and Cl concentrations of 15 to 40 mmol/L in other cells (93), suggest that the ionic gradients present in and across the tissue are consistent with the model.

The fluid in the lumen is isotonic with the interstitial fluid. Early measurements of the freezing point depression of luminal versus extracellular fluid by microcryoscopic techniques showed no differences between the two (94), although it is questionable whether a difference could have been detected with the technique. Later, micropuncture techniques were used to obtain samples of luminal fluid from the coil, and analysis showed Na, Cl, and K concentrations similar to those of plasma (95). More recently, after poisoning the intact gland with amiloride to selectively inhibit the reabsorptive process (84) or after removing the reabsorptive duct and collecting secretions directly from the isolated secretory coil (96), analysis of the collected samples again showed concentrations of major solutes similar to those in plasma.

These results show that water moves relatively freely across the epithelium of the secretory coil. Water movement analysis in the gallbladder (97) shows that the structure of the tight junction constrains it from allowing enough water flux to account for known volume flows. On the other hand, the hydraulic conductivity of the cell membrane is sufficient to account for these movements. We may surmise, again by analogy, that in the sweat gland secretory coil water moves freely and rapidly across the cells into the lumen to maintain osmotic equilibrium between the luminal compartment and the intersititial compartments. During secretion, solute (mainly NaCl) is pumped continuously into the lumen; water continuously follows, adding volume, and increases luminal hydrostatic pressure (Fig. 2). Since the proximal end of the secretory coil is closed, fluid must flow through the lumen into the reabsorptive duct to the open pore in the skin. Some insight into the forces involved in these processes is gained from Schulz's (95) observation that approximately 500 mm Hg pressure is required to block the flow of sweat from the gland.

VIII. MECHANISM OF SOLUTE REABSORPTION

As pointed out above, it is well known that sweat varies in its composition but is almost always hypotonic to plasma. Only in a few rare conditions such as cystic fibrosis or Addison's disease does the concentration of Na and Cl exceed 70–80 mmol/L (12). This fact, coupled with the knowledge that sweat is initially formed as an isotonic fluid with an Na concentration of about 145 mmol/L and a Cl concentration of about 110–115 mmol/L (HCO_3 and lactate make up most of the residual anions), requires two properties of the duct. First, it must actively reabsorb salt (mainly NaCl) and second, it must be relatively impermeable to water. Accordingly, NaCl can be actively transported in excess water so that the final sweat emerges as a significantly hypotonic fluid.

On the basis of recent evidence obtained by microperfusing isolated segments of the reabsorptive duct together with previous work, it is possible to present the following general picture of some of the major events in ductal absorption (Fig. 3). First, the driving force is supplied, as in secretion, by Na^+/K^+-ATPase. The enzyme is in the basal membrane as localized by autoradiography of ^3H-labeled ouabain, which has a high affinity for the enzyme (88). In this regard it is also significant that ouabain inhibits Na absorption only when applied to the basal, not the luminal, membrane (84,98). In this location the enzyme is situated to pump Na out of the cytoplasm in exchange

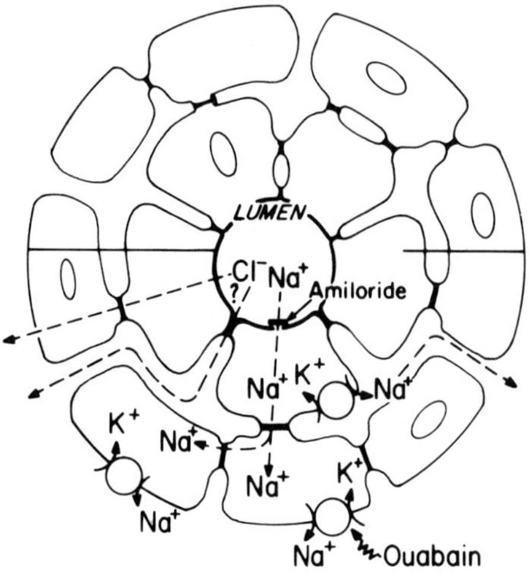

Figure 3 Diagram illustrating the multiple layers of epithelial cells that compose the reabsorptive duct of the eccrine sweat gland. The cells communicate via numerous gap junctions. The lumen of the duct is defined only by the apical membrane of the innermost cells. Functionally, sodium and chloride are moved out of the lumen primarily in response to a sodium gradient from the lumen into the cell cytoplasm. Sodium may move from cell to cell via the numerous gap junctions. The Na gradient is maintained by an Na/K pump located in all but the luminal membranes of the cells. Sodium is extruded via the Na/K pump from the cells into the intracellular spaces and interstitial fluid. By virtue of the electrical gradient set up by the movement of Na, Cl is placed at an electrochemical potential that favors its passive movement out of the lumen. This process may occur by simple diffusion directly across the tight junction or across the cell through both cell membranes and presumably through gap junctions into the intracellular space and interstitial fluid. The luminal entry of Na is blocked by the diuretic amiloride, and the Na/K pump, as in the secretory coil, is blocked by the glycoside ouabain.

for K pumped into the cytoplasm. Thus, the intracellular Na concentration is thought to be held relatively low (20–40 mmol/L). The basal membrane is permeable to K (99) so that K may recycle with the "pump." The high intracellular K concentration and its basal membrane permeability are also probably chiefly responsible for the cell electrical potential of about -40 mV (100). Thus a steep electrochemical gradient exists for the entry of Na into the cell. The luminal membrane of the duct cell contains Na conductive channels which can be blocked by amiloride. Luminal perfusion with amiloride-containing solutions completely inhibits Na reabsorption (84) and reduces the transepithelial electrical potential to zero (101).

The epithelium is also highly permeable to Cl ions (98,102). Substitution of impermeant ions such as SO_4 or gluconate for Cl ion in the perfusion causes the transepithelial potential to hyperpolarize from about -10 to about -100 mV. Such anion substitutions also increase the electrical resistance of the duct from about 7–8 to about 50–60 ohm-cm^2 (103). Such changes strongly indicate that most (~75%) of the electrodiffusive conductivity of the tissue is normally accounted for by Cl. After removing active transport by ouabain poisoning poisoning, NaCl diffusion gradients set up by reducing the luminal perfusion fluid to 50 mM show that the epithelium is 3–4 times more permeable to Cl than to Na. In sweat ducts taken from patients with cystic fibrosis, in which Cl permeability is abnormally low, the duct epithelium is 3–4 times more permeable to Na (98). It is not possible to clearly determine at present whether Cl moves across the cells through the cell membranes or between the cells through the tight junctions or both, but current evidence indicates that Cl moves largely, if not completely, according to electrochemical gradients set up by the active transport of Na (Fig. 3).

Thus, absorption of NaCl from the lumen of the sweat duct consists of Na moving passively from the luminal fluid across the luminal membrane into the cell cytoplasm. At this point we must recognize that the duct epithelium, as discussed above, is a multilayer of cells (Fig. 3). Only the innermost layer has a luminal membrane, and it is devoid of Na^+/K^+-ATPase; the cells of the other layers have only a single membrane, which contains Na^+/K^+-ATPase. Since all of the cells in the epithelium form numerous contacts with each other via gap junctions and desmosomes, it seems reasonable to assume that the cytoplasms of the cells communicate, at least for the movement of Na. In this configuration we may view the duct epithelium as a functional syncytium of cells with a single luminal membrane and a single basal membrane. The Na^+/K^+-ATPase in the basal membrane maintains a low cell concentration of Na by pumping it out across the basal membrane. The Cl apparently follows Na passively by moving either through or between the cells or both (Fig. 3).

The reabsorptive duct is so named because its most obvious action is the uptake of NaCl from sweat so that the final concentration is substantially below that of plasma. But, in addition to water, at least four other substances appear in final sweat at concentrations that are generally higher than those in plasma: K, H, lactate, and urea. All of these solutes increase in concentration as sweat rate decreases. Absorption of some of the water in the duct might explain some of the increase. Since the contact time of the fluid in the duct and the osmotic gradient increase at lower sweat rates, larger fractions of luminal water should be lost at lower rates because the duct is relatively impermeable to water. Such phenomena may well account for the increase in urea concentrations, which rise nonlinearly with decreasing rates (104).

However, additional phenomena must be involved in increasing the concentrations of K, H, and lactate. Lactate is a product of anaerobic glycolysis in both the secretory coil and the reabsorptive duct (77,105), so its appearance in sweat is probably a function of both gland metabolism and sweat pH (106). On the other hand, the fact that the pH of sweat may fall to 4.5 or so (107) indicates a concentration gradients of nearly 1000-fold. It seems unlikely that such a gradient could be established by completely passive mechanisms. In view of evidence (108,109) in support of a proton pump in the luminal membrane of urinary bladder, it is tempting to speculate that a similar mechanism may be present in the sweat duct.

IX. COMPOSITION

Inasmuch as the primary purpose of sweat is to provide evaporative cooling, the composition of human sweat may be thought of as the body's attempt to produce distilled water. The origin of the fluid forming sweat is the extracellular fluid, which has a composition in small solutes very similar to that of serum. Since sweat is excreted directly to the outside of the body, all solutes carried in sweat are lost to further use and therefore there is considerable economy in the body's efforts to reduce the sweat content of most solutes before excretion.

Still, sweat glands are not distillation devices, and sweat collected from the skin surface contains many of the same solutes as plasma. Furthermore, while these solutes generally appear in much smaller concentrations than in plasma, their concentrations are not constant but vary over substantial ranges even with the same gland.

Thus, the concentration of total proteins ranges from about 0.6 to 1.15 g/L (110,111) compared to 60–70 g/dL in plasma (112). Proteins in sweat have been poorly characterized, but Rubin and Penneys (113) report detection by two-dimensional gel electrophoresis of more than 100 different species, of which albumin and α-1-antitrypsin are presumed to be the major components. These investigators detected no globulins even though other workers have found at least small concentrations of IgA, IgE, and IgC (100). Proteins are also present as glycoproteins (114–116) and enzymes such as α-amylase (117), several esterases (118), a reninlike substance (119), and antibodies (120). Considering that sweat gland pores are potential sites for bacterial and viral invasion, it seems likely that protection against pathogens is a significant function of sweat proteins.

Numerous small organic solutes also appear in sweat over a range of concentrations. Although glucose concentrations are generally less than 1% of plasma values, sweat glucose does appear in proportion to blood glucose levels (121). On the other hand, lactate concentrations tend to exceed plasma values by as much as, or more than, an order of magnitude (122). Urea concentrations also slightly exceed those of plasma (123). Only about 3% of the total amino acids in sweat occur as proteins. The remainder appear (124) as free amino acids. Nearly all of the naturally occurring amino acids have been detected in sweat, but, interestingly, proline is reported only in sweat from women (125). Serine, glycine, and alanine occur in the highest concentration (125,126). Prostaglandin E_2-like substance (127) and cAMP (121) are also reported in low concentrations.

Except for lactate and bicarbonate, the major solutes present in sweat are inorganic electrolytes. All of the principal small electrolytes found in the plasma are also found in sweat. The concentrations of Na and Cl depend on sweat rate and normally range from about 10–15 mEq/L at low rates to 40–50 mEq/L at high rates (128). The concentration of Na is under the influence of aldosterone so that exercise and heat tolerance training tend to lower Na concentrations (129–131). The concentration of K appears slightly higher than that in plasma but, in contrast to Na and Cl levels, tends to increase with decreasing sweat rate (128,132). The concentration of Ca, usually between 1 and 2 mmol/L, also rises at very low secretory rates (133,134). The electrolytes Mg, PO_4, and SO_4 are present in concentrations of micromoles per liter (134,135).

At relatively high secretory rates, HCO_3 and lactate are present in concentrations of 20 mmol/L or slightly higher. But the concentrations of both solutes fall significantly with decreasing sweat rate. The HCO_3 disappears and sweat becomes acidic (pH < 5) at rates that are less than about 20% of maximal secretion. Lactate falls to between 5 and 10 mM as the sweat rate falls to less than about half-maximal but seems not to decrease much further at lower rates (106).

As a note of caution, it should be fully recognized that a principal difficulty in determining the true concentrations of solutes in sweat derives from the difficulty of collecting sweat uncontaminated by contact with substances on or in the skin surface. The problem is particularly relevant to substances appearing in small or trace concentrations, since solutes and debris may be added to sweat from the epidermis. Some improvements in collection techniques have been suggested (84,121,128).

REFERENCES

1. Jenkinson, D. McE. *Br. J. Dermatol. 88*: 397 (1973).
2. Quinton, P. M. In *Comparative Animal Nutrition* (M. Rechcigl, Jr., ed), Karger, Basel (1979).
3. Gisolfi, C. V., Sato, K., Wall, P. T., and Sato, F. *J. Appl. Physiol. 53*: 425 (1982).
4. Sato, K. *Rev. Physiol. Biochem. Parmacol., 79*: 51 (1977).
5. Kuno, Y. *Human Perspiration*, Charles C. Thomas, Springfield, Illinois (1956).
6. Quinton, P. M. In *Cystic Fibrosis* (L.Taussig, ed.), Thieme-Stratton, New York (1984) p. 338.
7. Randall, W. C., and Kimura, K. K., *Pharm. Rev., 7*: 365 (1955).
8. Sokolov, V. E., Shabadash, S. A., and Zelikina, T. I. In *Izv. Akad. Nauk SSSR Ser. Biol. 5*: 655 (1980).

9. Ellis, R. A. In *Normale und Pathologische Anatomie der Haut I* (O. Gans and G. K. Steigleder, eds.) (1968).
10. Hashimoto, K. In *The Physiology and Pathophysiology of the Skin* (A. Jarrett, ed.), Academic Press, London (1978).
11. Montagna, W., and Parakkal, P. F., eds. *The Structure and Function of Skin,* Vol. III, Academic Press, New York (1974).
12. Quinton, P. M. *Annu. Rev. Med., 34:* 429 (1983).
13. Sato, K. *Pharmacol. Ther., 24:* 435 (1984).
14. Hashimoto, K., Gross, B. G., and Lever, W. F. *J. Invest. Dermatol., 45:* 139 (1965).
15. Hashimoto, K., Gross, G. B., and Lever, W. F. *J. Invest. Dermatol., 46:* 513 (1966).
16. Dobson, R. L. *The Structure and Function of Skin,* Vol. III (W. Montagna and P. F. Parakkal, eds.), Academic Press, New York (1974) p. 336.
17. Verbov, J., and Baxter, J. *Br. J. Dermatol., 90:* 269 (1974).
18. Sato, K., and Sato, F. *Am. J. Physiol., 245:* R203 (1983).
19. Hibbs, R. G. *Am. J. Anat., 103:* 201–217 (1958).
20. Montagna, W., Ellis, R. A., and Silver, A. F. *Biology of Skin,* Vol. 3, Pergamon Press, Elmsford, New York (1962).
21. Dobson, R. L., Formisano, V., Lobitz, W. C., Jr., and Brophy, D. *J. Invest. Dermatol., 31:* 147 (1958).
22. Munger, B. L., Brusilow, S. W., and Cooke, R. E. *J. Pediatr., 59:* 497 (1961).
23. Montgomery, I., Jenkinson, D. M., Elder, H. Y., Czarnecki, D., and MacKie, R. M. *Br. J. Dermatol., 110:* 385 (1984).
24. Briggman, J. V., Bank, H. L., Bigelow, J. B., Graves, J. S., and Spicer, S. S. *Am. J. Anat., 162:* 357 (1981).
25. Constantine, V. S., and Mowry, R. W. *J. Invest. Dermatol., 46:* 536 (1966).
26. Hashimoto, K. *Ultrastruct. Res., 36:* 249 (1971).
27. Jenkinson, D. M., Nimmo, M. C., Jackson, D., McQueen, L., Elder, H. Y., MacKay, D. A., and Montgomery, I. *Tissue Cell, 15:* 573 (1983).
28. Spicer, S. S., Briggman, J. V., and Baron, D. A. In *Fluid and Electrolyte Abnormalities in Exocrine Glands in Cystic Fibrosis* (P. M. Quinton, R. J. Martinez, and U. Hopfer, eds.), San Francisco Press, San Francisco (1982).
29. Briggman, J. V., Bank, H., Graves, J. S., and Spicer, S. S. *Lab. Invest., 49:* 62 (1983).

30. Baron, D. A., Briggman, J. V., and Spicer, S. S. *Lab. Invest., 51:* 233 (1984).
31. Falanga, V., Zheng, P., and Lavker, R. M. *Br. J. Dermatol., 110:* 51 (1984).
32. Briggman, J. V., Tashian, R. E., and Spicer, S. S. *Am. J. Pathol., 112:* 250 (1983).
33. Sato, K., Nishiyama, A., and Kobayashi, M. *Am. J. Physiol., 237:* C177 (1979).
34. Puchtler, H., Waldorp, F. S., Carter, M. G., and Valentine, L. S. *Histochemistry, 40:* 281 (1974).
35. Bock, P. *Separatum Experientia, 35:* 538 (1979).
36. Korosumi, K., Kurosumi, U., and Tosaka, H. *Arch. Histol. Jpn., 45:* 213 (1982).
37. Ellis, R. A. In *Ultrastructure of Normal and Abnormal Skin* (A. S. Zelickson, ed.), Lea & Febiger, Philadelphia (1967).
38. Montgomery, I., Jenkinson, D. M., elder, H. Y., Czarnecki, D., and MacKie, R. M. *Br. J. Dermatol., 112:* 165 (1985).
39. Rick, R., Dorge, A., von Arnim, E., and Thurau, K. *J. Membr. Biol., 39:* 313 (1978).
40. Rick, R., Beck, F. X., Dorge, A., and Thurau, K. *Curr. Eye Res., 4* (4): 377 (1985).
41. Jones, C. J., and Kealey, T. *J. Physiol. (London),* 369: 121P (1985).
42. Hashimoto, K. *J. Ultrastruct. Res., 37:* 504 (1971).
43. List, C. F., and Peet, M. M. *Arch. Neurol. Psychiatry, 42:* 1098 (1939).
44. List, C. F., and Peet, M. M. *Arch. Neurol. Psychiatry, 40:* 269 (1938).
45. Foerster, O. In *Handbuch der Neurologie,* Vol. 2 (O. Bumke and O. Foerster, eds.), Springer, Berlin (1929).
46. Netsky, M. G. *Arch. Neurol. Psychiatry, 60:* 279 (1948).
47. Ellis, H. *Am. Surg., 45:* 546 (1979).
48. Shih, C. J., and Lin, M. T. *J. Neurosurg. 50:* 88 (1979).
49. Haimovici, H. J. *J. Appl. Physiol., 2:* 512 (1980).
50. Ranvier, L. *J. Micrographie, 10:* 544 (1887).
51. Langley, J. N. *J. Physiol. (London), 56:* 110 (1922).
52. Langley, J. N., and Vyeno, K. *J. Physiol. (London), 56:* 206 (1922).
53. Dale, H. H., and Feldberg, W. *J. Physiol. (London), 82:* 121 (1934).
54. Lobitz, W. C., and Campbell, C. J. *Arch. Dermatol. Syphilol. 67:* 575 (1953).

55. Rothman, S., and Coon, J. M. *J. Invest. Dermatol.*, *3:* 79 (1940).
56. Elliott, T. R. *J. Physiol. (London)*, *32:* 401 (1905).
57. Schilf, E., and Mandur, l. *Arch. Gesamte Physiol.*, *196:* 345 (1922).
58. Dale, H. H. *Br. Med. J.*, *1:* 835 (1934).
59. Benassi, P. *Boll. Soc. Ital. Biol. Sper.*, *28:* 723 (1953).
60. Haimovici, H. *Proc. Soc. Exp. Biol. Med.*, *68:* 40 (1948).
61. Sonnenschein, R. R., Kobrin, H., Janowitz, H. D., and Grossman, M. I. *J. Appl. Physiol.*, *3:* 573 (1951).
62. Uno, H., and Montagna, W. *Cell. Tissue Res.*, *158:* 1 (1975).
63. Yasuda, K., and Montagna, W. *J. Histochem. Cytochem.*, *8:* 356 (1959).
64. Chalmers, T. M., and Keele, C. A. *J. Physiol. (London)*, *114:* 510 (1951).
65. Allen, J. A., Jenkinson, D. J., and Roddie, I. C. *Br. J. Pharmacol.*, *47:* 487 (1972).
66. Gordon, N. F., Kruger, P. E., van Rensburg, J. P., van der Linde, A., Keilblock, A. J., and Cilliers, J. F. *J. Appl. Physiol.*, *58:* 899 (1985).
67. Sato, K., and Sato, F. *Am. J. Physiol.*, *241:* C113 (1981).
68. Sato, K., and Sato, F. *J. Clin. Invest.*, *73:* 1763 (1984).
69. Sato, K. *Am. J. Physiol.*, *247:* R646 (1984).
70. Rothman, S. *Physiology and Biochemistry of the Skin*, University of Chicago Press, Chicago (1954), p. 167.
71. Berridge, M. J. *Mol. Cell. Endrocrinol.*, *24:* 115 (1981).
72. Streb, H., Irvine, R. F., Berridge, M. J., and Schultz, I. *Nature (London)*, *306:* 67 (1983).
73. Williamson, J. R., Cooper, R. H., Joseph, S. K., and Thomas, A. P. *Am. J. Physiol.*, *248:* C203 (1985).
74. Petersen, O. H. *The Electrophysiology of Gland Cells*, Academic Press, New York (1980).
75. Prompt, C. A., and Quinton, P. M. *Nature (London)*, *272:* 171 (1978).
76. Sato, K., and Sato, F. *Am. J. Physiol.*, *242:* C353 (1982).
77. Sato, K. *Am. J. Physiol.*, *224:* 1149 (1973).
78. Sato, K., and Sato, F. *Pfluegers Arch.*, *390:* 49 (1981).
79. Kealey, T. *Biochem. J.*, *212:* 143–148 (1983).
80. Khullar, A. K., Schwarz, V., and Wilson, P. D. *Clin. Sci.*, *68:* 433 (1985).
81. Sato, K., and Sato, F. *Am. J. Physiol.*, *247:* C234 (1984).

82. Solack, S. D., Brengelmann, G. L., and Freund, P. R. *J. Appl. Physiol., 58(5):* 1546 (1985).
83. van Heyninger, R., and Weiner, J. S. *J. Physiol. (London), 116:* 404 (1952).
84. Quinton, P. M. *Pfluegers Arch., 391:* 309 (1981).
85. Skou, J. C. *Physiol. Rev., 45:* 596 (1965).
86. Sato, K., and Dobson, R. L. *J. Invest. Dermatol., 55:* 53 (1970).
87. Sato, K., Taylor, J. R., and Dobson, R. *J. Invest. Dermatol., 53:* 275 (1969).
88. Quinton, P. M., and Tormey, J. M. A question to the role of the enzyme in secretion, *J. Membr. Biol., 29:* 383 (1976).
89. Ernst, S. A., Riddle, C. V., and Karnaky, K. J. *Curr. Top. Membr. Transp. 13:* 355 (1980).
90. Krasny, E. J., and Frizzell, R. A. *Chloride Transport Coupling in Biological Membranes and Epithelia* (G. A. Gerencser, ed.), Elsevier Science Publishers, New York (1984).
91. Frizzell, R. A., Field, M., and Schultz, S. G. *Am. J. Physiol., 5:* F1 (1979).
92. Sato, K. *J. Membr. Biol., 42:* 123 (1978).
93. Greger, R., and Schlatter, E. In *Chloride Transport Coupling in Biological Membranes and Epithelia* (G. A. Gerencser, Ed.), Elsevier Science Publishers, New York (1984).
94. Slegers, J. F. G. In *Research on Pathogenesis of Cystic Fibrosis* (P. A. di Sant'Agnese, ed.), National Institute of Health, Bethesda, Maryland (1964).
95. Schulz, I. J. *J. Clin. Invest., 48:* 1470 (1969).
96. Sato, K. Mechanism of eccrine sweat secretion, *Fluid and Electrolyte Abnormalities in Exocrine Glands in Cystic Fibrosis* (P. M. Quinton, ed.), San Francisco Press, San Francisco (1982).
97. Persson, E.-E., and Spring, K. R. *J. Gen. Physiol., 79:* 481 (1982).
98. Quinton, P. M. *Nature (London), 301:* 421 (1983).
99. Bijman, J., and Quinton, P. M. Unpublished observations.
100. Reddy, M. M., and Quinton, P. M. *Biophys. J., 49*: 157a (1986).
101. Bijman, J., and Quinton, P. *Pediatr, Res., 18*: 1291 (1984.
102. Quinton, P. M., and Bijman, J. *N. Engl. J. Med., 308:* 1185 (1983).
103. Quinton, P. M. *Biophys. J., 49*: 399a (1986).

104. Quinton, P. M. Unpublished observation.
105. Wolfe, S., Cage, G., Epstein, M., Tice, L., Miller, H., and Gordon, R. S., Jr. *J. Clin. Invest.*, *49:* 1880 (1970).
106. Bijman, J., and Quinton, P. M. In *Secretion Mechanisms and Control* (R. M. Case, J. M. Lingard, and J. A. Young, eds.), Manchester University Press, Manchester, U.K. (1984).
107. Emrich, H. M., and Olert, H. *Pfluegers Arch.* *290:* 311 (1966).
108. Al-Awquati, Q., Gluck, S., Reeves, W., and Cannon, C. *J. Exp. Biol.*, *106:* 135 (1983).
109. Cannon, C., van Adelsberg, J., Kelly, S., and Al-Awqati, Q. *Nature (London)*, *314:* 443 (1985).
110. Cabau, M., Muller, O., and Levy, F. *C.R. Acad. Sci. (Paris)*, *275:* 297 (1972).
111. Boysen, T. C., Yanagawa, S., Sato, F., and Sato, K. *J. Appl. Physiol.*, *56:* 1302 (1984).
112. Keating, E. *J. Lab. Clin. Med.*, *73*: 825 (1969).
113. Rubin, R. W., and Penneys, N. S. *Anal. Biochem.*, *131:* 520 (1983).
114. Jirka, M., and Kotas, J. *Clin. Chim. Acta*, *2:* 292 (1957).
115. Pallavicini, J. C., Gabriel, O., di Sant'Agnese, P. A., and Buskirk, E. R. *Ann. N.Y. Acad. Sci.*, *106:* 330 (1963).
116. Seutter, E., Trijbels, J. M. F., Sutorius, A. H. M., and Urselmann, E. J. M. *Dermatologica*, *141:* 397 (1970).
117. Nikolajek, W. P., and Emrich, H. M. *Eur. J. Pediatr.*, *122:* 289 (1976).
118. Hermann, W. P., and Habbig, J. *Br. J. Dermatol.*, *95:* 67 (1976).
119. Emrich, H. M., and Dahlheim, H. *Klin. Wochenschr.*, *55:* 291 (1977).
120. Brodersen, M., and Wirth, M. *Acta Hepato-Gastroenterol.*, *23:* 194 (1976).
121. Boysen, T. C., Yanagawaw, S., Sato, F., and Sato, K. *J. Appl. Physiol.*, *56:* 1302 (1984).
122. Emrich, H. M., and Zwiebel, R. K. H. *Pfluegers Arch.*, *290:* 315 (1966).
123. Brusilow, S. W. *Nature (London)*, *214:* 506 (1967).
124. Jenkinson, D. M., Mabon, R. M., and Manson, W. *Br. J. Dermatol.*, *90:* 175 (1974).

125. Liappis, N., Kochbeck, E., Eckhardt, G., Hahne, H., Kesseler, K., and Bantzer, P. *Arch. Dermatol. Res., 269:* 311 (1980).
126. Gitlitz, P. H., Sunderman, W., Jr., and Hohnadel, D. C. *Clin. Chem., 20:* 1305 (1974).
127. Frewin, D. B., Easkins, K. E., Downey, J. A., and Bhattacherjee, P. *Aust. J. Exp. Biol., 51:* 701 (1973).
128. Bijman, J., and Quinton, P. M. *Am. J. Physiol., 247:* C3 (1984).
129. Conn, J. W. *J. Am. Med. Assoc., 183:* 775 (1963).
130. Conn, J. W. *J. Am. Med. Assoc., 183:* 871 (1963).
131. Ornstein, D. M., Henke, K. G., and Green, C. G. *J. Appl. Physiol., 57:* 408 (1984).
132. Emrich, H. M., and Ullrich, K. J. *Pfluegers Arch., 290:* 298 (1966).
133. Emrich, H. M., Stoll, E., Friolet, B., Colombo, J. P., Richterich, R., and Rossi, E. *Pediatr. Res., 2:* 464 (1968).
134. Prompt, C. A., and Quinton, P. M. *Nephron, 20:* 4 (1978).
135. Cole, D. E. C., and Landry, D. A. *J. Chromatograhy, 337:* 267 (1985).

5

The Mechanism of Antiperspirant Action in Eccrine Sweat Glands

Richard P. Quatrale* *Shulton Skin Care Center, American Cyanamid Company, Clifton, New Jersey*

I. INTRODUCTION

The structure and function of the eccrine sweat glands are eloquently described elsewhere in this volume. However, a brief review of those pertinent details of gland anatomy and physiology which are involved in the disruption of the sweating process will provide perspective for the understanding of antiperspirant mechanisms.

Credit for the discovery of the eccrine sweat glands is given to the 17th century anatomist Malpighi, but credit for the decision that it is aesthetically desirable to inhibit sweating, especially in the human underarm, is less readily rendered. Indeed, although these eccrine glands are now known to populate the palms and soles of most primates and the footpads and snouts of many mammals, their distribution in humans is unique. Human skin contains about 3 million sweat glands, which are distributed over virtually the entire body surface. We are also alone in that we rely predominantly on eccrine sweat for evaporative heat loss in response to thermal stress.

There are two functional types of eccrine sweat glands, but anatomically they are indistinguishable at this point in our

**Current affiliation:* Cosmair, Inc., Clark, New Jersey

understanding. Those found on the palms and soles respond exclusively to emotional stimulation, whereas those on almost the entire remainder of the body are predominantly thermally responsive. The striking exception to this broad classification is that the eccrine glands of the axilla can respond to both thermal and emotional stress.

The structure of an eccrine sweat gland is deceptively simple. It consists of a tubule which extends from the skin's surface down through the epidermis into the lower dermis. The tubule is composed of a secretory coil within the dermins, a resorptive duct region immediately adjacent to the coil and also in the dermis, and a more or less straight segment which enters the epidermis and courses to the exterior. In the stratum corneum layer of the epidermis, just as it empties onto the surface of the skin, the duct spirals in bedspring fashion. Lobitz et al. (1) first termed this most distal region of the duct, together with the surrounding epidermis, the epidermal sweat duct unit.

The secretory coil consists of three cell types—a large pale or clear cell, a small dark or mucoid cell, and a myoepithelium. Myoepithelial cells form a discontinuous layer along the basement membrane and contain myofilaments. These cells are capable of contracting, but it appears doubtful that active myoepithelial contractility plays a significant role in sweat formation or expulsion (2). Nor has there emerged any definitive evidence to implicate the process in an antiperspirant mechanism despite earlier speculation. Rather, it is more likely that the myoepithelium provides a supporting exoskeleton to maintain the structural, and functionally significant, integrity of the clear cell.

The large pale cells and the small dark cells occur in about equal numbers. Between adjacent clear cells only, there are intercellular canaliculi which empty into the lumen of the secretory coil. Clear cells contain numerous mitochondria and abundant glycogen. Because of its resemblance to other fluid-secretory cells, the clear cell is generally considered to be responsible for the production of sweat despite a lack of direct evidence.

Small dark cells contain numerous mucoprotein-containing granules of unknown function. A material with similar staining qualities often can be seen lining the lumen of the duct and it has been postulated that this is derived from dark cell granules. Whether this is correct is conjectural but, after profuse sweating, dark cell granules are depleted, suggesting an exocytotic process (3). Histologically, the resorptive duct region in the upper dermis resembles the deeper secretory coil region. In this segment of the sweat gland, the isomotic plasmalike sweat originally elaborated in the secretory coil may be modified.

From a physiological standpoint, the remainder of the tubule through which the sweat passes on its way to the surface of the skin is uneventful.

Excreted sweat is essentially a hypotonic solution of sodium chloride together with end products of metabolic activity such as lactate. In addition, sweat has been reported to contain a myriad of inorganic and organic constituents. It is likely that these casual components of sweat are of no physiological consequence but they may have some role in antiperspirancy. Numerous studies have described these minor components and their wide ranges in concentration. (for a review, see Ref. 4.)

Enumeration of sweat gland populations by anatomic methods has given strikingly different results from counts obtained after "maximal" thermal stimulation, as illustrated in Table 1. Particularly noteworthy is that the number of thermally stimulated glands consistently averages about half the number determined by direct anatomic visualization. It is not known whether these nonfunctioning glands remain so indefinitely or whether glands go through cycles of activity and quiescence. If such periodicy exists in the human underarm, then it certainly is a factor in our efforts to cause or improve antiperspirancy, our understanding of the process, and indeed our means of measuring it. According to Kuno (7), if all approximately 3 million sweat glands functioned at their maximum, an outpouring of about 10 liters per day would be achievable. There is, of course, considerable regional variability in sweat production in an individual, which is a function of the population density of the glands in a given area. Further, marked differences exist among individuals in the sweat response. These differences are influenced by many factors such as age, sex, race, acclimitization, and degree of hydration of the stratum corneum.

The effect of stratum corneum hydration on eccrine sweating is somewhat perplexing. For the general body surface, hydration due to soaking the skin in water, or to prevention of evaporation of excreted sweat, decreases sweat flows. This is probably due to poral occlusion resulting from the swelling of the surrounding stratum corneum. Yet, in the axillary vault, which is at least partially occluded by the nature of its anatomy, the sweat glands can continue to function in an admirably proficient manner despite the inhibitive environment.

The eccrine glands of the axilla are unusual in other respects as well, and these undoubtedly affect antiperspirant

Table 1 Sweat Gland Population by Region

Region	Glands/cm^2 ± SE	
	Thermal stimulation	Anatomic
Forehead	178 ± 6	360 ± 50
Cheek	—	320 ± 60
Chest	80 ± 2	175 ± 35
Abdomen	87 ± 3	190 ± 5
Scapular	80 ± 3	—
Lumbar	86 ± 3	160 ± 30
Arm	82 ± 3	150 ± 20
Forearm	106 ± 5	225 ± 25
Thigh	62 ± 7	120 ± 10
Calf	—	150 ± 15
Sole	—	620 ± 120

Source: Refs. 5 and 6

activity. Although sweat glands on the general body surface can be classified as either thermally or emotionally responsive, Ikeuchi and Kuno (8) first noted that in the underarm, sweating is provoked by either stimulus. Whether two distinct populations of glands exist or whether all glands are dually responsive remains unclear. Rothman (9) has categorically claimed that the same glands respond to both stimuli but supporting evidence is still forthcoming.

Rebell and Kirk (10) attempted to clarify this point. They described three patterns of axillary sweating in response to thermal stress but a single pattern in most subjects after nonthermal stress. These findings emphasize the complexity of axillary sweating responses. In some individuals thermal and emotional stimuli probably have an additive effect, whereas in others, different populations of glands seem to respond preferentially to either thermal or nonthermal stimulation. Inasmuch as these investigations also saw fit to detail a number of subtypes of response patterns for their male subjects and then concluded with the comment that the data from a preliminary study using females were at odds with the findings for the men, only the complexity of the issue is certain.

Nevertheless, despite our lack of complete understanding of the factors which govern the eccrine sweat glands of the axilla, vigorous efforts to control them persist.

II. MODE OF ACTION OF ANTIPERSPIRANTS

In a large segment of today's fastidious population, particularly in Western culture, there is a substantial demand for products which are capable of reducing or eliminating the wetness that originates from the activity of the eccrine sweat glands in the underarm. Numbering about 25,000 glands per axillary vault, this population of glands can produce a copious volume of sweat. As mentioned above, these glands not only secrete in response to heat as a mechanism for cooling the body, but also secrete in response to emotional stress. Sweat output caused by emotional stimuli can be particularly voluminous, as much as 1500 mg or more per axilla in only 10 minutes. It is likely that this potential for outpouring in emotionally charged social situations is a significant factor in the individual's desire for products that can effectively control sweating to avoid personal embarrassment (11). Further, the desire today for the ideally effective product can be viewed as particularly intense when one considers the growth of the antiperspirant market since Mum, a zinc oxide-based cream, was first introduced in 1888. Currently, the antiperspirant market in the United States is approaching $1.5 billion. The relative lack of brand loyalty by the consumer is still further testimony to the wish for a more effective product.

The major categories of agents which are recognized as capable of reducing eccrine sweating are the following: 1. metal salts, principally aluminum chlorohydrate and the aluminium zirconium chlorohydrate glycine complexes, 2. anticholinergic drugs, and 3. aldehydes such as formaldehyde and glutaraldehyde. In addition, a variety of other chemicals including antiadrenergic agents, metabolic inhibitors, various alcohols, and other organic acids have been reported to be capable of suppressing eccrine sweating.

A. The Metal Salts

1. *Aluminum Chloride*

By far the most widely used, best-known, and extensively studied antiperspirant materials are the salts of aluminum. The history of the use of aluminum salts for controlling perspiration is indeed long, with the first aluminum chloride solution being marketed in 1902. In 1914 product interest was substantial enough to support national advertising for ODO-RO-NO. Probably one of the first reasonably successful efforts was reported by Stillians in 1916 (12). With the finding that a 25% aqueous solution of aluminum chloride hexahydrate, applied for several days to the axillae, reduced excessive perspiration, the effectiveness of this aluminum salt as an antiperspirant was established. Stillians' discovery was not without attendant drawbacks, however. Whereas previously used materials, mostly "astringents" such as tannin, zinc sulfate, or alums, were marginally effective at best, aluminum chloride's effectiveness came at the expense of its deteriorating effect on clothing and, more important, its propensity to irritate the delicate axillary skin of many of the individuals who used it extensively.

Although aluminum chloride is not widely used commercially as an antiperspirant today, studies on its mechanism of antiperspirant action have been performed by numerous investigators. More than 30 years ago, Sulzberger et al. (13) proposed that a negative charge existed at the terminus of the sweat duct which somehow governed the transport of the secretions to the surface of the skin. They further reasoned that the application of electropositively charged metal salts, such as those of aluminum, neutralized that negative charge and accordingly inhibited sweating. This theory has scant supporting evidence.

Assignment of aluminum chloride's antiperspirant mechanism to its astringent action has been favored by many investigators in one form or another. For the most part, the researchers have set forth theories which center around the agent's ability to "shrink shut" the pore of the sweat gland, or to precipitate protein and form a plug which obstructs the flow of sweat to the skin surface. Shelley and co-workers (14,15) proposed that a histopathological condition, characterized by hyperkeratotic and parakeratotic plugs with the sweat duct, was produced in aluminum chloride-treated sweat glands. They based their profferings on the findings that the histology of the glands rendered anhidrotic was identical regardless of whether the miliaria was a result of the aluminum chloride application or of such injurious means as ultravoilet irradiation, heat, or freezing. But Kligman and Holzle (16) refuted this, contending that the plugs observed were merely a response to injury and an expected sequela of the healing process.

Papa and Kligman (17) had proposed a "leaky hose" hypothesis whereby the aluminum chloride damaged the sweat duct, causing the secreted sweat to leak back into the interstitial space rather than to reach the skin surface. The authors have since retracted their theory. Shelley and Hurley (18) more recently proposed that the aluminum chloride combines with intraductal keratin fibrils to cause closure or contraction of the duct and the formation of a horny plug which also serves to block sweat flow. Reller and Luedders (19), however, thought that an aluminum-containing mass forms at the level of the intraepidermal duct after aluminum chloride treatment. They further suggested that this mass was a hydroxide gel which resulted from the neutralization of the acidic metal and was capable of blocking the sweat from reaching the skin surface. Kligman and Holzle (20) provided evidence to substantiate the view that the physical obstruction of the duct opening, as professed by Reller and Luedders, together with cellular damage is the mechanism by which aluminum chloride causes antiperspirancy. It is this view which predominates today. The view is further supported by both transmission electron microscopy and fluorescence microscopy studies which demonstrate aluminum within the intradermal duct of $AlCl_3$-treated glands (21).

2. *Aluminum Chlorohydrate*

Without doubt, the overwhelming number of studies made and hypotheses offered by investigators relative to the mechanism of antiperspirant action by aluminum salts have been directed to aluminum chloride hexahydrate. When one considers that 1. aluminum chlorohydrate (ACH) was first introduced nearly 40 years ago as an antiperspirant having much less deleterious effects with respect to damage to fabric and irritation to human axillary skin (22), 2. aluminum chlorohydrate has been the predominantly used active ingredient in antiperspirant products for nearly that long (23), and 3. significant, highly germane differences between the chemistry of aluminum chloride and that of aluminum chlorohydrate exist (see Chapter 6 by J. Fitzgerald in this volume), then one must wonder why so much attention has been devoted by the students of antiperspirant mechanisms to

Cinderella's homely stepsister. Notwithstanding the claims (and apparent evidence) for superlative antiperspirant efficacy by such overpowering "formulations" as 25% aluminum chloride (in absolute ethanol applied under occulusion!), the acknowledged active antiperspirant ingredients of today, from the standpoint of both efficacy and consumer acceptance, are aluminum chlorohydrate and the aluminum zirconium chlorohydrate glycine complexes.

Attempts to define the mechanism of antiperspirant action by ACH have only recently been made. As stated earlier, Reller and Leudders (19) suggested that the means by which aluminum salts prevent the excretion of sweat is via their formation of an obstructive hydroxide gel within the sweat duct that blocks delivery of sweat. In support of their contention, the investigators offered histological evidence that such a mechanism held for aluminum chloride and for an aluminum-zirconium solution, but they did not comment on ACH.

Quatrale et al. (24) demonstrated that the site of action of ACH was relatively superficial compared to that of aluminum chloride. Others (17,25) had shown that Scotch tape stripping the stratum corneum layer of skin containing sweat glands which had been inhibited by the occlusive application of aluminum chloride did not restore those glands to active function. That finding indicated that the site of inhibition within these glands was deeper than the horny later. On the other hand, using the same approach, Quatrale et al. (26) found that stripping restored sweating in about half the glands in the forearm which had been inhibited by ACH, suggesting a more superficial primary site of action than that ascribed to $AlCl_3$. Similar tape-stripping studies of the axilla also established that ACH's site of action was closer to the skin surface than that of $AlCl_3$ (27). Histological studies of ACH-inhibited forearm sweat glands, using both transmission electron microscopy and fluorescence microscopy, demonstrated that as the duct leaves the stratum granulosum and passes through the stratum corneum to the surface, its lumen is filled with electron-dense amorphous material (Fig. 1). Similar material was frequently observed in the duct at the level of the upper layers of the viable epidermis. In all likelihood, the latter sweat glands were the counterparts of those for which stripping did not restore sweating. Fluorescense microscopy of similarly treated glands revealed substantial fluoresence, indicating the presence of aluminum within the sweat gland duct at the level of the stratum corneum (Fig. 2).

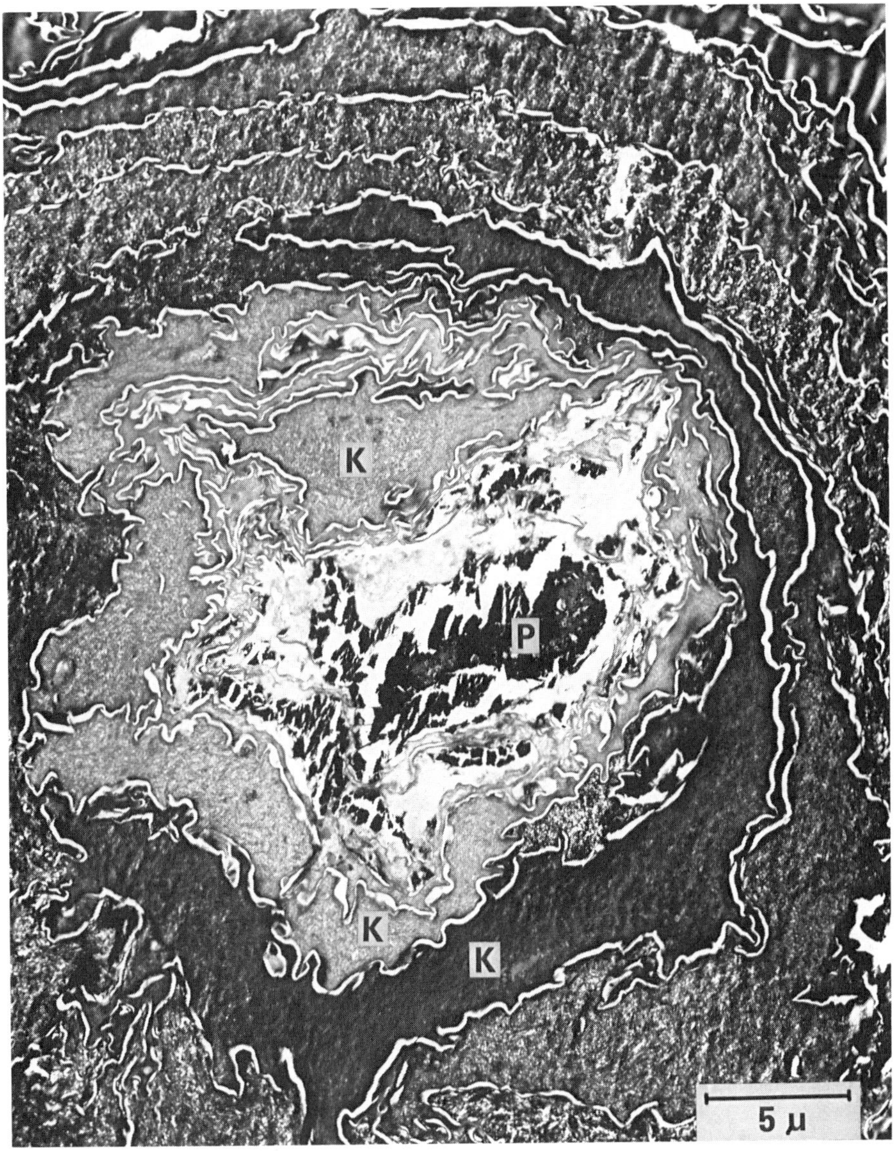

Figure 1 Transmission electron micrograph of an ACH-treated human forearm eccrine sweat gland duct at the level of the stratum corneum showing occluded lumen. P, plug; K, keratinized cells.

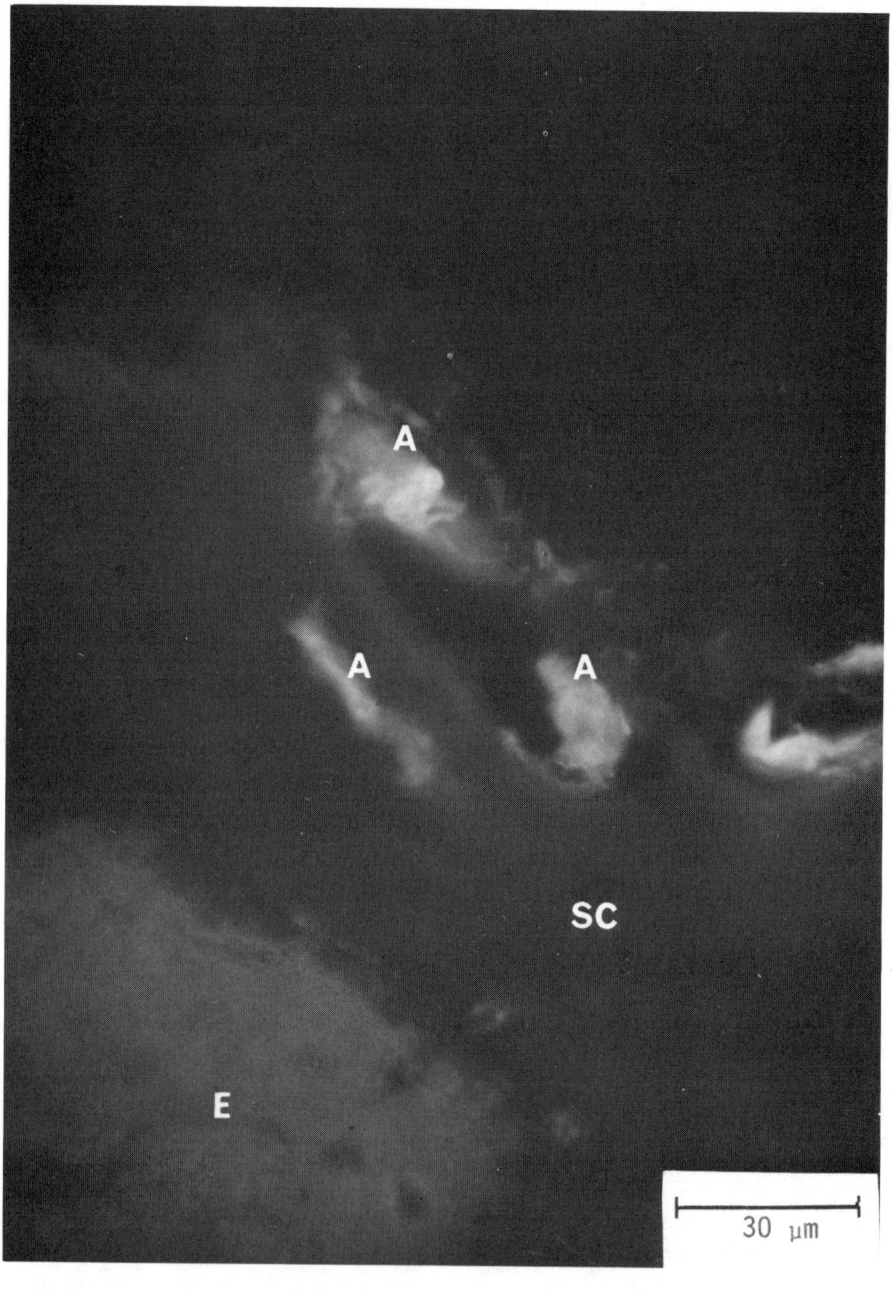

Figure 2 Serial section through an ACH-treated human forearm eccrine sweat gland depicting aluminum fluorescence in the duct at the level of the stratum corneum. E, viable epidermis; SC, stratum corneum; A, aluminum flourescence.

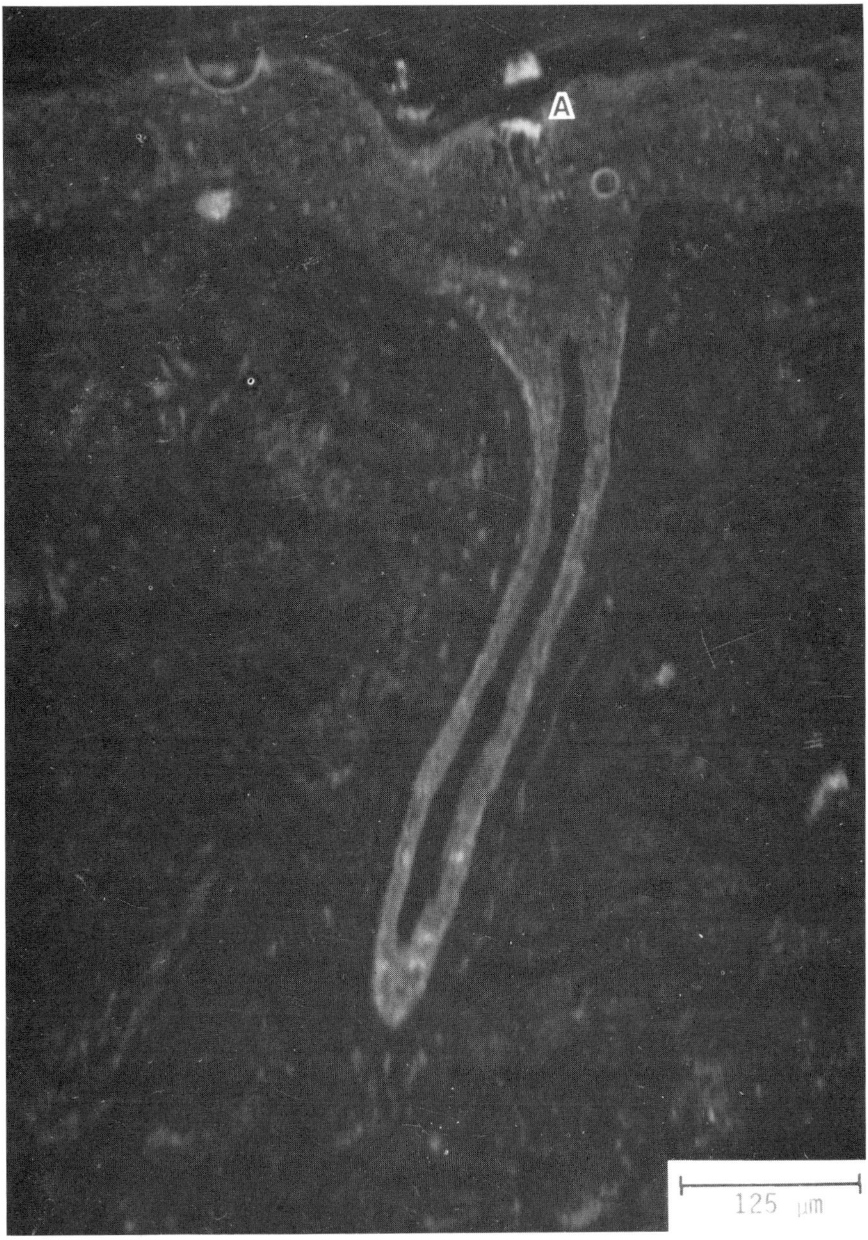

Figure 3 ACH-treated human forearm eccrine sweat gland depicting aluminum occlusion of the intracorneal and intraepidermal duct. A, aluminum flourescence.

In some instances, the fluorescence was also noted within the duct at the level of the stratum granulosum (Fig. 3). Again, these were taken as samples of some of the glands for which stripping did not restore sweating. Anatomically, the aluminum present within the duct as demonstrated by fluoresence was at the same site as the electron-dense material demonstrated with transmission electron microscopy. The mass, consisting of aluminum at least in part, is considered to be the plug responsible for the inhibition of sweating caused by ACH.

3. *Aluminum-Zirconium Complexes*

In recent years, aluminum zirconium chlorohydrate glycine complexes (AZAP) have gained in popularity as the active ingredient in antiperspirant products, principally because their sweat-inhibiting capability is demonstrably higher than that of ACH. Perhaps because of its recent appearance, even less is known about the antiperspirant action of AZAP than of ACH. Reller and Luedders (19) have shown that the effictiveness of aluminum-zirconium solutions against axillary sweating is concentration-dependent, producing an antiperspirant efficacy level ranging from about 25% for a 5% solution concentration to about 60% for a 20% concentration, whereas 20% aqueous ACH produces efficacy only to the level of about 40%. They further observed that AZAP's effectiveness was quite prolonged. From their data, a duration of antiperspirant effect as long as 78 days, after treatment had stopped, was projected. They also provided some histological evidence that AZAP's mechanism of antiperspirant action was via formation of an obstructive hydroxide gel in the lumen of the duct, and they commented that this mass was more superficially located than that caused by aluminum chloride.

More recently, Quatrale et al. (26) have shown that AZAP's site of action within the sweat gland is indeed superficial. Removal of the stratum corneum layer of forearm skin containing AZAP-inhibited glands resulted in the restoration of sweating for two-thirds of those glands, compared to restoration values of half for ACH and only 11% for aluminum chloride. Comparable tape-stripping studies in the underarm yielded a similar relationship with restoration values of 41% for AZAP-, 28% for ACH- and only 2% for $AlCl_3$-treated glands, respectively (27). Further, in histological studies of the forearm similar to those for ACH, morin-dye fluorescence indicative of the presence of aluminum and zirconium showed the antiperspirant metals to be located

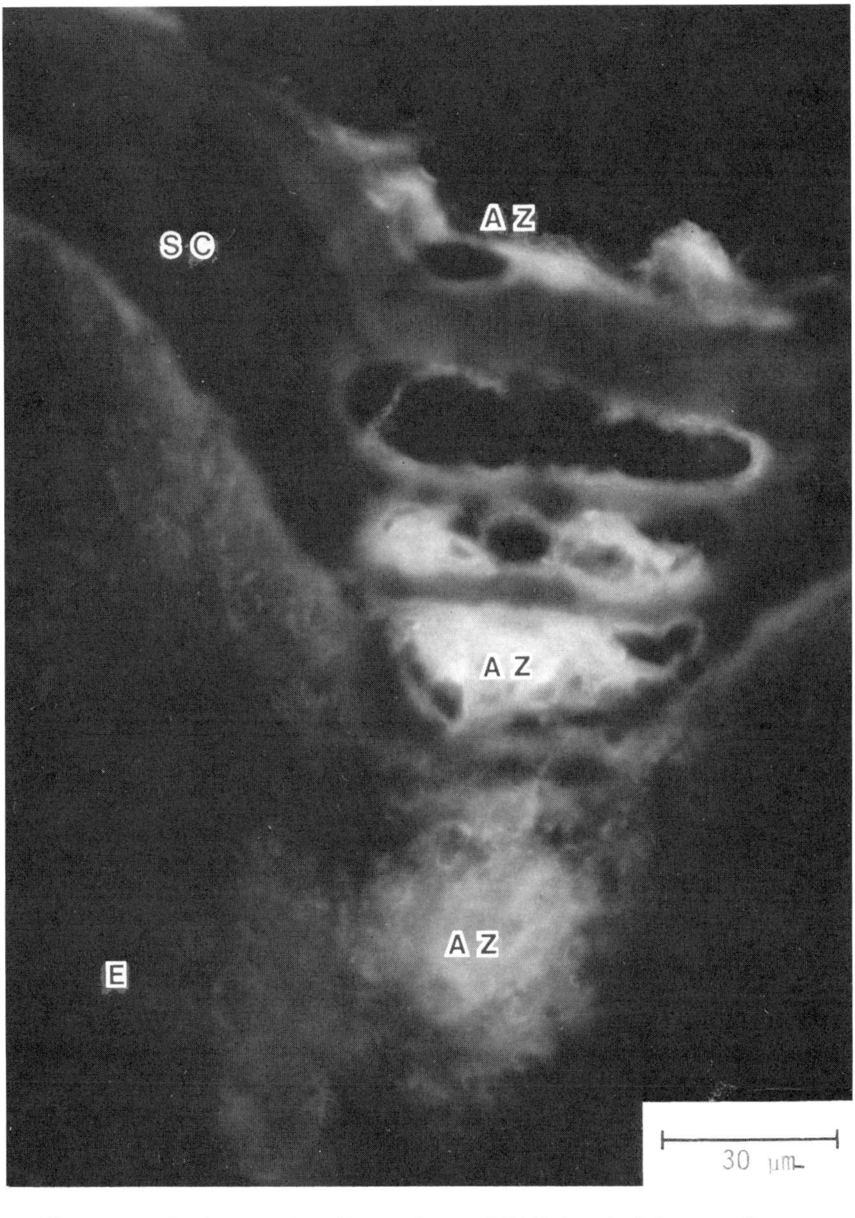

Figure 4 Serial section through an AZAP-treated human forearm eccrine sweat gland depicting aluminum and zirconium fluorescense in the duct at the levels of the stratum corneum and distal viable epidermis. SC, stratum corneum; E, viable epidermis; AZ, aluminum and zirconium fluorescence.

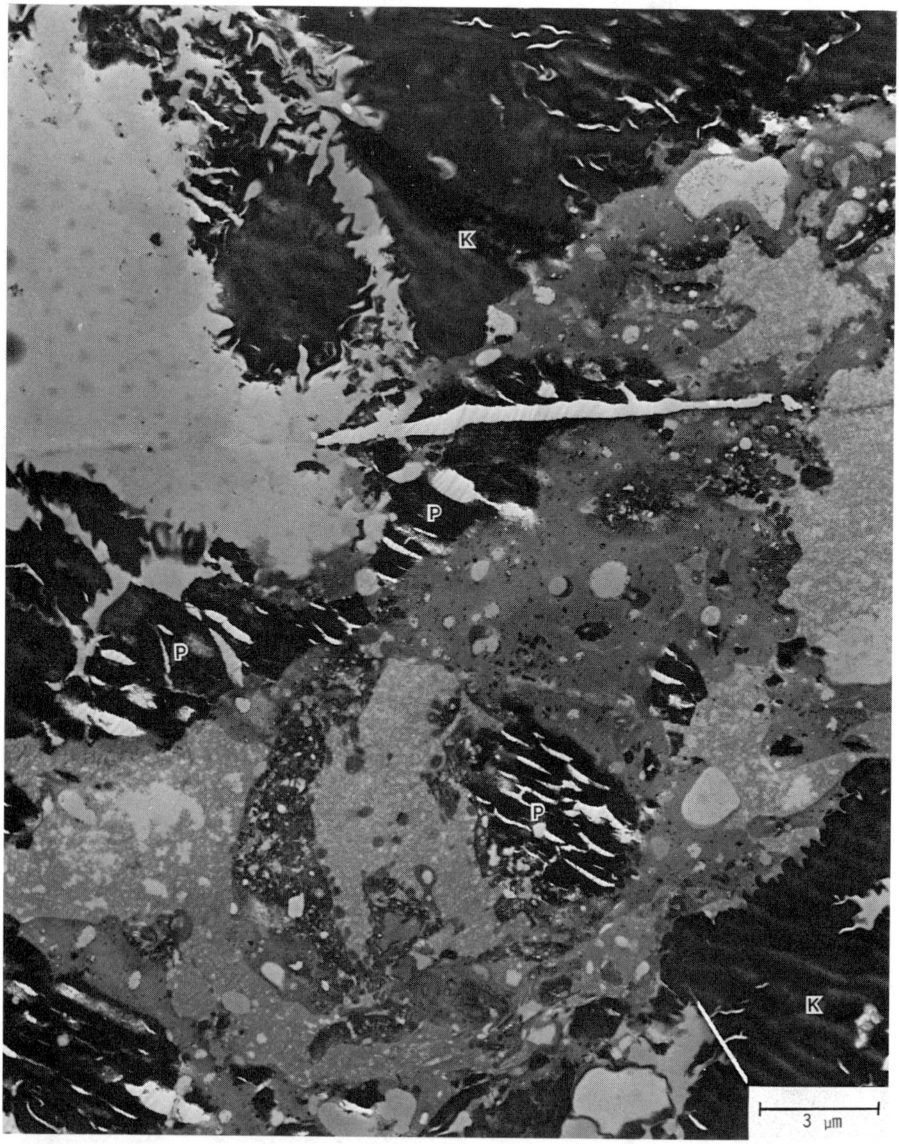

Figure 5 Cross section of an AZAP-treated human forearm eccrine sweat gland approximately 5 μm from the skin surface. Plug material is present throughout the area of the lumen. P, plug material; K, keratinized cells bordering the sweat gland ostium.

predominantly in the stratum corneum level of the duct (Fig. 4). However, fluorescence was not infrequently observed in the distal region of the intraepidermal duct. Companion transmission electron microscopy studies demonstrated the presence of an occlusive mass, amorphous, electron-dense, and charcoal-like in appearance, in the most distal region of the intracorneal duct but occasionally in the most distal region of the intraepidermal duct as well. As with ACH, the site of this material corresponded to the location of the fluorescence and the conclusion was drawn that AZAP, like ACH, acts by forming a plug which blocks delivery of sweat to the skin surface (Fig. 5) (28).

4. *Other Metal Salts*

ACH and AZAP are the two accepted aluminum salts which provide the greatest antiperspirant efficacy. Over the years, numerous other salts of aluminum have been examined for their potentially superior antiperspirant activity. Shelley and Hurley (18) have listed more than 40 salts of aluminum which have been studied; doubtless, many more have been examined but have gone unreported.

The search for metal salts which can inhibit the process of perspiration has not been limited to aluminum and zirconium. In fact, perusal of the available literature suggests that a reasonably thorough examination of Mendeleev's masterpiece has been made, ranging from the common (e.g., copper, iron, tin) to the obvious aluminum congeners (lanthanum, cerium) to the esoteric (samarium, praseodymium). Some of these, such as copper, zinc, nickel, chromium, and praseodymium, have been shown to be effective but have not found their way to the marketplace because of sensitization potential, cost, propensity to discolor skin, deleterious effect on clothing, etc., while others have simply been inactive. From a medical and aesthetic standpoint, aluminum chlorohydrate and, probably to a greater extent, aluminum zirconium chlorohydrate glycine complex are the most effective antiperspirants.

B. The Anticholinergics

By far the most effective antiperspirants known are the anticholinergic agents. Since eccrine sweat glands are innervated by cholinergic nerve fibers, and the sweat secretion process is stimulated by the release of acetylcholine at the neuroglandular junction, it is not unexpected that agents which block that

Table 2 Antiperspirant Activity of Scopolamine and Atropine Derivatives at 0.1% in Water

Very strong activity[a]	Strong activity
O-Acetylscopolamine HBr	Scopolamine MeBr (Pamine)
O-Pivalylscopolamine HBr	Scopolamine *N*-oxide
O-(2,2-Diethylacetyl)-scopolamine HBr	O-Acetylscopolamine MeBr O-(3-Cyclopentylpropionyl)-scopolamine HBr and MeBr Atropine $MeNO_3$

[a]Very strong activity means complete inhibition of sweating, spreading beyond treatment area, which lasted 24–48 hours.
Source: From MacMillan et al. (32).

release will efficiently prevent sweating. The phenomenon was first systematically studied by Shelley and Horvath in 1951 (29). The investigators reported that following topical administration of a number of anticholinergic compounds by iontophoresis, scopolamine HBr, atropine methyl nitrate, and atropine sulfate were among those more highly effective in preventing sweating. Hermann and Sulzberger (30) and Randall and Kimura (31) confirmed those early observations, and the findings have now been repeated on a number of occasions (for further references, see Ref. 19). To this point, however, it was invariably noted that although the capability of the anticholinergics to inhibit sweating was substantial, their poor skin penetration properties, i.e., their inability to reach the site of action, rendered them ineffective unless they were either injected or drive into the skin by iontophoresis. In 1964, MacMillian et al. (32) made the important observation that certain esters, particularly those of scopolamine, penetrate the skin much more readily and consequently inhibited sweating at very low concentrations. Although most esters exhibited greater activity than the unesterified form of scopolamine, the more efficacious esters were the straight and branched chains, cycloaliphatic, and aromatic forms (see Tables 2 and 3 for examples). In general, the most effective compounds were those which contained one or more elements of the scopolamine structure or the atropine structure (either tropane, a

Table 3 Antiperspirant Activity of Scopolamine Esters at 0.1% in Water

Very strong activity	Strong activity	Weak to moderate activity	Inactive
Butyrate	Acetate	Undecanoate	Oleate
Valerate	Propionate	Laurate	
Nonanoate	Hexanoate	Stearate	
Decanoate	Octanoate	Cyclopentyl-propionate	
Pivalate	Isopropylacetate		
Isovalerate	2-Ethylbutyrate		
Benzoate	2-Ethyl-3-methylbutyrate		
2,4-Dichloro-benzoate	4-Methylpentanoate Cyclohexylpropion-ate Naphthoate		

Source: From MacMillan et al. (32).

substituted phenylacetic acid related to tropic acid, or *N*-methyl-4 hydroxypiperidine). The authors further stated that the improved effectiveness of the scopolamine esters was due solely to their very high skin penetrability, rather than enhanced intrinsic activity of the modified compound itself. It was estimated that perhaps as much as 5–10% of the material applied was able to cross the horny later barrier.

From a commercial standpoint, MacMillian et al. made a second important observation. The influence of the vehicle on antiperspirant activity of the esters could be substantial. Activity was found to be greatly improved when a nonionic surfactant such as Pluronic F68 was added. Cationic surfactants such as cetyltrimethylammonium bromide also enhanced penetrability. On the other hand, anionic surfactants such as dodecylbenzene sulfonate detracted from it, probably because of the formation of an ionic complex between itself and the cationic scopolamine esters. Further, Carbowax mixtures and a cold cream base were also acceptable vehicles, but mineral oil, Solulan, and waxy fats were not. Finally, the authors reported that when as low a concentration as 0.0025% benzoylscopolamine HBr in 1% Pluronic F68 was applied to the axilla just once, sweating was reduced by 35%. Increasing the concentration of the anticholinergic to 0.05%, again using only a single application, virtually abolished axillary sweating (95% reduction).

Although scopolamine and its derivatives, and to a lesser extent atropine, have been the most extensively studied anticholinergic antiperspirants, numerous others had received early attention. Some of the more popular of these include diphemanil methylsulfate, the glycopyrrolate derivative AHR-483, poldine methylsulfate, and propantheline bromide (for references to those early studies, see Ref. 19). At present, basic research studies on the control of axillary sweating in vivo by anticholinergics are largely unreported.

It is indisputable that certain anticholinergics, probably those having the best skin penetration properties, are highly effective antiperspirants when applied topically. However, with the present state of the art, it appears unlikely that these compounds will find their way into antiperspirant product forms which are available to the general public. The reason for this is principally one of medical safety. The anticholinergics lack specificity of action for eccrine sweat glands and just as readily inhibit other functions which are under the mediation of acetylcholine. Thus, undesirable side effects such as dryness of mouth, urinary retention, and mydriasis (blurred vision) can be anticipated. Perhaps the most problematic adverse side effect is mydriasis. Although the delivered amounts of an anticholinergic may not necessarily be large enough to eventually reach demonstrable systemic levels, the potential for accidental instillation of the compound into the eyes, for example, by transfer from one's fingers, is a distinct danger, particularly to individuals who are predisposed to glaucoma. However, one interesting approach to circumventing this problem has been patented. By first converting the anticholinergic to an inactive form (in this case the β-glucuronide ester) and then reactivating it in situ via enzymatic hydrolysis (e.g., with β-glucuronidase), Felger and co-workers (33,34) claimed to have minimized untoward side effects. Finally, one anticholinergic-based antiperspirant product, Ercoril, a roll-on formulation which contains 5% propantheline bromide, has been available for over-the-counter sale in Denmark for about 10 years.

C. The Antiadrenergics

It is generally held that the innervation of eccrine sweat glands is by postganglionic sympathetic cholinergic fibers and that the principal mediator of sweating is acetylcholine released by these fibers. However, it is equally well known that catecholamines, particularly adrenaline and noradrenaline are capable of eliciting the sweating response in humans when they are injected intradermally, as first shown by Haimovici (35,36) and Barnett (37). As stated earlier, histological studies have indeed demonstrated the presence of catecholamine-containing nerve fibers, indicating the occurrence of dual innervation. However, the phenomenon of adrenergically mediated eccrine sweating is complex and at best poorly understood.

Nevertheless, with the observation established that catecholamines do evoke perspiration, several investigators have examined the possibility that antiandrenergic compounds can inhibit sweating just as their anticholinergic counterparts do. More than 30 years ago (36,37) it was noted that *N*,*N*-dibenzyl-β-chloroethylamine hydrochoride (Dibenamine) inhibited spontaneous sweating on the palms when the drug was systematically administered. The phenomenon was ascribed to an adrenergic blocking effect. The original observations for Dibenamine were confirmed and extended to show that Priscol (tolazoline), another antiandregeric compound, had a similar effect (38). Goodall (39) examined a number of antiadrenertic agents for their capability to inhibit epinephrine-induced axillary sweating and found them all to have at least some antiperspirant activity. Among the more potent were dibenzyline, reserpine, Priscol, $NaNO_3$, tetraethylammonium chloride, hexamethonium bromide, and trimethaphan camsylate (Arfonad). Finally, using isolated sweat glands from the monkey palm, Sato (40) clearly demonstrated that secretion could be stimulated by the additon of norepinephrine, but this stimulatory effect could be markedly suppressed by the antiadreneric propanol. (For a further discussion of these studies, refer to Quinton's review in Chapter 4 of this volume.)

It is likely that sweating which results from adrenergic stimulation is not as important as that which is of cholinergic origin. Thus, it is not surprising that the attention directed toward antiadrenergics as antiperspirants has been substantially less extensive and more academically oriented. Indeed, there is only one published study which is concerned with the effect of these agents on axillary sweating (39). However, since it is well known that eccrine sweat glands in the axilla respond to emotional as well as thermal stress and that adrenaline is certainly involved in emotional stress, future studies—for example, to elucidate the innervation of the axillary glands—may lead to increased interest in this class of compounds.

D. The Aldehydes

In 1936, Ichihaski (41) reported that the iontophoretic application of 5 or 10% formaldehyde solutions to various areas of the human body, including the palms of the hands, the soles of the feet, and the axillae, resulted in prolonged suppression of sweating. He further observed that when formaldehyde was similarly applied to the eccrine sweat gland-containing footpads of the cat, histological examination showed that those glands had many cells which were shrunken or flattened but not destroyed. Ichihaski concluded that formaldehyde acted selectively on the sweat glands, causing inhibition of sweating by direct action against them. Since then, aldehydes, particularly formaldehyde and glutaraldehyde, have enjoyed the periodic attention of students of antiperspirancy. In 1966, Papa and Kligman (42) observed that the occlusive application of 10% formaldehyde to the forearm was highly effective in reducing sweating. They reported that the effect lasted from a few days up to a few weeks (as Ichihaski had also noted). The investigators further showed that normal sweating could be restored if the stratum corneum layer of skin was removed by stripping with plastic tape. This finding indicated a blockage of delivery of sweat to the skin surface which was located in the outermost layers of the epidermis. Gordon and Maibach (43) obtained similar results when the formaldehyde was applied using multiple topical doses. The repeated applications apparently resulted in deeper penetration of the material, however, since tape-stripping away the stratum corneum did not completely remove the anhidrosis. Zahejsky and Rovensky (44) found that the antisudorific activity of formaldehyde on the forearm was not readily removable by tape stripping, but their method of application was relatively aggressive.

Most investigators have postulated that the mechanism by which formaldehyde (or aldehydes in general) causes suppression of sweating is through a modification of the sweat duct orifice, perhaps by a denaturing of the proteins of the stratum corneum which leads to a mechanical obstruction of the pore itself. In support of that view, Papa (45) provided scanning electron micrographs which show formaldehyde-treated sweat duct pores obstructed by amorphous noncellular plug material.

Despite its effectiveness in preventing perspiration, formaldehyde enjoys no favor as a commercial antiperspirant, at least in the United States. The drawback of its potent sensitizing capability has restricted its use largely to laboratory investigations. On the other hand, glutaraldehyde has seen at least limited clinical usage in that it effectively controls palmar and plantar hyperhidrosis. Its mechanism of action, like that of formaldehyde, is believed to be via blockage of the sweat pore. Beyond its sensitizing property, however, glutaraldehyde stains the skin brownish-yellow (43,46,47).

The ability of two other aldehydes to inhibit sweating was examined by Hunziker et al. (48). Acetaldehyde was ineffective, whereas propionaldehyde was a properspirant.

E. Metabolic Inhibitors

As described above, the process of sweat secretion (and reabsorption) is dependent on a supply of energy and probably uses the same biochemical pathways as other secretory organs for that energy production. Consequently, it is to be expected that metabolic inhibitors can interfere with the biochemistry of sweat gland function and can cause antiperspirancy. One of the better-known metabolic inhibitors, ouabain, whose effects have been studied in innumerable biological systems, has been examined for its action in sweat glands as well.

Ouabain is an inhibitor of Na^+/K^+-ATPase, the enzyme responsible for energy production required to transport sodium and potassium (and consequently water) across cell membranes during secretion. The antimetabolite has been shown to inhibit a variety of secretory processes. Emrich and Ulrich (49) injected ouabain into human fingertips and examined its effect on sweating. They found that shifts in the concentrations of various sweat constituents resulted. Most notably, there was an increase in sodium concentration, which they ascribed to the ouabain inhibiting the ductal mechanism responsible for reabsorbing the cation. They found no change in the glands' ability to secrete sweat, however. The studies of Sato and co-workers (50–52), on the other hand, convincingly demonstrated that ouabain inhibits the secretion of sweat in a variety of systems both in vitro and in vivo. The antimetabolite was shown to act by inhibiting Na^+/K^+-ATPase either in the secretory coil region of the gland to prevent the secretion of sweat or in the resorptive duct region, depending on its concentration.

Carbonic anhydrase is another enzyme involved in the secretion of electrolytes. Acetazolamide (Diamox) is an antimetabolite known to interfere specifically with this enzyme's activity, and accordingly it is reasonable to suggest that it can inhibit sweating. However, whereas Emrich and Ulrich (49) did demonstrate the antiperspirant capability of Diamox, Richterich and Friolet (53) did not.

Because of the process of sweating, like all physiological processes, is energy-dependent, one might logically expect that any agent which can interfere with energy production would be an antiperspirant. The classical inhibitors of the Embden-Meyerhof pathway and the Krebs cycle, such as cyanide, fluoride, sodium azide, malonic acid, and dinitrophenol, are all probably "physiological" antiperspirants, but for obvious reasons, their use will remain of academic interest only.

III. MODELS FOR THE STUDY OF ANTIPERSPIRANT ACTIVITY

A. In Vitro Systems

To study the processes and factors involved in eccrine sweat gland function and to search for and evaluate agents which can safely but more effectively inhibit human axillary sweating, model systems are highly desirable. They not only provide the opportunity to perform experiments which are not feasible or practical to conduct in humans, but also offer the advantage of screen-testing potential new antiperspirants for efficacy more rapidly and less expensively. To that end, a variety of in vitro systems have been described. For perspective, two of them are mentioned here. Christian and co-workers (54–56) developed an in vitro model using the frog skin permeability system in which they could study the effect of aluminum salts and other astringents. They were able to relate the changes in permeability of the frog skin, as measured by increase in diffusion rates, to the increasing concentrations of the salts studied. However, as additional understanding of the mechanism of antiperspirant action by proprietary aluminum salts such as ACH and AZAP is gained, the usefulness of models based on permeability changes may diminish. More recently, Reller and Luedders (19) described a Millipore filter system which they used to measure flow obstruction caused by the neutralization of various metal salts. They then compared their in vitro findings with those obtained for antiperspirant efficacy of the salts which had been tested in humans and found reasonable correlations.

B. Animal Models

Although such in vitro systems have their value, an in vivo animal model system is much preferred in that the investigator is now examining the effect of potential commercial antiperspirants on actual functioning sweat glands. As mentioned earlier, a variety of animals are known to have eccrine sweat glands and several of them have been proposed as models for the evaluation of prototype antiperspirant products. Kaszynski and Frisch (57) described a simple, rapid, inexpensive, quantitative screening procedure to determine the degree of sweat inhibition by anticholinergic agents using the eccrine glands in the footpads of mice. The presence or absence of sweating after injection of the injection of the test material was noted using the starch/castor oil/iodine colorimetric procedure originally developed by Wada and Takagaki (58). The parameters of sweating by the eccrine glands in the footpads of rats and the methods for collecting that sweat to measure its volume have also been described (59–62). With this technique, a quantitative determination of antiperspirant effects after either topical application or intradermal injection of test materials has been made. Lansdown (63) also described the use of the rat as a model for studying the sweat-inhibiting effects of aluminum salts and various other agents, but he did not employ quantitative means of measuring those effects. Although the footpads of cats have also been used by some investigators (64,65), this approach has disadvantages. As larger animals, cats are more difficult to work with, and the accumulation of sufficient data proceeds much more slowly. The same remarks pertain to an even greater extent if one considers using the palms of monkeys.

Alternatively, in recent years sophisticated techniques for studying isolated eccrine sweat glands in vitro have been developed by Sato (66) and further refined by Quinton and Tormey (67) and Vitale et al. (68). The application of this study system to the search for the subtleties of antiperspirant mechanisms can be a most rewarding spproach.

Nevertheless, given the specificity of function of the eccrine sweat glands of humans, prototype antiperspirants must of course ultimately be evaluated in the human axilla, with no guarantee that positive results found in models will be duplicated therein. Indeed, the uniqueness of the human axillary organ can even make direct extrapolation of test results obtained using the human forearm or the back tentative. In the final analysis, although data derived from model systems can usually be regarded as indicative and directional, the dimensions of antiperspirant efficacy can be determined with certainty only using the human underarm.

REFERENCES

1. Lobitz, W. C., Jr., Holyoke, J. B., and Montagna, W. B. The epidermal eccrine sweat duct unit. A morphologic and biologic entity, *J. Invest. Dermatol., 22:* 157–158 (1954).
2. Sato, K. Pharmacology and function of the myoepithelial cell in the eccrine sweat gland, *Experientia, 33:* 631–633 (1977).
3. Dobson, R. L. The effects of repeated episodes of profuse sweating in the human eccrine glands, *J. Invest. Dermatol., 35:* 195–198 (1960).
4. Robinson, S., and Robinson, A. H. Chemical composition of sweat, *Physiol. Rev., 34:* 202–220 (1954).
5. Roberts, D. F., Salzano, F. M., and Wilson, J. O. C. Active sweat gland distribution in Caingang indians, *Am. J. Phys. Anthropol., 32:* 395–400 (1970).
6. Szabo, G. The number of eccrine sweat glands in human skin, *Adv. Biol. Skin, 3:* 1–5 (1962).
7. Kuno, Y. *Human Perspiration,* Charles C. Thomas, Springfield, Illinois (1956).
8. Ikeuchi, K., and Kuno, Y. On the regional differences of the perspiration on the surface of the human body, *J. Orient. Med., 7:* 67–89 (1927).
9. Rothman, S. *Physiology and Biochemistry of the Skin,* University of Chicago Press, Chicago (1954).
10. Rebell, G. and Kirk, D. Patterns of eccrine sweating in the human axilla, *Adv. Biol. Skin, 3:* 108–126 (1962).
11. Quatrale, R. P., Stoner, K. L., and Felger, C. B. A method for the study of emotional sweating, *J. Soc. Cosmet. Chem., 28:* 91–101 (1977).

12. Stillians, A. W. The control of localized hyperhidrosis, *J. Am. Med. Assoc., 67:* 2015 (1916).
13. Sulzberger, M. B., Hermann, F., Keller, R., and Pisha, V. B. Studies of sweating. III. Experimental factors influencing the function of the sweat ducts, *J. Invest. Dermatol., 14:* 91–111 (1950).
14. Shelley, W. B., and Horvath, P. N. Experimental miliaria in man. II. Production of sweat retention anhidrosis and miliaria crystallina by various kinds of injury, *J. Invest. Dermatol., 1:* 9–20 (1950).
15. Shelley, W. B., Horvath, P. N., and Pillsbury, D. M. Anhidrosis. An etiologic interpretation, *Medicine (Baltimore), 29:* 195–224 (1950).
16. Kligman, A. M., and Holzle, E., Factors influencing the antiperspirant action of aluminum salts, *J. Soc. Cosmet. Chem., 30:* 357–367 (1979).
17. Papa, C. M., and Kligman, A. M. Mechanisms of eccrine anhidrosis. II. The antiperspirant effect of aluminum salts, *J. Invest. Dermatol., 49:* 139–145 (1967).
18. Shelley, W. B., and Hurley, H. J., Jr., Studies on topical antiperspirant control of axillary hyperhidrosis, *Acta Dermato.-Venereol., 55:* 241–260 (1975).
19. Reller, H. H., and Luedders, W. L. Pharmacologic and toxicologic effects of topically applied agents on the eccrine sweat glands, *Mod. Toxicol., 4:* 1–54 (1977).
20. Kligman, A. M., and Holzle, E. Mechanism of antiperspirant action of aluminum salts, *J. Soc. Cosmet. Chem., 30:* 279–295 (1979).
21. Quatrale, R. P., Stoner, K. L., Coble, D. W., and Felger, C. B. Unpublished observations.
22. Govette, T., deNavarre, M. G. Aluminum chlorohydrate, new antiperspirant ingredient, *Am. Perfum. Essential Oil Rev., 49:* 365–368 (1947).
23. Carson, H. Deodorants/antiperspirants, *HAPPI, 18:* 33–48 (1981).
24. Quatrale, R. P., Coble, D. W., Stoner, K. L., and Felger, C. B. The mechanism of antiperspirant action by aluminum salts. II. Histological observations of human eccrine sweat glands inhibited by aluminum chlorohydrate, *J. Soc. Cosmet. Chem., 32:* 107–136 (1981).
25. Gordon, B. I., and Maibach, H. I. Studies on the mechanism of aluminum anhidrosis, *J. Invest. Dermatol., 50:* 411–413 (1968).

26. Quatrale, R. P., Waldman, A. H., Rogers, J. G., and Felger, C. B. The mechanism of antiperspirant action by aluminum salts. I. The effect of cellophane tape stripping on aluminum salt-inhibited eccrine sweat glands, *J. Soc. Cosmet. Chem., 32:* 67–73 (1981).
27. Quatrale, R. P., Thomas, E. L., and Birnbaum, J. E. The site of antiperspirant action by aluminum salts in the eccrine sweat glands of the axilla, *J. Soc. Cosmet. Chem., 36:* 435–440 (1985).
28. Quatrale, R. P., Coble, D. W., Stoner, K. L., and Felger, C. B. The mechanism of antiperspirant action of aluminum salts. III. Histological observations of human eccrine sweat glands inhibited by aluminum zirconium chlorohydrate glycine complex, *J. Soc. Cosmet. Chem., 32:* 195–222 (1981).
29. Shelley, W. B., and Horvath, P. N. Comparative study of the effect of anticholinergic compounds on sweating, *J. Invest. Dermatol., 16:* 267–274 (1951).
30. Hermann, F., and Sulzberger, M. B. Some aspects of therapy of sweat disturbances, *Arch. Dermatol., 66:* 162–179 (1952).
31. Randall, W. C., and Kimura, K. K. The pharmacology of sweating, *Pharmacol. Rev., 7:* 365–397 (1955).
32. MacMillan, F. S. K., Reller, H. H., and Snyder, F. H. The antiperspirant action of topically applied anticholinergics, *J. Invest. Dermatol., 43:* 363–377 (1964).
33. Felger, C. B. Anticholinergic glucuronide compounds and antiperspirant use thereof, U.S. Patent 4,517,176 (1985).
34. Herlihy, W. C., Epstein, D. M., and Felger, C. B. Anticholinergic glucuronides and antiperspirant use thereof, U.S. Patent, 4,546,096 (1985).
35. Haimovici, H. Evidence for an adrenergic component in the nervous mechanism of sweating in man, *Proc. Soc. Exp. Biol. Med., 68:* 40–41 (1948).
36. Haimovici, H. Evidence for adrenergic sweating in man, *J. Appl. Physiol., 2:* 512–521 (1950).
37. Barnett, A. J. Sweating in man from the intradermal injection of noradrenalin, *Nature (London), 167:* 482–483 (1951).
38. Chalmers, I. M., and Keele, C. A. The nervous and chemical control of sweating, *Br. J. Dermatol., 64:* 43–54 (1952).
39. Goodall, M. C. Innervation and inhibition of eccrine and apocrine sweating in man, *J. Clin. Pharmacol., 10:* 235–246 (1970).

40. Sato, K. Stimulation of pentose cycle in the eccrine sweat gland by adrenergic drugs, *Am. J. Physiol., 224:* 1149–1154 (1973).
41. Ichihaski, T. Effect of drugs in the sweat glands by cataphoresis and effective method for suppression of local sweating, *J. Orient. Med., 25:* 103–104 (1936).
42. Papa, C. M., and Kligman, A. M. Mechanism of eccrine anhidrosis. I. High level blockade, *J. Invest. Dermatol., 47:* 1–9 (1966).
43. Gordon, B. I., and Maibach, H. I. Eccrine anhidrosis due to glutaraldehyde, formaldehyde, and iontophoresis, *J. Invest. Dermatol. 53:* 436–439 (1969).
44. Zahejsky, J., and Rovensky, J. A comparison of the effectiveness of several external antiperspirants, *J. Soc. Cosmet. Chem., 23:* 775–789 (1972).
45. Papa, C. M. Mechanism of eccrine anhidrosis. III. Scanning electron microscopic study of poral occlusion, *J. Invest. Dermatol., 59:* 295–298 (1972).
46. Sato, K., and Dobson, R. L. Mechanism of antiperspirant effect of topical glutaraldehyde, *Arch. Dermatol., 100:* 564–569 (1969).
47. Juhlin, L., and Hansson, H. Topical glutaraldehyde for plantar hyperhidrosis, *Arch. Dermatol., 97:* 327–330 (1968).
48. Hunziker, N., Brun, R., Vidmar, B., and Laugier, P. Zur Wirkung der lokalen Antihydrotika, *Hautarzt, 24:* 301–304 (1973).
49. Emrich, H. M., and Ulrich, K. J. Ausscheidung verschiedener Stoffe im Schweiss in Abhangigkeit von der Schweissflussrate, *Pfluegers Arch., 290:* 298–310 (1966).
50. Sato, K., Taylor, J. R., and Dobson, R. L. The effect of ouabain on eccrine sweat gland function, *J. Invest. Dermatol., 53:* 275–282 (1969).
51. Sato, K., and Dobson, R. L. Glucose metabolism of the isolated eccrine sweat gland. I. The effect of mecholyl, epinephrine and ouabain, *J. Invest. Dermatol., 56:* 272–280 (1971).
52. Sato, K., Dobson, R. L., and Mali, J. W. H. Enzymatic basis for the active transport of sodium in the eccrine sweat gland. Localization and characterization of Na-K-ATPase, *J. Invest. Dermatol., 57:* 10–16 (1971).
53. Richterich, R., and Friolet, B. The effect of acetazolamide on sweat electrolytes in mucoviscidosis, *Metabolism, 12:* 1112–1121 (1963).

54. Lux, R. E., and Christian, J. E. Permeability of frog skin by means of radioactive tracers, *Am. J. Physiol. (London), 162:* 193–197 (1950).
55. Lux, R. E. and Christian, J. E. A study of the effect of certain astringents on the permeability of frog skin using radioactive tracer techniques, *J. Am. Pharm. Assoc., 40:* 160–162 (1951).
56. Urakami, C., and Christian, J. E. An evaluation of the effectiveness of antiperspirant preparations using frog skin membrane and radioactive tracer techniques, *J. Am. Pharm. Assoc., 42:* 179–180 (1953).
57. Kaszynski, E., and Frisch, S. B. Mouse foot screen for the inhibition of sweating by anticholinergic drugs, *J. Invest. Dermatol., 62:* 510–513 (1974).
58. Wada, M., and Takagaki, T., A simple and accurate method for detecting the secretion of sweat, *Tohoku J. Exp. Med., 49:* 284 (1948).
59. Quatrale, R. P., and Laden, K. Solute and water excretion by the eccrine sweat glands of the rat, *J. Invest. Dermatol., 51:* 502–504 (1968).
60. Quatrale, R. P., and Speir, E. H. The effect of ADH on eccrine sweating in the rat, *J. Invest. Dermatol., 55:* 344–349 (1970).
61. Brusilow, S. W., and Gordes, E. H. Ammonia secretion in sweat, *Am. J. Physiol., 214:* 513–517 (1968).
62. Bruislow, S. W., Ikai, K., and Gordes, E. H. Comparative physiological aspects of solute secretion by the eccrine sweat gland of the rat, *Proc. Soc. Exp. Biol. Med., 129:* 731–732 (1968).
63. Lansdown, A. B. G The rat foot pad as a model for examining antiperspirants, *J. Soc. Cosmet. Chem., 24:* 677–684 (1973).
64. Alphin, R. S., Saunders, D., and Ward, J. W. Method for the evaluation of anhidrotic substances in the anesthetized cat, *J. Pharm. Sci., 56:* 449–452 (1967).
65. Alphin, R. S., Vocac, J. A., Saunders, D., and Ward, J. W. Effects of some lignosulfonates on sweat gland activity, *J. Pharm. Sci., 58:* 902–903 (1969).
66. Sato, K. Sweat induction from an isolated eccrine sweat gland, *Am. J. Physiol., 225:* 1147–1151 (1973).
67. Quinton, P. M., and Tormey, J. M. Localization of Na-K-ATP-ase sites in the secretory and reabsorptive epithelia of perfused eccrine sweat glands. A question to the role of

the enzyme in secretion, *J. Membr. Biol.*, *29:* 383–399 (1976).
68. Vitale, G. I., Quatrale, R. P., Giles, P. J., and Birnbaum, J. E. Electrical field stimulation of isolated primate sweat glands, *Br. J. Dermatol.*, in press.

6

Chemistry of Basic Commercial Aluminum Hydrolysis Complexes

John J. Fitzgerald *South Dakota School of Mines and Technology Rapid City, South Dakota*

I. INTRODUCTION

The history of commercial utilization of aluminum compounds as antiperspirants extends from the early application of a simple aluminum salt, $AlCl_3 \cdot 6H_2O$, in 1902 through the development of the basic aluminum salts such as aluminum chlorohydrate, $Al_2(OH)_5Cl$, in the 1940s (Huehn and Haufe, 1940; Govett and deNavarre, 1947) to the introduction of various aluminum/zirconium complexes in the 1960s (Daley, 1957b; Grad, 1958). Most recently, a spectrum of activated basic aluminum systems (Gosling et al. 1978; Fitzgerald, 1981; Nelson, 1985; Callaghan and Phipps, 1985) has been developed. The number of unique water-soluble basic aluminum hydrolysis complexes of pharmaceutical utility has grown enormously since the original patent (Huehn and Haufe, 1940) on the redox synthesis of basic aluminum complexes. Present-day basic aluminum hydroxide complexes include aluminum dichlorohydrate (ADCH), aluminum sesquichlorohydrate (ASCH), and aluminum chlorohydrate (ACH), the major complex salt produced commercially from the 1940 patent of Huehn and Haufe. In addition to the basic aluminum chloride (BAC) salt complexes, aluminum hydrolysis complexes containing sulfate, bromide, iodide, and nitrate may be synthesized by this process. The addition of various other chemical entities to ACH, including $AlCl_3$ solutions, chelating and buffering agents such as glycine, urea, and phenolsulfonate, alcohol solubilizing agents such as propylene glycol (Jones and Rubino, 1969), and other hydrolyzed metal salt solutions such as basic zirconium complexes (Daley, 1957a, 1957b; Grad, 1958), is now part of standard procedures for preparing a wide variety of different basic aluminum antiperspirant systems. Furthermore, alterations in the synthesis conditions, starting materials, and solution heating processes as well as various solubilization, spray-drying, and freeze-drying techniques have been used by various workers and manufacturers to modify the chemical and biological properties of these systems.

The chemistry of aluminum antiperspirant (A/P) salts in the solution and the solid state has been the subject of intense scientific, product development, and manufacturing interest for well over 20 years. While reports of the general chemical and physical characteristics of antiperspirant salts have appeared in patents, manufacturing bulletins, and a limited number of technical publications and presentations, the results of intense and detailed research studies by various industrial and academic research groups have led to a depth of knowledge of aluminum antiperspirant chemistry which has gone largely unreported in the scientific literature. This chapter represents the first major scientific summary describing the current state of understanding of aluminum antiperspirant systems in both the solution and the solid state.

The almost unlimited solubility of all major classes of solid aluminum antiperspirant salts in aqueous solution, the relevance of the solution state to their pharmacologic activity, and the significant number of chemical characterization approaches available to examine aqueous aluminum hydrolysis chemistry have led investigators to emphasize research studies of aqueous solutions of aluminum antiperspirant salts. While the solid-state properties of aluminum A/P salts are significant in the manufacturing, formulation, and stability of commercial products, the majority of solid-state chemical characterization techniques have been of limited value in aiding our understanding of the synthetic and structural chemistry of aluminum hydrolysis complexes contained in aluminum A/P systems. This chapter will, therefore, focus primarily on the chemical behavior of aluminum A/P systems in aqueous solution. Solid-state investigations of aluminum A/P salts will be considered where investigative results show relevance in supplementing our understanding of the body of knowledge now available on their aqueous chemistry.

The review of aluminum antiperspirant chemistry in this chapter will cover six primary areas:

1. The major classes of aluminum compounds, their nomenclature, and their chemical compositions
2. The hydrolysis chemistry of Al(III) ion in dilute aqueous solution

3. Approaches to examining concentrated aluminum hydrolysis systems
4. The solution chemistry of basic aluminum halides, particularily basic aluminum chloride systems such as ACH
5. Solid-state chemistry of basic aluminum halides, especially the basic aluminum chlorides
6. The solution chemistry of basic aluminum/zirconium systems

In addition, a brief discussion of other aluminum- and non-aluminum-containing systems consisting of hydrolyzable metal ions other than zirconium(IV) ion will be given.

This chapter will emphasize chemical investigations of aqueous aluminum systems which are presently well defined and of major commercial utility. Our understanding of the chemical nature of these systems as derived from specific physical-chemical techniques and approaches will be reviewed in depth. A major goal of this chapter will be to describe experimental studies which have provided concrete, interpretable information about the structure, macromolecular nature, and stability of aluminum hydrolysis species contained in a wide range of aluminum A/P salts over a broad concentration range. It is anticipated that this chapter will enable chemists to understand and appreciate the complexity of the various aluminum A/P salts and will stimulate further research on their chemistry and the chemistry of related hydrolysis systems.

II. CLASSES OF ALUMINUM ANTIPERSPIRANT SALTS

A unique group of inorganic salts of aluminum(III) ion are the basis for all commercial antiperspirant metal salts utilized today. The majority of these aluminum salt systems are contained in the tentative final FDA-OTC Antiperspirant Monograph (1982) summary of antiperspirant active ingredients. The four major classes of aluminum antiperspirant salts are

1. Simple aluminum salts
2. Basic aluminum salts, including aluminum chlorohydrates or BAC salts
3. Basic aluminum/zirconium salts
4. Miscellaneous complex basic aluminum salts

Table 1 summarizes a range of compounds contained in these four classes of aluminum antiperspirant salts. This table notes their general chemical formulas and the FDA-OTC classification (categories I, II, and III) for these salts as contained in either nonaerosol or aerosol antiperspirant formulations.

Table 1 Classes of Aluminum Salts

Type of aluminum salt	FDA-OTC classification	
	Nonaerosol	Aerosol
Simple aluminum salts		
$AlCl_3 \cdot 6H_2O$ (15% aq.)	I	III
$AlCl_3 \cdot 6H_2O$ (alcohol)	II	II
$Al_2(SO_4)_3 \cdot 18H_2O$	III	III
$Al_2(SO_4)_3$ (buffered)	I	III
$KAl(SO_4)_2 \cdot 12H_2O$ (potassium alum)	III	III
Basic aluminum salts		
$[Al_2(OH)_m(X)_n]$, where $m + n = 6$		
Aluminum chlorohydrates ($X = Cl^-$)	I	I
Aluminum bromohydrates ($X = Br^-$)	II	II
Aluminum iodohydrates ($X = I^-$)	—[a]	—[a]
Aluminum nitrohydrates	—[a]	—[a]
Basic aluminum/zirconium salts		
Aluminum/zirconium chlorohydrates (without glycine)	I	II
Aluminum/zirconium chlorohydrexes (with glycine)	I	II
Miscellaneous basic aluminum salts (complexed)		
Aluminum chlorohydrexes—PG and PEG[b]	I	I
Sodium aluminum chlorohydrate lactate $[(NaAl(OH)_m(lactate)_n]$	III	III
Aluminum phenolsulfonatohydrate	—[a]	—[a]

[a]Not classified by FDA-OTC Monograph (1982).
[b]PG and PEG refer to propylene glycol and polyethylene glycol, respectively.

A. Simple Aluminum Salts

The simple aluminum salts include aluminum chloride hexahydrate, $AlCl_3 \cdot 6H_2O$, aluminum sulfate, $Al_2(SO_4)_3$, which is soluble to ca. 50% weight in water but insoluble in alcohols, and aluminum sulfate octadecahydrate, $Al_2(SO_4)_3 \cdot 18H_2O$, also called alum or potassium alum (solubility of 12% weight in water). Dissolution of aluminum chloride hexahydrate in water at 1.0 M Al produces an acidic solution which contains the aluminum hexaaquo species, $Al(H_2O)_6{}^{3+}$ ion, whereas dissolution of the various aluminum sulfate compounds produces aqueous solutions containing this species and a range of coordinated sulfate complexes (Stryker and Matijevic, 1969; Akitt, 1972,1985a) depending on concentration and the original salt used.

B. Basic Aluminum Halides

The aluminum salts termed basic aluminum halides have the general empirical formulas $Al_2(OH)_m(X)_n$, where $m + n = 6$. The dominant series of commercial antiperspirants is the aluminum chlorohydrate series, $Al_2(OH)_m(Cl)_n$, which includes three widely used compounds, aluminum chlorohydrate, $Al_2(OH)_5Cl$, also referred to as "5/6 basic," aluminum sesquichlorohydrate or "3/4 basic," $Al_2(OH)_{4.5}Cl_{1.5}$, and aluminum dichlorohydrate or "2/3 basic," $Al_2(OH)_4Cl_2$. The BAC salts are very complex systems in aqueous solution, containing varying compositions of polymeric, oligomeric, dimeric, and monomeric species, and will be described in detail in this chapter. Similar series of partially hydrolyzed basic aluminum systems exist where the chloride ion is replaced by bromide, iodide, or nitrate, e.g., $Al_2(OH)_5Br$ or $Al_2(OH)_5NO_3$. None of these systems has been employed in commercial use. The basic aluminum bromide, aluminum bromohydrate, has been produced commercially and assessed for antiperspirant activity (Bretschneider et al., 1975).

C. Basic Aluminum/Zirconium Systems

Basic aluminum/zirconium salts represent extremely complex antiperspirant salt systems and are usually produced by reactions of aluminum chlorohydrate with various hydrolyzed basic zirconium chlorohydrates such as zirconium oxydichlorohydrate or zirconyl chloride, $ZrOCl_2$, and zirconium oxyhydroxychlorohydrate or zirconyl hydroxychloride, $ZrO(OH)Cl$. The synthesis of commercial mixed aluminum/zirconium chlorohydrates is usually carried out using varying amounts of glycine (Daley, 1957b; Grad, 1958), the resulting complexes being referred to as zirconium/aluminum/glycine (ZAG) complexes or aluminum/zirconium chlorohydrexes. The FDA-OTC monograph (1982) identifies specific Al/Cl and Al/Zr mole ratios to designate or describe a specific salt system. For example, the ZAG salt of the composition $2Al_2(OH)_5Cl \cdot ZrO(OH)Cl \cdot$glycine has a 4:1 Al/Zr ratio, a 4:3 or 1.5:1 Al/Cl ratio, and a 1:1 Zr/glycine ratio. This salt is referred to as an aluminum/zirconium tetrachlorohydrex (Al/Zr = 2.0–6.0, and Al/Cl = 1.5–0.9). The zirconium/glycine mole ratios are not pertinent to the designation of the aluminum/zirconium chlorohydrexes but do influence their solution chemistry.

D. Basic Complexed Aluminum Salts

Finally, a variety of miscellaneous basic complexed aluminum salts such as aluminum chlorohydrex PG containing aluminum chlorohydrate complexed with propylene glycol (PG), aluminum chlorohydrex PEG containing aluminum chlorohydrate complexed with polypropylene glycol (PEG), and sodium aluminum chlorohydroxy lactate have been patented (Jones, 1969, 1975) as antiperspirants and received an OTC designation. These systems will not be examined further in this chapter because of the lack of reported investigations of them.

III. HYDROLYSIS CHEMISTRY OF Al(III) ION IN SOLUTION

The chemistry of aluminum(III) ion in aqueous solution, particularly its hydrolysis reactions, has been extensively investigated for over 30 years by virtually every applicable physical-chemical technique. An understanding of Al(III) ion hydrolysis reactions in dilute solution and in the presence of simple anions such as Cl^-, SO_4^{2-}, PO_4^{3-}, and OAc^- ions is of importance in its own right in order to more fully explain the geological, mineralogical and environmental behavior of Al(III) ion in natural water systems (Hem and Robinson, 1967; Hem 1968a). From the viewpoint of understanding and investigating the aqueous chemistry of the more complex and highly concentrated aqueous aluminum A/P salt systems, the chemistry of Al(III) ion in dilute solutions provides an important reference point to 1. establish the applicability of various techniques to study a wide range of these hydrolysis systems in general, 2. gain insight into the fundamental hydrolysis reactions in the acidic, neutral, and basic pH regions, and 3. provide evidence for the existence and structures of various stable hydrolysis species which may best describe a given hydrolysis system of interest.

A. pH Region of Al(III) Hydrolysis

The dilute aqueous chemistry of trivalent aluminum (Al^{3+}), the only known oxidation state of aluminum in aqueous media, is roughly divided into three major regions as depicted by the following neutralization reaction:

$$Al(H_2O)_6^{3+}(aq) \xrightarrow{3OH^-} Al(OH)_3(s) \xrightarrow{OH^-} Al(OH)_4^-(aq) \quad (1)$$

In acidic media below pH 3, the Al^{3+} ion has a hydration number of six, is of octahedral coordination geometry around the Al^{3+} ion with six water molecules coordinated by oxygen donors, and exists as the cationic hexaaquo species (see Fig. 1). Neutralization of this species by 3 moles of OH^- ion in the pH range 4–8 leads to the formation of various amorphous or crystalline solid aluminum hydroxides, $Al(OH)_3$ or trihydrated aluminum oxides, $Al_2O_3 \cdot 3H_2O$ (Hsu, 1964; Hsu and Bates, 1964; Schoen and Roberson, 1970), the nature of the precipitate being dependent on the method of neutralization and conditions of synthesis (particularly concentration and temperature). In addition to various transient amorphous precipitates, two crystalline forms of the hydroxide are prominent: α-$Al(OH)_3$ (gibbsite) and γ-$Al(OH)_3$ (bayerite). Gibbsite is thermodynamically the most stable aluminum hydroxide phase at room temperature (Baes and Mesmer, 1976). The solubility of gibbsite in acid media,

$$Al(OH)_3 \text{ (gibbsite)} + 3H^+ \rightleftharpoons Al^{3+}(aq) + 3H_2O \quad (2)$$

has been reported (Baes and Mesmer, 1976) to be log $K_{sp} = 10^{-8.5}$. In the basic pH region, the hydrolysis chemistry of aluminum is dominated by the aluminate ion, $Al(OH)_4^-$, an anionic tetrahedrally coordinated hydroxide species.

B. Mononuclear Hydrolysis Equilibrium

While the simple three-component aluminum hydrolysis scheme shown in Eq. 1 depicts the three major regions of aluminum hydrolysis chemistry, extensive potentiometric pH measurements (Brosset, 1952; Brosset et al., 1954; Sillen, 1954, 1959; Frink and Peech, 1963; Frink and Sawhney, 1967) have established that a variety of mononuclear species are in stepwise reversible equilibria shown below:

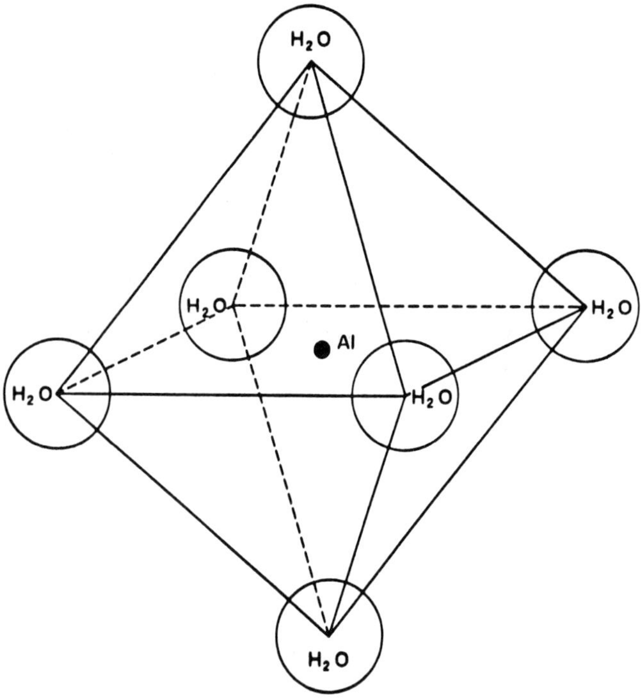

Figure 1 Schematic Representation of the hexaaquo aluminum(III) ion, $Al(H_2O)_6^{3+}$. (Reproduced by permission of Hem and Roberson, 1967.)

$$Al(H_2O)_6^{3+} \rightleftarrows Al(H_2O)_5(OH)^{2+} + H^+ \qquad pK_1 = 5.0 \tag{3}$$

$$Al(H_2O)_5(OH)^{2+} \rightleftarrows Al(H_2O)_4(OH)_2^{+} + H^+ \qquad pK_2 = 4.8 \tag{4}$$

$$Al(H_2O)_4(OH)_2^{+} \rightleftarrows Al(H_2O)_3(OH)_3 + H^+ \qquad pK_3 = 5.7 \tag{5}$$

$$Al(H_2O)_3(OH)_3 \rightleftarrows Al(OH)_4^{-} + 2H_2O + H^+ \qquad pK_4 = 7.6 \tag{6}$$

While these mononuclear species are shown to form rapidly and reversibly upon hydrolysis of Al(III) ion from best-fit analysis of potentiometric data, hydrolysis schemes involving both mononuclear and polynuclear hydrolysis species must be used to account for the other physical-chemical data for these systems,

C. Polynuclear Hydrolysis Species

Baes and Mesmer (1976) have summarized a number of important conclusions concerning hydrolyzed aluminum solutions up to $\bar{n}$ = 2.5 (OH/Al ratio) based on potentiometric, small-angle x-ray scattering, ultracentrifugation, light scattering, and ferron spectrophotometric analysis of extensively hydrolyzed and aged aluminum solutions:

1. Mononuclear species such as those described above are formed rapidly and reversibly.
2. Small polynuclear species, such as $Al_2(OH)_2^{4+}$ and $Al_3(OH)_4^{5+}$ ions, are formed less rapidly.
3. At least one large stable polymeric ion, $Al_{13}O_4(OH)_{24}^{7+}$, is formed even slower.
4. The precise conditions of hydrolysis, including $\bar{n}$, synthesis, method, form and rate of OH^- addition, and temperature, determine the nature and amounts of transient polymeric species, colloidal particles, or amorphous solid phases formed.
5. The rates at which transient (or metastable) soluble species or solid precipitates are converted to other polymers in solution or crystalline gibbsite is dependent on temperature, $\bar{n}$, pH, the amounts of initially formed gibbsite, as well as the presence of various anions.

The latter two conclusions of Baes and Mesmer indicate that equilibria descriptions of hydrolyzed aluminum solutions represent unique species distributions relevant to only a particular set of these solutions and usually derived from a single experimental approach. Kinetic effects, therefore, can significantly affect the species distribution of hydrolyzed aluminum solutions, particularly where these solutions contain large metastable polymeric species or transient precipitates. In the latter cases, such soluble or solid hydroxide complexes are metastable with respect to gibbsite. Furthermore, reactions to form large polymeric or colloidal species are quite slow at room temperature, and the reestablishment of equilibria following perturbation of these systems is likewise slow (on the order of months at room temperature).

Whereas early potentiometric studies (Brosset, 1952; Brosset et al., 1954; Sillen, 1954, 1959) of aluminum hydrolysis solutions generally proposed that such data most readily fit the formation of an infinite series of polynuclear complexes of the form $[Al(OH)_3Al]_n^{3+}$ in the acid range and $[Al(OH)_3]_nOH^-$ series in the basic range, the application of new techniques such as ultracentrifugation, small-angle x-ray scattering, and light scattering first provided workers with strong evidence for the formation of a limited number of polynuclear aluminum hydroxide complexes in the $\bar{n}$ = 1.0–2.5 range, pH 3–4. Rausch and Bale (1964), using small-angle x-ray scattering, first demonstrated that freshly hydrolyzed (1 M Al) aluminum solutions were metastable and contained polynuclear species from 10 to 300 Å in size over the $\bar{n}$ = 0–2.5 range. Equilibrated solutions prepared by aging at $\bar{n}$ = 2.5 were found to contain polymeric species with a radius of gyration of 4.3 Å, consistent with the radius of gyration of 4.15 Å for the $Al_{13\text{-mer}}$ complex from Johansson's x-ray diffraction work (1962b, 1962c). The classical studies of Aveston (1965) first bridged the gap between pH measurements and macromolecular techniques by examining freshly hydrolyzed aluminum solutions and equilibrated solutions (hydrolyzed at 50°C up to $\bar{n}$ = 2.43 and allowed to reach equilibrium for 2 weeks at 25°C) using pH and ultracentrifugation measurements. Aveston interpreted his data as supporting the presence of two major stable polynuclear species, the hydroxy bridged dimer, $Al_2(OH)_2{}^{4+}$ ion, and the polynuclear $Al_{13\text{-mer}}$ species, $Al_{13}O_4(OH)_{24}{}^{7+}$ ion. He reported the least-squares refined formation constants for these two complexes as given below:

$$2Al^{3+} + 2H_2O \rightleftarrows Al_2(OH)_2{}^{4+} + 2H^+ \quad \log{}_{2,2} = -7.07 \tag{7}$$

$$13Al^{3+} + 32H_2O \rightleftarrows Al_{13}(OH)_{32}{}^{7+} + 32H^+ \quad \log_{13,32} = -104.5 \tag{8}$$

It was suggested that the polymerization of Al^{3+} up to $\bar{n}$ = 1.0 involves only the polynuclear $Al_2(OH)_2{}^{4+}$ ion, whereas hydrolysis in the $\bar{n}$ = 1.0–2.5 range results in increasing amounts of only one polynuclear species, the $Al_{13\text{-mer}}$ cation. The choice of these species as "best fit" complexes was based largely on the availability of x-ray diffraction studies reported by Johannson and co-workers (1960, 1962a, 1962b, 1962c) for the basic dimeric sulfate complex $[Al_2(OH)_2(H_2O)_8[(SO_4)_2$ and the basic $Al_{13\text{-mer}}$ sulfate and selenate complex containing the ion $Al_{13}O_4(OH)_{24}(H_2O)_{12}{}^{7+}$.

The dimeric ion $Al_2(OH)_2(H_2O)_4{}^{4+}$, as depicted in Fig. 2, was shown by Johansson to consist of two Al^{3+} octahedra doubly bridged by OH^- ions via edge sharing. Formation of this species is favored at $\bar{n}$ = 1.0 as a result of the hydrolysis reaction sequence involving the hexaaquo species and the monohydroxy Al^{3+} ion as shown below:

$$H_2O + Al(H_2O)_6{}^{3+} \rightleftarrows Al(H_2O)_5(OH)^{2+} + H_3O^+ \tag{9}$$

$$Al(H_2O)_6{}^{3+} + Al(H_2O)_5(OH)^{2+} \rightleftarrows Al_2(OH)_2(H_2O)_8{}^{4+} + H_2O + H_3O^+ \tag{10}$$

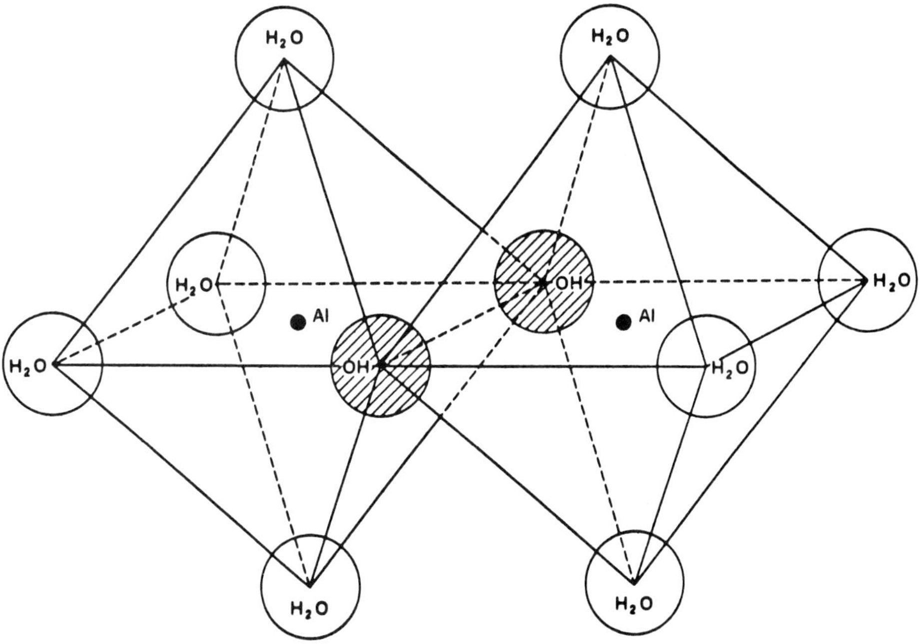

Figure 2 Schematic representation of u-bis(hydroxy)bis(tetraaquo-aluminum(III)) ion, $Al_2(OH)_2(H_2O)_8^{4+}$. (Reproduced by permission of Hem and Roberson, 1967.)

as proposed by Grunwald and Fong (1969) and suggested by Akitt et al. (1969, 1969b, 1972). The structure of the $Al_{13\text{-mer}}$ species from Johansson's x-ray diffraction studies (1962b/c) of crystals obtained from highly hydrolyzed solutions ($\bar{n}$ = 2.43) shows that this ion consists of 12 aluminum octahedra and 1 aluminum tetrahedron. The ion is built up from 12 AlO_6 octahedra with shared edges and apices arranged around an AlO_4 tetrahedron. Details of the structure of this polynuclear species in the solid state, depicted using a number of structural models in Fig. 3, indicate that this species consists of a regular arrangement of three distorted AlO_6 octahedra grouped around each apex of the highly symmetrical AlO_4 tetrahedron. These trimeric AlO_6 units are interconnected by their edges and vertices with Al--O distances which vary from 1.85 to 2.05 Å and interoxygen edge distances which vary from 2.34 to 2.85 Å.

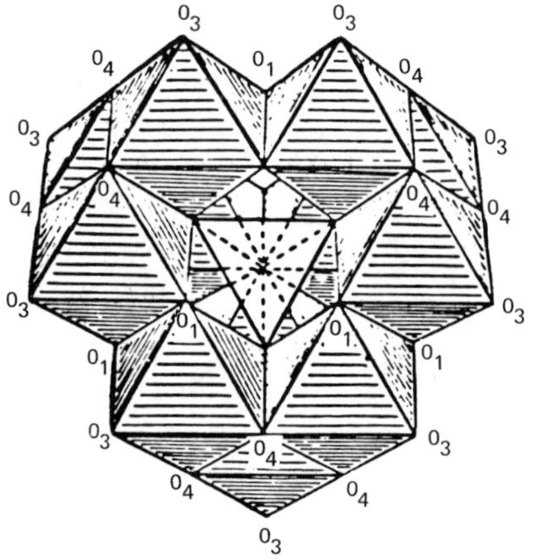
O_3
O_3
O_4
O_1
O_4
O_3
O_3
O_4
O_4
O_4
O_4
O_3
O_3
O_1
O_1
O_1
O_1
O_3
O_3
O_4
O_4
O_4
O_3

(a)

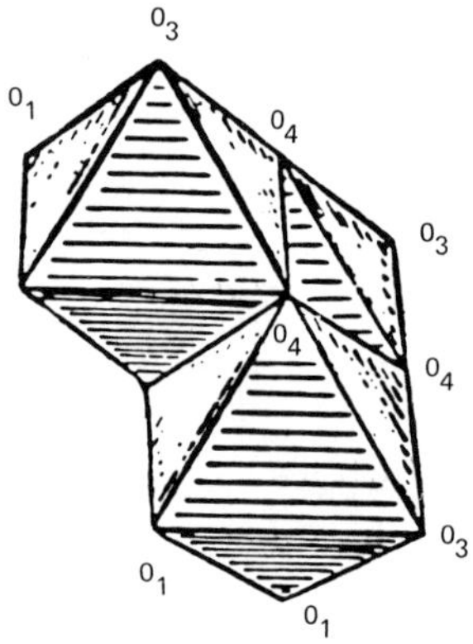
O_3
O_1
O_4
O_3
O_4
O_4
O_3
O_1
O_1

(b)

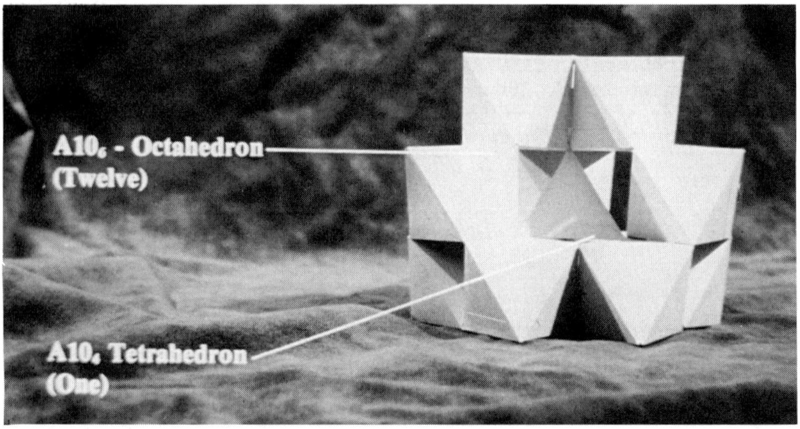
AlO_6 - Octahedron
(Twelve)
AlO_4 Tetrahedron
(One)

(c)

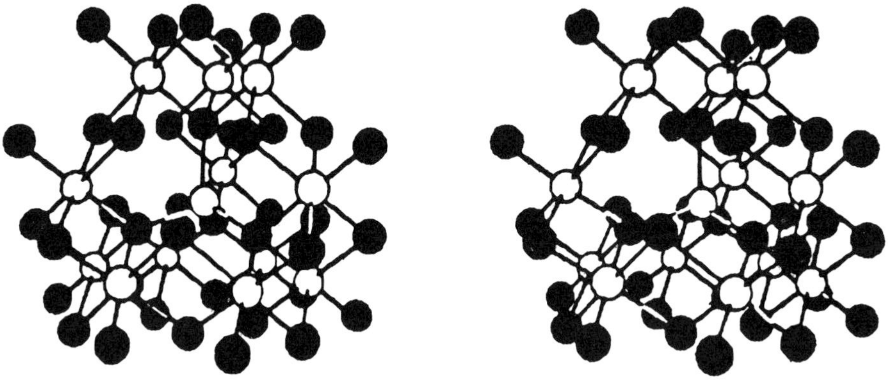

(d)

Figure 3 Various representations of the $[Al_{13}O_4(OH)_{24}(H_2O_{12}]^{7+}$ ion and the trimeric unit, Al_3O_{13}: (a) The $Al_{13\text{-mer}}$ ion according to Johansson (1962a), (b) the Al_3O_{13} unit of the tridecameric ion, according to Johansson (1962a), (c) photograph of $Al_{13\text{-mer}}$ model according to Fitzgerald (1981), and (d) structural model of $Al_{13\text{-mer}}$ according to Waters and Henty (1977). (Reproductions by permission of authors.)

In addition to the ultracentrifugation studies of Aveston (1965), the light scattering work of Patterson and Tyree (1972, 1973) on hydrolyzed aluminum solutions provides evidence for the dimer and $Al_{13\text{-mer}}$ species. The earliest work on hydrolyzed aluminum nitrate solutions (Ruff and Tyree, 1958) showed that the average molecular weight of hydrolyzed solutes increased with OH/Al ratio from 138 at $\bar{n} = 0$ to approximately 1430 at $\bar{n} = 2.25$, as shown in Table 2. Later work (Patterson, and Tyree, 1972, 1973) on hydrolyzed aluminum perchlorate solutions indicated that the presence of the dimer at $\bar{n} = 1.0$ and of species up to the $Al_{13\text{-mer}}$ at $\bar{n} = 2.4$ was consistent with the Rayleigh turbidity measurements.

While these physical-chemical measurements provided valuable evidence for the presence of only two stable polynuclear species, the dimer and the $Al_{13\text{-mer}}$, a number of workers have suggested additional polynuclear species. The very extensive potentiometric measurements of Baes and Mesmer (1971) at low $\bar{n}$ values up to 150°C suggest that early hydrolysis of Al^{3+} can be best fit by assuming two polynuclear species at low $\bar{n}$, the dimeric ion $Al_2(OH)_2{}^{4+}$ and the trimeric ion $Al_3(OH)_4{}^{5+}$ or $Al_3(OH)_5(H_2O)_9{}^{5+}$. The latter species could, in fact, have the structure shown in Fig. 3 for the trimer units of the $Al_{13\text{-mer}}$. Mesmer and Baes' work (1971) at high $\bar{n}$ values, like that of Aveston and Tyree, was consistent with the formation of only one larger polynuclear cation, the $Al_{13\text{-mer}}$. A number of additional intermediate-sized polynuclear species such as $Al_6(OH)_{15}{}^{3+}$, $Al_6(OH)_{12}(H_2O)_{12}{}^{6+}$, $Al_7(OH)_{17}{}^{4+}$, $Al_8(OH)_{20}{}^{4+}$, and $Al_{14}(OH)_{34}{}^{8+}$ have been suggested as principal species in hydrolyzed aluminum solutions. The majority of these species have been proposed on the basis of best-fit analysis of potentiometric data, with the exception of two complexes. The $Al_6(OH)_{12}(H_2O)_{12}{}^{6+}$ ion was proposed (Hem and Roberson, 1967; Hem 1968a; Roberson and Hem, 1969; Hsu, 1968, 1977) on the basis of its potential as a plausible structural precursor of the various crystalline forms of $Al(OH)_3$. The $Al_8(OH)_{20}{}^{4+}$ ion was proposed from the potentiometric work by Hayden and Rubin (1973) and the silver halide sol coagulation experiments of Matijevic et al. (1961, 1964) for solutions of $\bar{n} = 2.5$.

The presence of various mononuclear and polynuclear hydrolysis species in the acidic pH range in dilute aqueous Al^{3+} solutions has also been extensively investigated by ^{27}Al nuclear magnetic resonance (NMR) spectroscopy as well as by infrared (IR) and Raman spectroscopy. In acidic media, infrared studies (Fripiat et al., 1965) of hydrolyzed aluminum solutions have identified the OH stretching frequency of bridging hydroxide, presumably in the dimer. Raman spectra (Waters and Henty, 1977) have shown a number of unique spectral bands associated with the $Al_{13\text{-mer}}$ complex in both the

Table 2 Summary of Average MW for Various Hydrolyzed Aluminum Solutions

$[OH^-]/[Al^{3+}]$	Average molecular weight (MW_w)
0.0	138
0.5	256
1.0	308 (dimer MW = 232)
1.5	526
1.75	715
2.0	1000
2.25	1430 ($Al_{13\text{-mer}}$ MW = 1039)

Source: Patterson and Tyree (1973).

solution and the solid state. The use of ^{27}Al NMR (Akitt et al., 1969a, 1969b, 1972, 1985b; Fitzgerald, 1984c) has provided the most direct spectroscopic evidence for the existence of four of the hydrolysis species in aqueous solution: $Al(H_2O)_6{}^{3+}$, $Al(H_2O)_5(OH)^{2+}$, $Al_2(OH)_2(H_2O)_8{}^{4+}$, and the $Al_{13\text{-mer}}$ complex. In addition, indirect evidence for the $Al_8(OH)_{20}{}^{4+}$ ion was given by Akitt (1972). The relative concentration of the four species idenitfied by ^{27}Al NMR resonances was estimated for solutions hydrolyzed over the $\bar{n}$ = 0.0–2.5 range, although their NMR signal intensities did not account for all of the soluble aluminum species present.

In the highly basic pH range in dilute solution, the aluminate ion, $Al(OH)^{4-}$, has been unequivocally identified from IR, Raman, and ^{27}Al NMR spectroscopic studies. Infrared and Raman measurements by Lippincott et al. (1952) and Moolenaar et al. (1970) have all shown that the aluminate ion predominates in the pH 8–12 range for dilute aluminate solutions. Solutions of higher concentration (greater than 1.5 M Al) at pH values greater than 13 are believed to contain condensed aluminate ions such as $Al_2O(OH)_6{}^{2-}$, for which x-ray diffraction information (Johansson, 1966) is available on the crystalline salt, $K_2[Al_2O(OH)_6]$, or polyaluminate species of unknown structure. The presence of the aluminate species in alkaline solutions has been established by the ^{27}Al NMR experiments of Akitt and Farthing (1978) and others, although NMR spectral evidence for the existence of polyaluminate species is at present less certain.

Table 3 Summary of Aluminum Hydrolysis Species and Techniques to Support Species Existence

Hydrolysis species	Complex formula	Techniques used[a]
Mononuclear complexes		
Al^{3+}	$Al(H_2O)_6^{3+}$	All
$Al(OH)^{2+}$	$Al(H_2O)_5(OH)^{2+}$	P, NMR
$Al(OH)_2^+$	$Al(H_2O)_4(OH)_2^+$	P
$Al(OH)_3(aq)$	$Al(H_2O)_3(OH)_3$	P, UC, XR, NMR, F
$Al(OH)_4^-$	$Al(OH)_4^-$	P, IR, R, NMR
Polynuclear complexes		
$Al_2(OH)_2^{4+}$	$Al_2(OH)_2(H_2O)_8^{4+}$	P, UC, LS, XR, IR, R
$Al_3(OH)_4^{5+}$	$Al_3(OH)_4(H_2O)_9^{5+}$	P
$Al_6(OH)_{15}^{3+}$	—	P
$Al_6(OH)_{12}^{6+}$	$Al_6(OH)_{12}(H_2O)_{12}^{6+}$	P, F
$Al_7(OH)_{17}^{4+}$	—	P
$Al_8(OH)_{20}^{4+}$	—	P, CG, NMR
$Al_{13}(OH)_{24}^{7+}$	$Al_{13}(OH)_{24}(H_2O)_{12}^{7+}$	P, UC, LS, XR, R, SXR, GFC, IR
$Al_{14}(OH)_{34}^{8+}$	—	P
Series of polynuclear complexes		
$Al_{10}(OH)_{22}(H_2O)_{16}^{8+}$ to $Al_{54}(OH)_{144}(H_2O)_{38}^{18+}$		F (Hsu, 1964)
$Al_6(OH)_{12}^{6+}$ to $Al_{24}(OH)_{60}^{12+}$		F (Hem and Roberson, 1967)
Al_{20}–Al_{200}		P (Stol, 1976)

[a]P, potentiometric; UC, ultracentrifugation; LS, light scattering; XR, x-ray diffraction crystallography; SXR, small-angle x-ray diffraction; GFC, gel filtration chromatography; NMR, ^{27}AL NMR; CG, coagulation; F, ferron or other spectrophotometric methods; IR, infrared spectroscopy; R, Raman spectroscopy.

D. Summary of Aluminum Hydrolysis Species in Dilute Solutions

The number of discrete mononuclear and polynuclear aluminum(III) ion hydrolysis species for which sufficient evidence has been reported, particularly from a number of physical-chemical approaches, to propose their existence in aqueous solution is summarized in Table 3. This summary includes the abbreviated formulas denoting their aluminum and hydroxide contents, their complete formulas including the composition of coordinated water molecules where it has been proposed, and a listing of the techniques which have been used to support their existence in aqueous media. A number of species which are part of a series of complexes proposed in various nucleation schemes derived from either potentiometric or timed spectrophotometric complexation methods have been included here also, although they will be discussed in more detail shortly. For a large number of the complexes given in Table 3, calculated concentration quotients Q_{xy} have been reported by Baes and Mesmer (1976) for the reaction

$$xAl^{3+} + H_2O \rightleftarrows Al_x(OH)_y^{(3x-y)+} + yH^+ \tag{11}$$

where $Q_{xy} = [Al_x(OH)_y^{(3x-y)+}][H^+]^y/[Al^{3+}]^x$. In addition, Mesmer and Baes (1971) have reported the Q_{xy} values over the temperature range 25–150°C for the polynuclear dimer, trimer, and Al_{13}-mer and the thermodynamic parameters ΔH_{xy} and ΔS_{xy} for the species $Al(OH)^{2+}$, the dimer, trimer, and Al_{13}-mer at 25°C. Table 4 summarizes the stability constants at 25°C for various soluble mononuclear and polynuclear species and solid gibbsite, as derived from the equation

$$\log K_{xy} = \log Q_{xy} - \frac{aI^{1/2}}{1 + I^{1/2} + bm_x} \tag{12}$$

where I, a, b, and m_x are defined in the work of Baes and Mesmer (1976).

Various workers have utilized these or other calculated K_{xy} values and the solubility quotient K_{s10} to generate species distribution diagrams for the soluble hydrolysis species as a function of pH and to generate soluble species distribution diagrams for solutions which are saturated with respect to gibbsite. For example, Morgan (1967) reported the species distribution diagrams shown in Fig. 4a for various mononuclear and polymeric aluminum species assumed to be in equilibrium with gibbsite and in Fig. 4b for similar species assumed to be in equilibrium with freshly precipitated $Al(OH)_3$. Note that these species distribution diagrams include the $Al_7(OH)_{17}^{4+}$ ion, a species which has not been proposed by other workers. The species distribution diagrams derived by Baes and Mesmer (1976), by contrast, utilize their K_{xy} values at 25°C for the species summarized in Table 4. Figures 5a and 5b summarize the soluble species distribution content for various mononuclear and polynuclear species

Table 4 Summary of Stability Constants (K_{xy}) for Various Mononuclear and Polynuclear Species

Species or phase	Species abbreviation	log K_{xy}
$Al(OH)^{2+}$	1,1	- 4.97
$Al(OH)_2^+$	1,2	- 9.3
$Al(OH)_3(aq)$	1,3	-15.1
$Al(OH)_4^-$	1,4	-23.0
$Al_2(OH)_2^{4+}$	2,2	- 7.7
$Al_3(OH)_4^{5+}$	3,4	-13.94
$Al_{13}O_4(OH)_{24}^{7+}$	13,32	-98.73
$Al(OH)_3(s)$	Gibbsite	8.5 log Q_{s10}

Source: Baes and Mesmer (1976).

as a function of solution pH for 10^{-1} and 10^{-5} M Al(III) ion solutions, respectively. Figure 5c presents a species distribution diagram for these similar species in equilibrium with solutions saturated with gibbsite [species $Al(OH)_3$, where $x = 1$ and $y = 3$].

The species distribution diagrams of Baes and Mesmer (1976) include a range of mononuclear species (the "1,0," "1,1," "1,2," "1,3," and "1,4" species) and three polynuclear species (the "2,2," "3,4," and "13,32" species). Of these species, Baes and Mesmer have concluded that only four, the monomeric hexaaquo species, the monomeric monohydroxy species, the dihydroxy dimer, and the Al_{13}-mer, are present in sufficient quantity in equilibrated solutions which do not contain gibbsite. The ^{27}Al NMR work of Akitt et al. (1969a, 1969b, 1972, 1985b) provides direct spectroscopic evidence for these four species, lending support to the conclusions of Baes and Mesmer.

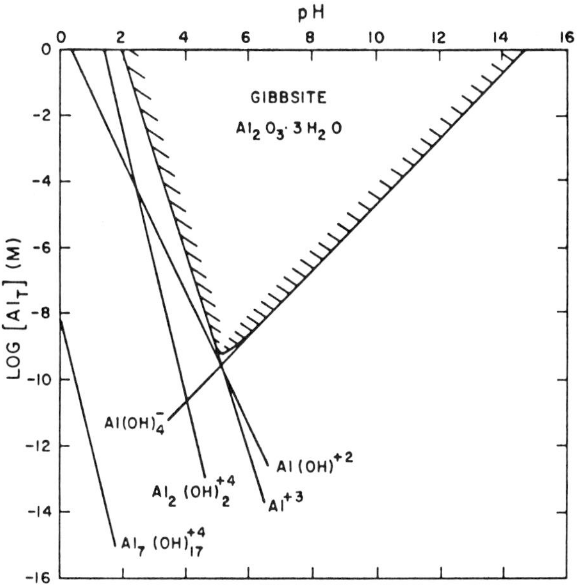

(a)

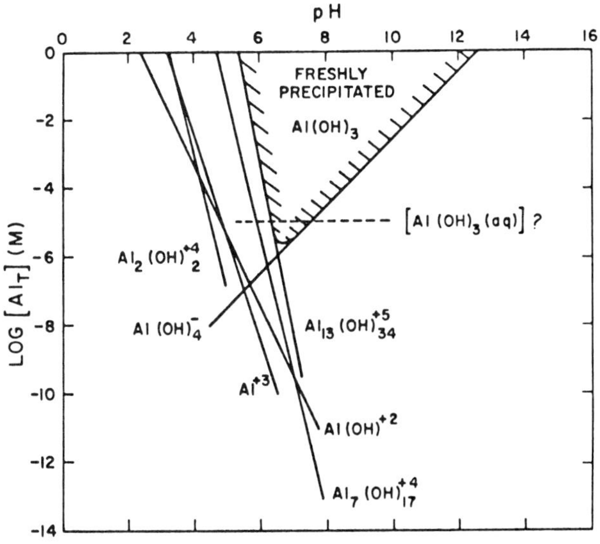

(b)

Figure 4 Species composition profile (concentration versus pH) for aluminum species in equilibrium with (a) gibbsite and (b) freshly precipitated $Al(OH)_3$. (Reproduced by permission of Morgan, 1967.)

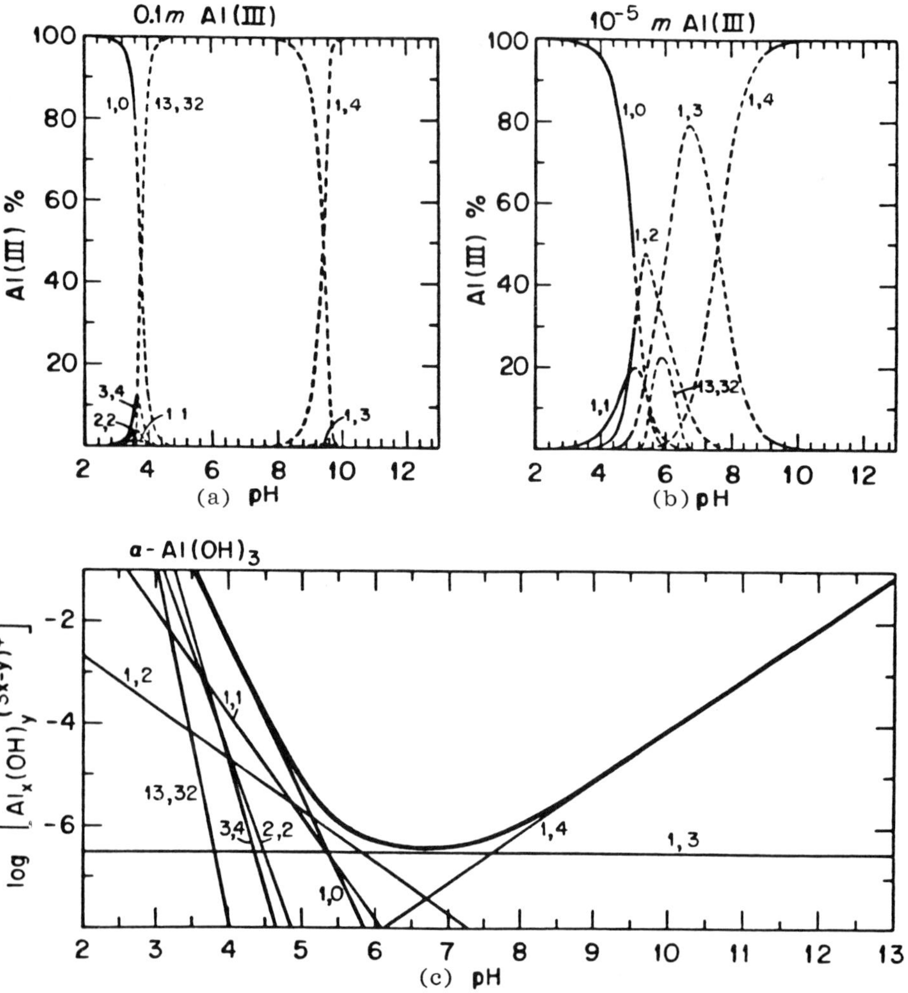

Figure 5 Distribution of aluminum(III) hydrolysis species (a) at 0.1 M Al(III), (b) at 10^{-5} M Al(III), and (c) in solutions saturated with gibbsite (α-$Al(OH)_3$). (Reproduced by permission of Baes and Mesmer, 1976.)

E. Mechanisms of Nucleation of Al(III) Ion in Dilute Solutions

While the existence of various stable solution hydrolysis species in equilibrium with gibbsite or other $Al(OH)_3$ solids has been of paramcunt importance in understanding the hydrolysis of Al^{3+} in the pH range 3–5, the rates of formation of such species and the mechanisms of species interconversions have attracted increasing attention. Kinetic effects in aluminum hydrolysis chemistry have been well recognized in potentiometric, light scattering, small-angle x-ray scattering, ultracentrifugation, and ^{27}Al NMR studies. However, the development of a number of timed spectrophotometric compleximetric methods of analysis has provided a means of readily examining kinetic changes occurring in hydrolyzed aluminum solutions over periods of days to months. The research groups who have utilized these spectrophotometric approaches, the various complexation reagents, and the range and nature of hydrolyzed aluminum solutions studied are summarized in Table 5. The studies of these four groups have defined aluminum hydrolysis products in terms of three kinds of aluminum(III) complexes:

1. Fast-reacting (up to 2 minutes) Al^{3+} ion forms consisting of mononuclear and possibly dimeric species, termed Al^a species by Turner
2. Slower-reacting (up to 3 hours) polynuclear hydrolysis species, termed Al^b species
3. Very slowly reacting high-molecular-weight polymers, microcrystalline gibbsite, or various amorphous solids, termed Al^c species

The various definitions of the three forms of Al(III) hydrolysis products and the interpretation of where to include the various proposed mononuclear, polynuclear, and solid complexes in each of the reaction schemes have differed among these workers. Nevertheless, in the various approaches reported, each method is based on the relative rates of reaction of the various hydrolysis complexes with the compleximetric agent used.

The work of Turner and co-workers (Turner, 1965, 1968a, 1968b, 1969, 1971, 1975, 1976a, 1976b; Turner and Ross, 1969, 1970; Turner and Sulaiman, 1971), using hydroxyquinoline, involved comparative analysis of neutralized and aged $AlCl_3$ solutions at 10^{-2} to 10^{-3} M Al. These studies have revealed that various hydrolyzed aluminum solutions contain varying levels of Al^a, Al^b [proposed to be a single polynuclear ion, possibly the $Al_6(OH)_{15}{}^{3+}$ ion], and Al^c complexes. Turner proposed that polymerization on aging at $\bar{n}$ = 1.5–3.0 involved two pathways: an $Al^a \rightarrow Al^b$ and an $Al^a \rightarrow Al^c$ (an initial amorphous aluminum hydroxide solid containing Cl^- ion) pathway. These two pathways are summarized in the following scheme:

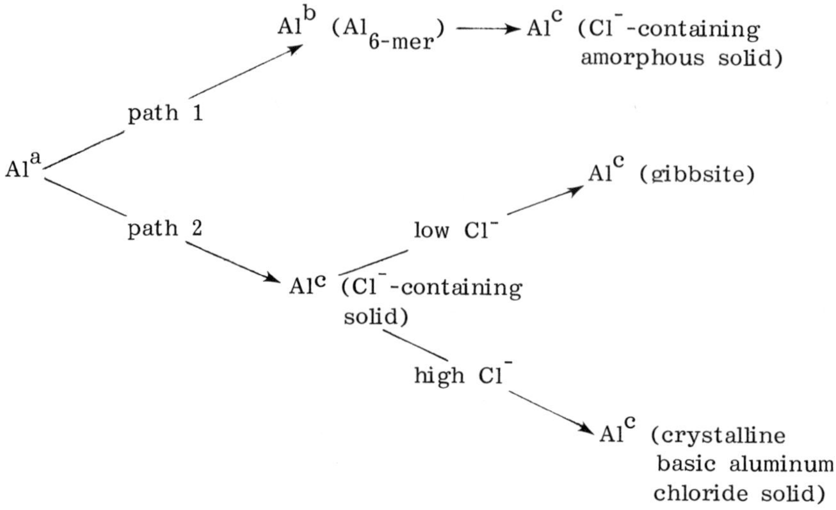

Reaction Scheme (13)

Table 5 Summary of Spectrophotometric Studies of Aluminum Hydrolysis Systems Using Various Related Complexation Reagents

Workers	Spectrophotometric reagent	Solutions examined
Turner et al.	8-Hydroxyquinoline	Neutralized and aged $AlCl_3$ solutions, 10^{-2} to 10^{-3} M Al ($\bar{n}$ = 0.5–3.0)
Hem and Smith	Ferron (8-hydroxy-7-iodo quinoline-5-sulfonic acid)	Neutralized and aged $Al(ClO_4)_3$ solutions, 4×10^{-3} M Al, up to 2 years ($\bar{n}$ = 0.5–3.0)
Hsu et al.	Aluminon	Neutralized and aged $Al(NO_3)_3$ solutions, 2×10^{-2} M Al, up to 2 years ($\bar{n}$ = 0.3–3.0)
Gessner et al.	Ferron	Neutralized and aged $AlCl_3$, $Al(ClO_4)_3$, and $Al_2(SO_4)_3$ solutions, 4.5×10^{-4} M Al, up to 40 days ($\bar{n}$ = 0.5–2.5)

Turner proposed that path 1 produced Al^c (containing Cl^-) during the initial neutralization but that these complexes redissolved on aging. Only path 2 produced gibbsite, at a much lower rate, in solutions with Cl^- ion concentrations below 0.6 M, whereas solutions with higher Cl^- concentrations produced insoluble crystalline basic aluminum chloride Al^c-type solids. Turner (1976a, 1976b) has suggested the presence of a second soluble Al^b species, which is 100-fold less reactive than the original Al^b and is formed on aging of hydrolyzed aluminum solutions.

Hem (1968a, 1968b), Hem and Roberson (1967), (Roberson and Hem, 1969), Smith (1969, 1970), and Smith and Hem (1972), using the ferron method, provided the most extensive studies of neutralized and aged $Al(ClO_4)_3$ solutions in the $\bar{n} = 0.5-3.0$ range at 4.5×10^{-4} M Al. Hem and Smith also studied the formation of gibbsite in these aged solutions over a period of 2 years in the absence of Cl^- ion to clarify its mechanism of nucleation from soluble precursors above $\bar{n} = 1.36$. Based on determination of the percentages of Al^a, Al^b, and Al^c (termed microcrystalline gibbsite by these workers) content changes, they proposed that the formation of gibbsite involved the simple pathway

$$Al^a \rightarrow Al^b \rightarrow Al^c \text{ (gibbsite)} \tag{14}$$

where the Al^b species was assumed to be an $Al_{6\text{-mer}}$ species, a solution precursor of gibbsite. The initial formation of the Al^b species involved stepwise hydrolysis from $Al(H_2O)_6{}^{3+}$ to $Al(H_2O)_5(OH)^{2+}$ to significant quantities of $Al_2(OH)_2(H_2O)_8{}^{4+}$, which then combined to yield a ring structure of six octahedrally coordinated aluminum ions with the formula $Al_6(OH)_6(H_2O)_{12}{}^{6+}$ (Al^b). They proposed that the $Al_{6\text{-mer}}$ ion, consisting of six-membered aluminum hydroxide rings, further coalesced on aging to form larger, slower-reacting Al^b soluble complexes which ranged in size from 24 to 200 aluminum ions per molecular ion. The structure of the $Al_{6\text{-mer}}$ species and a coalesced network containing this species are shown in Fig. 6. Eventually, these complexes were proposed to grow in size according to the diagram shown in Fig. 7, reaching sufficient organization to be converted to Al^c, microcrystalline gibbsite. The formation of gibbsite precipitate was envisioned to involve ordering of these soluble two-dimensional complexes into well-ordered three-dimensional layered solid particle structures of crystalline gibbsite. Smith and Hem (1972) estimated that $\bar{n} = 2.47$ solutions contained Al^b species corresponding to the $Al_{54}(OH)_{144}{}^{8+}$ ion.

The aluminom studies of Hsu (1966, 1968, 1977) carried out on hydrolyzed and aged (up to 2 years) 2×10^{-2} M Al solutions provided information regarding the mechanism of formation of gibbsite

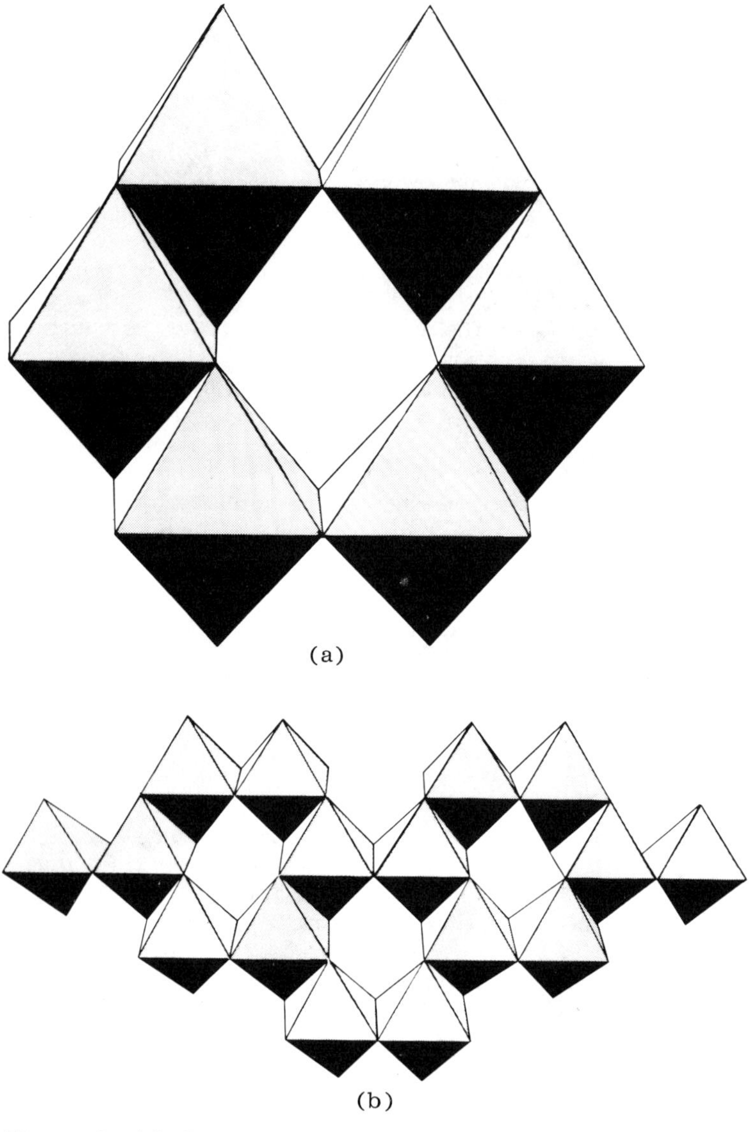

Figure 6 (a) Structure of $Al_{6\text{-mer}}$ cation and (b) schematic representation of coalescence of networks of aluminum hydroxide octahedra. (Reproduced by permission of Hem and Roberson, 1967.)

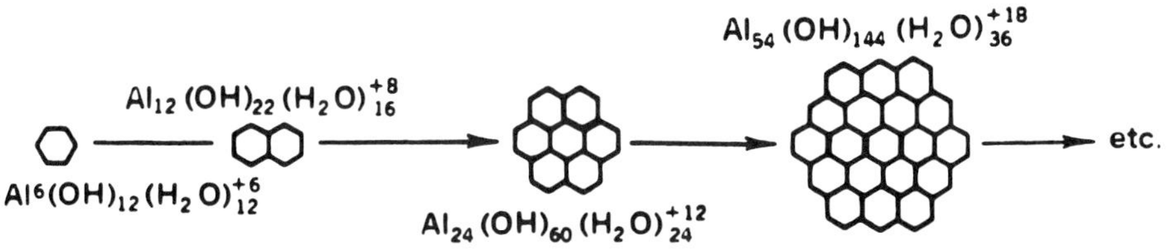

Figure 7 Proposed growth of polynuclear aluminum hydrolysis species from the $Al_{6\text{-mer}}$. (Reproduced by permission of Smith and Hem, 1972.)

similar to that proposed by Hem and Smith in Eq. 14. In addition, they reached a number of additional conclusions: first, solutions below an $\bar{n}$ of 1.5 were stable with respect to gibbsite precipitation, second, solutions from $\bar{n}$ = 1.8 to 2.7 produced gibbsite slowly in the course of 6 months to 1 year, and third, bayerite was formed rapidly at $\bar{n}$ = 3.0. In addition, their work supported the conclusions of Turner that extensive hydrolysis at low Cl^- levels leads to the initial formation of Al^b and Al^c species containing Cl^- which, on aging, produce gibbsite. At Cl^- ion concentrations greater than 1.0 M, the solution Cl^- ion inhibited the formation of gibbsite.

The work of Gessner and Winzer (1979) and Schonherr et al. (1979), using the ferron technique of Smith and Hem (1972), involved the study of various hydrolyzed solutions of 4.5×10^{-4} M $AlCl_3$, $Al(ClO_4)_3$, and $Al_2(SO_4)_3$ over $\bar{n}$ = 0.5–2.5 values. The quantitative ferron measurements showed that the % Al^a species contents decreased linearly with increases in $\bar{n}$ values, e.g., 73% Al^a at $\bar{n}$ = 1.0 to 12% Al^a at $\bar{n}$ = 2.5 for neutralized $Al(ClO_4)_3$ solutions. While extensive quantitative ferron analysis data were presented for the $\%Al_{monomer}$, $\%Al_{oligomer}$, and $\%Al_{polymer}$ contents (similar to the Al^a, Al^b, and Al^c terminology) of a range of hydrolyzed solutions, the aging studies led these workers to propose two parallel pathways for the formation of gibbsite as shown below:

$$\begin{array}{lll} & Al^b \longrightarrow Al^c & \\ Al^a \nearrow & & \\ \quad\searrow & & (15) \\ & Al^c & \end{array}$$

One of the most interesting results of Gessner and Winzer (1979) is the determination that the mononuclear species, $Al(H_2O)_6{}^{3+}$ and $Al(H_2O)_5(OH)^{2+}$ ions, react completely in minutes with the ferron reagent, whereas solutions containing the $Al_2(OH)_2{}^{4+}$ and $Al_{13\text{-mer}}$ ions, prepared by dissolution of their respective crystalline salts, react completely in approximately 14 and 60 minutes, respectively; thus, the dimer may be classified as either an Al^a or Al^b species, and the $Al_{13\text{-mer}}$ may be classified as an $Al_{polymer}$ or Al^b-type solutes on the basis of the Smith and Hem terminology. Larger soluble polymeric species as well as microcrystalline gibbsite, by contrast, take from 10 to 48 hours from complete reaction in the ferron reagent assay system used by Gessner and Winzer (1979) and Smith and Hem (1972).

IV. APPROACHES TO EXAMINING COMMERCIAL ALUMINUM HYDROLYSIS SYSTEMS

The chemical literature describing the nature of aluminum hydrolysis species contained in dilute (10^{-4} to 10^{-1} M) hydrolyzed aluminum(III) ion solutions provides the theoretical and experimental background and direction necessary for studying concentrated basic aluminum chloride and mixed aluminum/zirconium chlorohydrates. The research reported from 1960 to 1975 showed that dilute aluminum hydrolysis solutions can be described on the basis of a narrow range of unique mononuclear and polynuclear complexes of varying contents. The relative concentration of these species is markedly dependent on pH. A given aqueous aluminum system may be characterized on the basis of four principal parameters:

1. The $[OH^-]/[Al^{3+}]$ ratio, or $\bar{n}$
2. The solution concentration, ionic stength, temperature, pH, and presence and concentration of various anions
3. The method of synthesis and "history of solution"
4. The equilibrium status and hydrolyzed solutions

Species distribution diagrams have shown that the existence and amounts of different species are controlled to the first approximation by pH.

Although many of these soluble hydrolysis species have been proposed over the years, there is definitive evidence for the existence of only a few complexes based on experiments on similar solutions by at least two or three different physical-chemical methods or approaches. The use of macromolecular techniques, particularly ultracentrifugation, light scattering, and more recently gel filtration chromatography, together with spectroscopic methods such as ^{27}Al NMR, IR, and Raman spectroscopy, has greatly complemented the work using potentiometric measurements. These approaches, together with the x-ray diffraction results for the crystalline dimer and $Al_{13\text{-mer}}$ salts, have greatly aided our understanding of base-neutralized aluminum solutions at equilibrium.

Aged hydrolyzed aluminum solutions of various $\bar{n}$ values are more easily examined by rapid techniques such as the ferron technique and related timed spectrophotometric methods. This work has provided quantitative information about the mononuclear and polynuclear species contents of various dilute solutions. Studies of the mechanisms of formation of gibbsite and the mechanisms of nucleation of soluble polymeric species have consistently stressed that the $Al_6(OH)_{12}^{6+}$ ion is a convenient structural unit which may act as a precursor of larger polymeric species and eventually well-ordered crystalline gibbsite. Organization of these larger, essentially two-dimensional polymeric sheetlike species into ordered three-dimensional layered microcrystalline $Al(OH)_3$ or gibbsite particles is an attractive mechanism leading to $Al(OH)_3$ precipitation from hydrolyzed aluminum solutions of $\bar{n}$ greater than 2.0.

A. Commercial Basic Aluminum Complexes

The experimental study of commercial basic aluminum chlorides and mixed aluminum/zirconium complexes in solution require the use of characterization methods similar to those used to study dilute Al^{3+} ion hydrolysis systems. In the 1970s, Reheis Chemical Company (1977) summarized two major complications which represented significant barriers in the scientific study of commercial basic aluminum hydrolysis systems: first, these systems are highly concentrated, and second, the role of synthesis conditions is exceedingly important. Regarding the effects of concentration, the following ideas were summarized:

> Appreciable factors such as temperature, age, and particularly concentration contribute to equilibrium reactions in dilute aluminum chlorhydrol solutions. At any instance, there are competing equilibria giving rise to a spectrum of different basicity aluminum salts being present in disequilibrium in solution. One cannot deny the formidable impact of concentration in all its implications in fostering association and polymerization.

In discussing the effects of synthesis methods, additional ideas were presented:

In the manufacture of 5/6 basic aluminum chloride, the atomic ratio of Al to Cl cannot be controlled at exactly 2 to 1, hence ... changes in degree of polymerization might be expected. It is unlikely, however, that only one species exists in solution. Evidence points to the fact that a wide spectrum of basic aluminum inorganic polymeric salts exists [and of different properties] undergoing slow and continuing changes as a function of method of preparation, temperature, age and pH. Hence, factors such as the history of preparation, ratio of Al to Cl, age, temperature, concentration, storage conditions, among others, determine the ultimate average ionic weight and structure of these species in solution.

In essence, a number of important distinctions must be considered in studies aimed at characterizing commercial basic aluminum hydroxide salt systems:

1. Commercial ACH and other basic aluminum salt solutions can be initially examined without regard to studies of their synthesis, since their aqueous solutions are well-defined analytically from manufacturing specifications. Such studies provide a reference point from which to further examine dilution and other influences due to methods of synthesis.
2. Dilution of concentrated (6–2 M Al) basic aluminum solutions is required to make these systems amenable to examination by various characterization approaches. Since dilution alters their chemical state to produce undefined changes in species equilibria, systematic studies of dilution must be carried out to aid the understanding of these kinetic effects.
3. The addition of various chemical entities, including anions, $AlCl_3$, HCl, complexing agents such as propylene glycol, and even more acidic basic zirconium-glycine complexes will significantly change the various species equilibria.
4. The selection of applicable physical-chemical techniques requires that these approaches be developed sufficiently with regard to the types of information obtainable by each technique and the relationship of this information to previously reported studies using that technique for dilute Al^{3+} systems. In addition, the simultaneous use of a number of approaches is necessary to correlate their characterization results for similar solutions.

Based on these guidelines, extensive characterization studies of ACH, other BAC salts, and ZAG salts have been carried out over the past decade. These studies have sought to provide a well-defined physical-chemical picture of the species composition of these systems at equilibrium. They have provided both macromolecular and structural information about these systems on dilution. In addition, efforts are under way to more fully understand the role of synthesis conditions in producing a range of commercial aluminum hydrolysis systems which differ in their chemical behavior and pharmacological activity.

Chemical techniques for examining commercial aluminum(III) ion hydrolysis systems may be conveniently divided into four major approaches:

1. *Analytical methods* of analysis of solutions and solids before and after dissolution
2. *Macromolecular techniques* to provide a perspective on the polymeric characteristics of aqueous solutions
3. *Spectroscopic methods* to define the structural nature of solutes contained in these systems
4. *Differential kinetic reactivity methods* such as the ferron method

B. Analytical Analyses

The methods most pertinent to chemical analysis of both solution and solid antiperspirant materials include elemental analysis of Al and Cl contents, solution pH, density, conductivity, and viscosity. In addition, measurements of the critical coagulation concentration (Teagarden et al., 1981) and relative solution solute charge by a streaming current detector (Bretschneider et al., 1975) have been reported. The most important analysis for concentrated basic aluminum chloride systems is the accurate determination of Al and Cl contents. Aluminum analysis methods for BAC solutions include atomic absorption analysis or compleximetric titration methods with EDTA or other compleximetric reagents. Chloride analysis of solutions is most readily performed using Cl^- ion-specific electrodes with acidified BAC solutions or the Volhard titrametric method. For BAC solids, aluminum and chloride contents can be determined by the above solution methods on solutions prepared gravimetrically. Thermogravimetric approaches involving furnace treatment of BAC salts to produce Al_2O_3 at temperatures greater than 950°C have also been used. It is important to note that the degree of basicity ($\bar{n}$ or OH/Al ratio) of bulk BAC solutions or solids is determined indirectly from experimentally derived Al/Cl ratios using the relationship $Al_2(OH)_x(Cl)_y$.

C. Macromolecular Techniques

The polymeric nature of highly concentrated aluminum hydroxide systems has been characterized using four principal techniques, which are summarized in Table 6 together with the specific information

Table 6 Summary of Various Macromolecular Techniques Used to Study Commercial Aluminum Hydrolysis Solutions

Method	Types of information[a]	Investigators
Ultracentrifugation (UC)	Molecular weight range, MW_n, MW_w, and MW_z values, N_z, weight-average degree of polymerization, and polydispersity index	Aveston (1965), Langerman (1986)
Light scattering (LS)	Average molecular weight, N_z', average number of aluminums per aggregate, and z', charge per aggregate	Tyree (1958, 1973), Riesgraf et al. (1979)
Vapor phase osmometry (VPO)	MW_n at infinite dilution	Bretschneider et al. (1975)
Gel filtration chromatography (GFC)	MW distribution profiles, MW_n and MW_w values, polydispersity	Fitzgerald and Rand (1975)

[a]For definitions of MW_n, MW_w, and MW_z, see Poole and Schuette (1984), p. 299.

attainable by each method. The physical-chemical basis and general experimental approaches of these classical techniques have been well described in the literature (Flory, 1953; Rollins et al., 1983; Bombaugh, 1971).

The *ultracentrifugation technique* (Ansevin et al., 1970; Johnson et al., 1959; Yphantis, 1974) involves examination of the sedimentation or diffusion of solute species in intense gravitational fields induced by centrifuging samples at high rotational speeds up to 60,000 rpm. Two basic experiments are commonly used: sedimentation velocity and sedimentation equilibrium. In a sedimentation velocity experiment, the rate of sedimentation of solute species is measured at high rotational speeds and the solute species observed in a rotating cell using Schlieren optical systems. Sedimentation velocity experiments provide information on sample heterogeneity or polydispersity since solute molecules have different sedimentation rates (denoted by sedimentation constants in units of svedbergs, S, where 1 S is 10^{-13} sec) depending on their molecular size. If the diffusion constants and partial specific volumes of solutes can be measured, one can also obtain molecular weight information. A sedimentation equilibrium experiment is carried out at lower speeds to avoid sedimentation of larger molecules to the bottom of the sample cell. Centrifugation is done at speeds necessary to achieve an equilibrium distributions of solutes throughout the rotating centrifugation cells. Analysis of the solute distribution throughout the centrifuge cell by various optical measurements (Johnson et al., 1959) at equilibrium (sometimes requiring days) allows one to determine the sample heterogeneity and to calculate the weight-averaged molecular weight values for various solutes in polydisperse solutions. This technique does not require measurements of solute diffusion constants; however, MW_w values generally overemphasize larger molecules.

Light scattering studies of macromolecules (Flory, 1953; Patterson and Tyree, 1973; Ruff, 1959) require measurement of the scattering or diffraction of light incident on solute-containing solutions. The light scattering by a solution is usually expressed by its Rayleigh turbidity at 90° to the incident light. If the solute-containing solutions are polydisperse, weight-average molecular weight values can be calculated. The technique of Patterson and Tyree (1973) has been utilized effectively to examine various metal ion hydrolysis systems. This technique permits the determination of two important parameters for charged polyelectrolytes: N_z, the average number of metal ions per aggregate, and z', the average charge per aggregate.

Vapor phase osmometry (Flory, 1953) utilizes measurement of the osmotic pressure, a colligative property, of various solutes using a semipermeable membrane apparatus. These measurements provide a means of determining solute species molecular weights, specifically number-average molecular weights, MW_n. The MW_n values generally overemphasize contributions from molecules of relatively small size. In the case of polyelectrolyte solutes, solute charges reduce the observed osmotic pressure, necessitating measurements in media of high ionic strength such as 0.2 M salt solutions.

Gel filtration chromatography (GFC) or *gel permeation chromatography* (GPC) is one of the most rapid and useful techniques (Kirkland, 1971; Bombaugh, 1971; Rollins et al., 1983; Poole and Schuette, 1984) applicable to the study of the macromolecular nature of polymeric species in aqueous or nonaqueous solutions. Its major advantage is speed of analysis, allowing the direct determination of both qualitative and quantitative information about the species composition of polydisperse polymer systems. Chemical characterization of both aqueous and nonaqueous polymer-containing solutions by this technique provides a diverse range of characterization information, including

1. Quantitative and qualitative molecular weight distribution information including molecular weight profiles, MW_w and MW_n values, and polydispersity data
2. Estimates of molecular size of individual species
3. Molecular weight distribution profiles of kinetic phenomena, including polymerization and depolymerization reactions

The use of the technique to define the species composition and molecular weight characteristics of aluminum hydrolysis complexes as a function of $\bar{n}$, solution media, temperature, aging, and methods of synthesis has been described (Fitzgerald, 1975, 1977, 1981; Fitzgerald and Johnson, 1980). Gel filtration chromatography, which is based on molecular size separation, has received widespread application in polymer and biological chemistry. It has been applied specifically to the study of iron(III) and silicon(IV) hydrolysis systems, in addition to aluminum systems.

In a gel filtration experiment involving the separation of various polymeric hydrolysis systems, it has been established that molecular size, chiefly solute hydrodynamic radius, is the controlling separation mechanism under optimum elution and eluent conditions. Polymeric solutes are separated by differential migration and partitioning between the mobile liquid phase and a steric-exclusion solid stationary phase (gel).

The GFC method for separating aluminum hydrolysis solutes has concentrated on the use of Sephadex gels under hydrostatic flow conditions. Most recently, the use of high-pressure liquid chromatographic (HPLC) approaches, using a range of aqueous steric exclusion gels as the stationary phase for studies of commercial basic aluminum hydrolysis systems, has been reported (Gosling et al., 1978a, 1982a; Fitzgerald, 1981; Nelson, 1985). Pertinent properties of the various Sephadex gels used to date to study aluminum systems are given in Table 7. It should be noted that Sephadex G-25, with a fractionation range of 1500–5000 daltons, is most applicable to basic aluminum salts, whereas Sephadex G-50, with an exclusion limit of 30,000–50,000 daltons, is more applicable to mixed aluminum/zirconium salt systems.

Various theoretical and experimental relationships have been derived to explain the GFC mechanism of separation and the relationship between the elution behavior of macromolecules and their molecular weights (Bombaugh, 1971). Experimentally, a gel filtration chromatogram such as the one depicted in Fig. 8A for the four solutes labeled C, D, E, and F is obtained. The elution profile of these solute species is usually measured by any of a number of detectors, including refractive index (RI) or ultraviolet (UV)-visible

Table 7 Pertinent Sephadex® Gel Properties

Gel type	Dry particle diameter (μm)	Water regain, W_r (ml/g)	Fractionation range (MW)
G-10	40–120	1.0 ± 0.1	0–700
G-15	40–120	1.5 ± 0.1	0–1500
G-25			
Coarse	100–300	2.5 ± 0.2	1000–5000
Medium	50–150	2.5 ± 0.2	1500–5000
Fine	20–150	2.5 ± 0.2	1500–5000
Superfine	10–40	2.5 ± 0.2	1500–5000
G-50 medium	50–100	5.0 ± 0.3	1500–30,000

Source: Determan (1969) and Pharmacia (1968). Sephadex® is a registered trademark of Pharmacia Inc.

detectors. The intensity of the measured parameter in the eluted stream is generally related to the concentration of various solute species, often by correlation via metal analysis for metal hydrolysis systems.

The differential elution of a solute contained in the sample is described by its elution volume, V_e, and three volumes which are constant for a given gel column (Figs. 8A and 8B). The total liquid column volume, V_t, is equal to the sum of the liquid volume inside the gel matrix, V_i (imbibed volume), and the liquid volume external to the gel matrix, V_0 (void volume). Molecules larger than the pores of the gel bed matrix, above the exclusion limit (C), cannot penetrate the gel particles, pass through the liquid phase, are excluded, and are eluted first at the void volumn. Molecules much smaller than the gel pore size completely penetrate the gel and are eluted last at the total volume (F). Selective permeation occurs for intermediate-sized molecules (D and E), which are eluted between V_t and V_0. The classical chromatography equation,

$$V_e = V_0 + K_d V_i \quad \text{or} \quad K_d = \frac{V_e - V_0}{V_i} \tag{16}$$

where V_e = elution volume of solute
V_0 = column void volume (K_d = 0, solute C)
V_i = imbibed volume

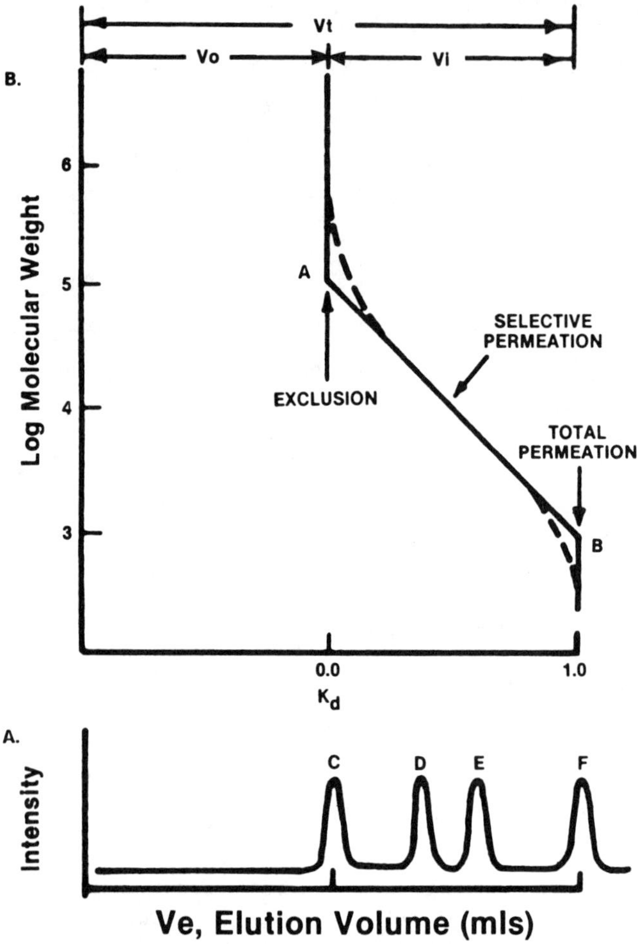

Figure 8 Depiction of (A) idealized gel filtration chromatogram and (B) relationship of GFC peak elution volume to various column parameters, K_d values, and the log molecular weight of solute species. (Modified by permission from Kirkland, 1971.)

V_t = total column volume (K_d = 1, solute F)
K_d = distribution of solute in imbibed volume ($0 < K_d < 1$)

best describes the relationship between the eluted volume of a solute and the column parameters. Here K_d represents the distribution coefficient of the solute in the imbibed volume. Experimental chromatograms, therefore, may be described mathematically by peak K_d

values and the relative intensity of the observed elution peaks containing the variable-sized solutes.

To obtain molecular weight data (both individual peak distributions and statistical molecular weight parameters) from GFC data, a logarithmic relationship between molecular size (either MW or hydrodynamic radius) and elution volume (expressed as K_d) is used. Calibration plots of log MW versus K_d may be constructed as shown in Fig. 8B. Ideally, the region of selective permeation is a straight line given by the equation

$$\log MW = A \log K_d + B \qquad (17)$$

where A and B are constants determined experimentally using calibration standards of known hydrodynamic radius or MW.

D. Spectroscopic Techniques

Although the use of this broad range of macromolecular techniques to study and characterize the polymeric properties of commercial aluminum hydrolysis systems has been highly successful, the structural nature of individual species is most readily approached using a number of applicable spectroscopic techniques, such as *Fourier transform infrared spectroscopy (FTIR)*, *Raman spectroscopy*, and *^{27}Al NMR*. The technique of ^{27}Al NMR has received the most extensive attention in the past decade. A number of excellent reviews of ^{27}Al NMR have been published (Hinton and Briggs, 1978; Akitt, 1972; Delpuech, 1983), including its use in solution and the solid state. The pertinent characteristics of ^{27}Al NMR spectra of basic aluminum hydrolysis systems will be discussed later for some of the most important work reported using this approach.

E. Kinetic Reactivity Methods (Ferron)

The ferron reactivity assay, by contrast, is a timed spectrophotometric analysis of the rate of complexation of aluminum(III) ion complexes with the ferron reagent in aqueous solution. The ferron reagent refers to a buffered mixture (pH 5.0) of the aluminum specific ligand, 7-iodo-8-hydroxyquinoline-5-sulfonic acid, sodium acetate, and hydroxylamine. The ferron kinetic assay has been extensively used by Hem and Smith, Gessner, and Schonherr to examine a range of aluminum hydrolysis solutions. Although the technique has been critized by Gildea et al. (1977) on a number of kinetic grounds, it permits one to describe the differences in the rate of reaction of aluminum solutes in terms of concentrations of three independent forms of aluminum hydroxy complexes. As originally defined by Smith and Hem (1972), the technique quantifies

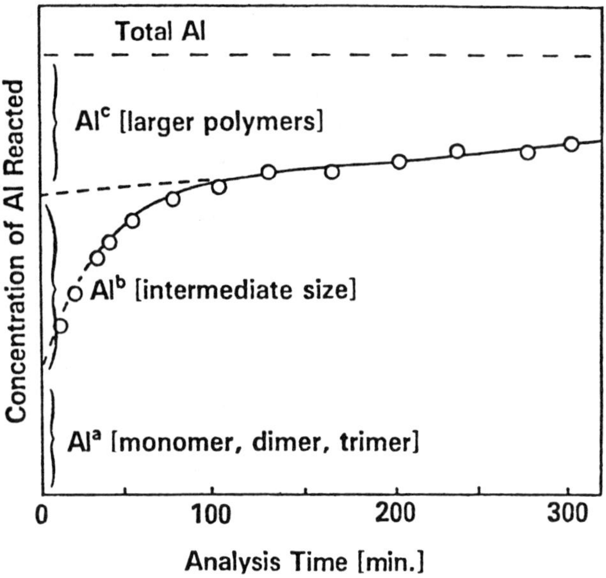

Figure 9 Idealized reaction profile of various aluminum species with the ferron reagent. (Reproduced by permission of Riesgraf, 1979, and Smith and Hem, 1972.)

aqueous aluminum hydrolysis complexes on the basis of three species types: Al^a, monomeric Al^{3+} species; Al^b, first-order reacting polymeric species; and Al^c, microcrystalline filterable gibbsite or high-molecular-weight polymeric species. This reaction scheme and the various hydrolysis complexes which are included in each of the different types of aluminum forms are shown in Fig. 9. In essence, Al^a includes soluble monomeric species, Al^b includes slower-reacting polynuclear species such as the dimer and $Al_{13\text{-mer}}$, and Al^c includes soluble aluminum hydroxide complexes which are unreactive in a 3-hour period with the ferron reagent system. Aqueous solutions of $AlCl_3 \cdot 6H_2O$ contain principally (90–95%) Al^a. The monomeric Al $(H_2O)_6{}^{3+}$ ion is instantaneously reactive to ferron; i.e., it is readily available for ligand complexation forming a 3:1 ferron:Al^{3+} complex. Polymer solutes, on the other hand, require a slow dissociative reaction to monomeric units prior to complexation,

$$Al_n \text{ (polymer)} \xrightarrow{\text{slow}} Al_{n-1} \text{ (polymer)} + Al(H_2O)_6{}^{3+} \qquad (18)$$

$$Al(H_2O)_6^{3+} + \text{ferron} \xrightarrow{\text{fast}} \text{Al-ferron complex} \quad (19)$$

As such, Al^b- and Al^c-containing solutions undergo "slow" and "very slow" reactions with ferron, respectively. Studies of the rate of reaction of the ferron reagent with aluminum hydrolysis complexes utilize timed spectrophotometric measurements of the absorbance of the aluminum-ferron complex at 368 nm in a pH 5.0 acetate buffer system. The results of this timed spectral analysis are typically reported on the basis of two major parameters:

1. Quantitative determination of the % Al^a, % Al^b, and % Al^c contents of the solutions assayed.
2. Calculation of the pseudo-first-order rate constants of the Al^b species and Al^c species. In addition, some workers have also reported the rates of reaction of Al^c complexes in terms of the micrograms of Al reacted per minute (μg Al/min).

The determination of the % Al contents of the various Al^a, Al^b, and Al^c species is usually carried out by graphical or computer methods of analysis of ferron reaction profiles such as those shown in Fig. 9. The % Al^a species is determined by extrapolation of the reaction profiles to zero time to obtain the Al^a reacted in order to calculate % Al^a = Al^a(reacted)/Al_{total}(reacted) × 100. The % Al^b is calculated from the equation % Al^b = Al^b(reacted)/Al_{total}(reacted) × 100, where the Al^b(reacted) is determined from a reaction profile such as that shown in Fig. 9 by extrapolating the linear region of the profile to zero time. The % Al^c content of aluminum solutions is determined from the equation % Al^c = Al^c(reacted)/Al_{total}(reacted) × 100, where Al^c(reacted) is determined by difference, i.e., Al_{total}(reacted) - [Al^b(reacted) + Al^a(reacted)]. Generally, the Al^b species are completely reacted with the ferron reagent in the first 3 hours of a reaction assay. The determination of the pseudo-first-order rate constants for the Al^b and Al^c species is usually carried out by computational methods using standard first-order kinetic treatment. Similarly, if the rate of reaction of Al^c species is expressed by the quantity μg Al(reacted)/min, this parameter is usually calculated directly from the reaction profiles using various computational techniques. To obtain good rate of reaction data for Al^c species, reaction profiles on the order of 6–24 hours are needed.

Baseline studies of the ferron reaction kinetics of unique aluminum hydrolysis species such as the aquated monomer, the dimer, and the $Al_{13\text{-mer}}$ have been reported by Gessner and Winzer (1979).

These workers used aluminum solutions prepared from the soluble crystalline salts of the following complex ions: $Al(H_2O)_6^{3+}$, $Al_2(OH)_2(H_2O)_8^{4+}$ and $Al_{13}O_4(OH)_{24}(H_2O)_{12}^{7+}$. A summary of the various reaction kinetics reported by Gessner and Winzer, including their observed pseudo-first-order rate constants, half-lives of reaction, and estimates of the time for complete reaction, is given in Table 8.

Table 8 Summary of Various Aluminum Hydrolysis Species on the Basis of Ferron[c] Reaction Assay

Form[a]	Complexes	Description of Kinetics			
		Relative Rate[a]	k[b] (min)	$t_{1/2}$[b] (min)	Total reaction time (min)
Al^a	$Al(H_2O)_6^{3+}$ $Al(H_2O)_5(OH)^{2+}$	Fast	2.3	0.3	2
Al^b	$Al(OH)_2(H_2O)_8^{4+}$	Slow	0.22	3.0	14
	$Al_{13}O_4(OH)_{24}(H_2O)_{12}^{7+}$	pseudo-first-order	7.5×10^{-2}	9.5	60
Al^c	High-molecular-weight soluble polymers[b] or micro-crystalline gibbsite	Very slow	3.1×10^{-3}	125	days

[a]Notation Al^a, Al^b, and Al^c is based on terminology of Smith and Hem (1972); Al^c refers to microcrystalline gibbsite.
[b]Specific rate constants and other kinetic data reported by Gessner and Winzer (1979). Designation of high-molecular-weight species as Al^c-type aluminum complexes noted by these workers.
[c]Ferron refers to a buffered mixture (pH 5.0) of the aluminum specific ligand, 7-iodo-8-hydroxyquinoline-5-sulfonic acid, sodium acetate, and hydroxylamine.

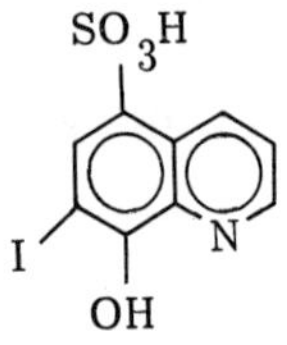

The aquated monomer (and some of the hydroxy monomer $Al(OH)(H_2O)_5^{2+}$ ion) reacts completely in less than a minute to form the ferron complex. The dimer and the $Al_{13\text{-mer}}$ take about 14 and 60 minutes, respectively, to react completely. These two species and other intermediate polynuclear complexes may be classified as Al^b species according to the Smith and Hem notation. Larger, less reactive polymeric species, classified as Al^c complexes according to the Smith and Hem notation, require upward of days to completely react with the ferron reagent.

V. SOLUTION CHEMISTRY OF BASIC ALUMINUM HALIDE SYSTEMS

The solution chemistry of commercial basic aluminum halide complexes of the general empirical formula $Al_2(OH)_m(X)_n$, where $m + n = 6$ and X is an anion of the type Cl^-, Br^-, I^-, or NO_3^-, includes a broad range of basic aluminum hydroxide solutions. Although the patent literature pertaining to basic aluminum halide systems includes this whole family of anion complexes, the research literature consists primarily of studies to characterize the basic aluminum chloride (BAC) series of complexes. A very limited number of experimental measurements have been reported on aluminum bromohydrate, $Al_2(OH)_5Br$, aluminum iodohydrate, $Al_2(OH)_5I$, or aluminum nitrohydrate, $Al_2(OH)_5NO_3$. These systems and their less basic homologs will not be discussed in this chapter.

A. Basic Aluminum Chloride Systems—Nomenclature

Basic aluminum chloride systems comprise a series of related complexes of any fractional basicity such as shown in Table 9. The use of

Table 9 Various Basic Aluminum Chloride Systems

BAC salts	Empirical formula	$\bar{n}$ or $[OH^-]/[Al^{3+}]$
1/6 basic	$Al_2(OH)Cl_5$	0.5
2/6 basic (1/3)	$Al_2(OH)_2Cl_4$	1.0
3/6 basic (1/2)	$Al_2(OH)_3Cl_3$	1.5
4/6 basic (2/3)	$Al_2(OH)_4Cl_2$	2.0
5/6 basic	$Al_2(OH)_5Cl$	2.5

Table 10 Nomenclature for BAC Salts

Chemical name	Common name	Abbreviation
Aluminum chlorohydrate	5/6 basic	ACH
Aluminum sesquichlorohydrate	3/4 basic	ASCH
Aluminum dichlorohydrate	2/3 basic	ADCH

the empirical formula $Al_2(OH)_m(Cl)_n$ is convenient for relating aqueous or solid BAC salt systems to the commercial 5/6 basic aluminum chloride system, $Al_2(OH)_5Cl$, referred to as aluminum chlorohydrate. Studies of commercial basic aluminum chlorides have most clearly focused on three distinct commercially available complexes, referred to as 5/6 basic, 3/4 basic, and 2/3 basic. The relationships of the various fractional basicities, their accepted CTFA nomenclature, and common abbreviations, e.g., ACH, ASCH, and ADCH, respectively, are given in Table 10.

B. Basic Aluminum Chloride Systems—Analytical Analyses

BAC salts are readily available both in commercial aqueous solutions and as spray-dried powders of various particle sizes. The chemical analysis of powders relies primarily on measurements of their Al/Cl ratios as determined by elemental analysis, measurements of their water contents, and various solution analyses following dissolution of these powders in aqueous solution. The % Al and % Cl contents of powders vary from 27 to 21% Al and 17 to 27% Cl for the three most common salts. Typical examples of the elemental composition of these three salts are given in Table 11 together with corresponding analytical data for concentrated BAC solutions. Solution elemental analyses for % Al and % Cl contents of concentrated BAC solutions are also given in Table 11, together with calculated Al/Cl ratios, $\bar{n}$ or $[OH^-]/[Al^{3+}]$ ratios, concentrated solution densities, and comparative pH values for 5% (wt/wt) solutions. The % Al and % Cl contents of concentrated BAC solutions vary from 10 to 12% Al and 14 to 8% Cl over the OH/Al ratios 1.00–2.00. The most basic system, ACH, is the most concentrated (12.5% Al or approximately 5–6 M Al) and has the highest $\bar{n}$ value of 2.50, highest solution density of 1.34 g/ml, and highest solution pH of 4.3 for a 5% solution. Conductivity measurements for ACH and ADCH solutions (Bret-

Table 11 Nomenclature, Elemental Composition, and Relevant Solution Parameters for BAC Powders and Solutions

Nomenclature	Elemental composition						Solution parameters	
	Powders		Solution					Density
Chemical name	%Al	%Cl	%Al	%Cl	Al/Cl[a]	$\bar{n}$(OH/Al)	pH(5%)	(g/ml)
Aluminum chlorohydrate	26.4	17.6	12.5	8.2	2.00	2.50	4.3	1.34
Aluminum sesquichlorohydrate	22.6	22.5	11.5	11.5	1.33	2.33	3.9	1.30
Aluminum dichlorohydrate	21.0	26.2	10.6	13.8	1.00	2.00	3.8	1.25

[a]According to recent FDA-OTC definitions, the following Al/Cl ranges are applicable: ACH, 2.1 down to, but not including 1.9:1; ASCH, 1.9 down to, but not including 1.25:1; ADCH, 1.25 down to, but not including 0.9:1.

Schneider et al., 1975) reveal that dilute ADCH solutions have a higher degree of dissociation than ACH solutions.

C. Basic Aluminum Chloride Systems—Solution pH and Density

pH measurements (Ftizgerald et al., 1980) of a series of ACH, ASCH, and ADCH solutions of different concentrations prepared from concentrated solutions from the same supplier are shown in Fig. 10. These measurements illustrate the concentration dependence of "freshly" diluted BAC solutions. The range of pH values for ACH is 1.4 units, for ASCH 2.2 pH units, and for ADCH 2.3 pH units over the concentration range examined. The largest pH changes occur in the lower concentration range from 5 to 0.025%. The pH is noted to increase with concentration as expected. Solution H^+ ion concentrations are highest for ADCH solutions over the entire concentration range for comparable solution concentrations. The order of basicity of these complexes is in agreement with their $\bar{n}$ values noted in Table 11. Aging of these BAC salts over 2 months has a pronounced influence on the solution pH values for ACH, ASCH, and ADCH solutions (25 to 2%) as shown in Fig. 11. All

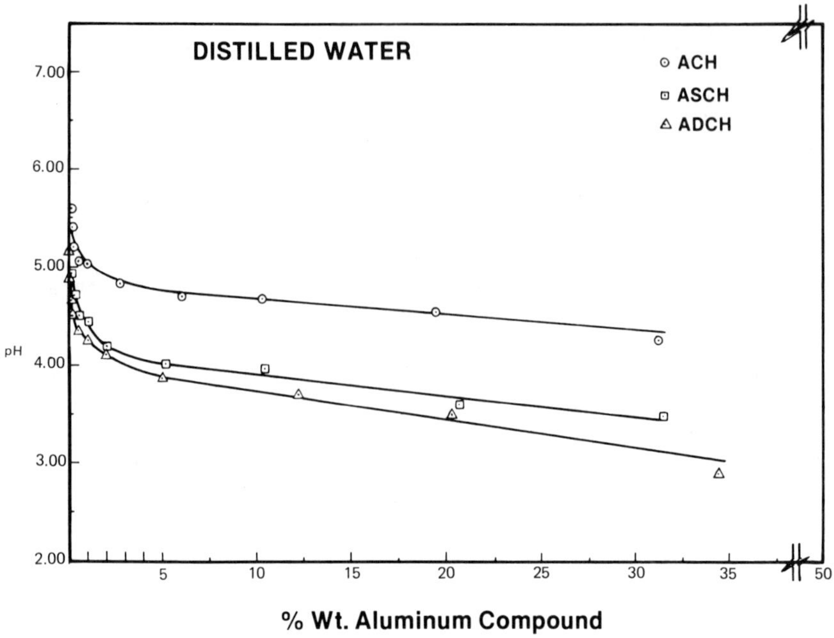

Figure 10 Plot of pH versus concentration for various freshly diluted basic aluminum chloride solutions. (Fitzgerald, unpublished results.)

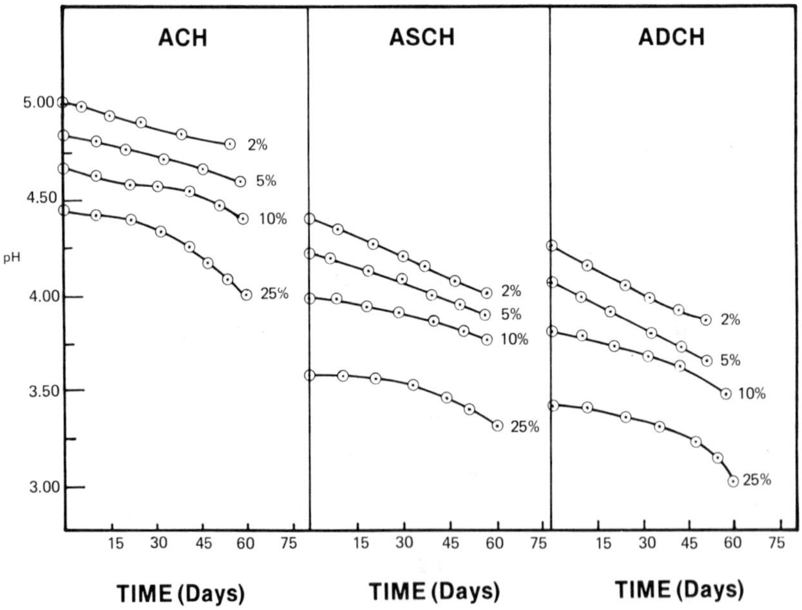

Figure 11 Profiles of pH for BAC solutions from 25 to 2% concentration over 2 months of aging. (Fitzgerald, unpublished results.)

dilute BAC solutions decrease in pH on aging, the largest decreases being noted for the 25% solutions. Aging of BAC solutions produces a net increase in $[H^+]$, which must arise from reorganization of aluminum solutes during the aging process, concomitant with release of H^+ ion. At comparable concentrations, a minimal pH change corresponding to the release of H^+ ion is observed for ACH, in contrast to the less basic systems. Larger pH changes as a function of aging for ASCH and ADCH imply that these systems undergo more extensive reorganization or hydrolysis reactions on dilution and aging.

Solution densities for ACH as well as ASCH and ADCH are also found to be concentration-dependent, as expected. The density of a 50% ACH solution [based on the formula $Al_2(OH)_5Cl \cdot 2H_2O$] is 1.34 g/ml at 25°C, in comparison with 1.04 g/ml for a 10% ACH solution. Solution density is a very sensitive physical-chemical parameter for BAC solutions.

D. Basic Aluminum Chloride Systems—Methods of Synthesis

The synthesis methods used to prepare commercial BAC solutions involve three approaches: 1. the oxidation-reduction reaction used to prepare the parent system, ACH, 2. the preparation of ASCH and ADCH solutions using ACH as a starting reagent, and 3. the preparation of activated ACH (ACH') using ACH solutions as a starting reagent. Since ACH solutions are used as starting reagents in the latter two synthetic approaches, the redox synthesis of ACH is considered first.

The oxidation of aluminum metal in highly acidic aqueous aluminum chloride solutions was first reported in the patent by Huehn and Haufe (1940). This method of synthesis is used at present to produce highly concentrated aluminum chlorohydrate as given by the following balance reaction equation:

$$10Al(s) + 2AlCl_3(aq) + 30H_2O \rightarrow 6Al_2(OH)_5Cl(aq) + 15H_2 \qquad (20)$$

This reaction is envisioned to involve two fundamental reaction steps: 1. the oxidation of aluminum metal to Al^{3+} ion,

$$Al(s) \rightarrow Al^{3+}(aq) + 3e^- \qquad (21)$$

and the reduction of water,

$$2H_2O + 2e^- \rightarrow H_2 + 2OH^-(aq) \qquad (22)$$

to produce hydrogen gas and liberate hydroxide ion in solution, and 2. the hydrolysis of Al^{3+} ion to solutions containing polymeric aluminum hydrolysis species of a given Al/Cl ratio as shown below:

$$Al(H_2O)_6^{3+} + mOH^- + nCl^- \rightarrow Al_2(OH)_m(Cl)_n \quad (23)$$

The redox synthesis of ACH thus produces an aqueous solution of Al/Cl ratio 2:1 starting with a heterogeneous reaction system including an aqueous $AlCl_3$ solution of Al/Cl ratio 0.33. The reaction progresses from an initial basicity of $\bar{n} = 0.00$ to $\bar{n} = 2.50$, thus progressing through the various basicity BAC solutions are defined by their analytical Al/Cl values. Production of a 50% ACH solution necessitates carrying out this reaction with an initial aluminum content in the reaction mixture that will eventually produce a concentrated ACH solution near 6.5 M Al. According to the Huehn and Haufe patent (1940), the choice of Al metal—the use of HCl/Al or $AlCl_3$/Al mixtures—is not as critical as the reaction temperature of 70–90°C.

The synthesis of commercial less basic ASCH and ADCH solutions is carried out by two approaches: 1. HCl neutralization of ACH solutions, and 2. aqueous $AlCl_3$ neutralization of ACH solutions. These two reactions may be generically summarized using the two equations shown below:

Reaction 1: $$Al_2(OH)_5Cl(aq) + HCl(aq) \rightleftarrows Al_2(OH)_mCl_n(aq) \quad (24)$$

Reaction 2: $$Al_2(OH)_5Cl(aq) + AlCl_3(aq) \rightleftarrows Al_2(OH)_m(Cl)_n(aq) \quad (25)$$

Balanced equations representing the reaction stoichoimetry using these two reactions are given below for the synthesis of ADCH ($\bar{n} = 2.00$) solutions:

$$HCl(aq) + Al_2(OH)_5Cl(aq) \rightleftarrows Al_2(OH)_4Cl_2(aq) + H_2O \quad (26)$$

$$2AlCl_3(aq) + 4Al_2(OH)_5Cl(aq) \rightleftarrows 5Al_2(OH)_4Cl_2(aq) \quad (27)$$

The synthesis of activated ACH systems, by contrast, has been carried out using a number of approaches involving the thermal treatment of dilute ACH solutions.

The preparation of commercial ACH, ASCH, ADCH, and ACH' solid systems involves spray-drying, lyophilization, or other concentrating methods which are standard techniques in the industrial preparation of powdered materials.

VI. CHEMISTRY OF ALUMINUM CHLOROHYDRATE (ACH)

A. Macromolecular Nature of ACH

The macromolecular nature of aqueous solutions of ACH is exceedingly complex. Concentrated aqueous solutions are approximately 50% (wt/wt) on the basis of the formula $Al_2(OH)_5Cl \cdot 2H_2O$, when prepared commercially. Even higher ACH solution concentrations may be prepared by dissolution of powdered forms. Examination of these concentrated solutions directly is not achievable by any of the macromolecular techniques which have been used to examine the polymeric species compositions of this system. Therefore, studies of ACH solutions by gel filtration chromatography, light scattering, ultracentrifugation, and vapor phase osmometry have used solutions in the 2.5–0.05 M Al range. Although the latter three techniques have commonly used solutions of 4–0.1% (0.4–0.01 M Al), ACH samples of 25–1% (2.5–0.1 M Al) may be conveniently studied by GFC techniques.

Gel filtration chromatography studies of ACH solutions (Fitzgerald, 1975) were initially carried out using Sephadex G-25 columns with mixed KCl/HCl eluents. A typical gel filtration chromatogram of a 10% aqueous sample of ACH obtained by elution on a Sephadex G-25 column is shown in Fig. 12. The experimental ACH chromatograms are reported as refractive index profiles or as corresponding aluminum content profiles as shown in Figs. 12a and 12b, respectively. The experimental RI profile consists of four major peak distributions, labeled peaks 1, 2, 2b, and 4. Peak distribution 4 is essentially a non-aluminum-containing "artifact" or "ghost" peak as evidenced by the aluminum content profile. The GFC profile of ACH on Sephadex G-25 normally exhibits three aluminum species distributions of K_d values and relative peak intensities (normalized peak heights) as summarized in Table 12. The peak K_d values are invariant for all ACH samples prepared from commercial powders or concentrated solutions; however, there is a dependence on relative peak intensities. The GFC profiles of ACH readily illustrate the polydisperse nature of these hydrolyzed systems. Peak distributions 1, 2, and 2b at K_d values of 0.00, 0.40, and 0.69 contain polymeric aluminum species of varying molecular size from 1100 to 8000 daltons. In addition, a shoulder peak at K_d = 0.25, on the lower elution volume side of peak 2, is sometimes observed. This additional polymeric peak distribution has been label peak 2a.

The molecular size range of polymeric solutes in ACH has been obtained (Fitzgerald, 1975) by the examination of ACH samples

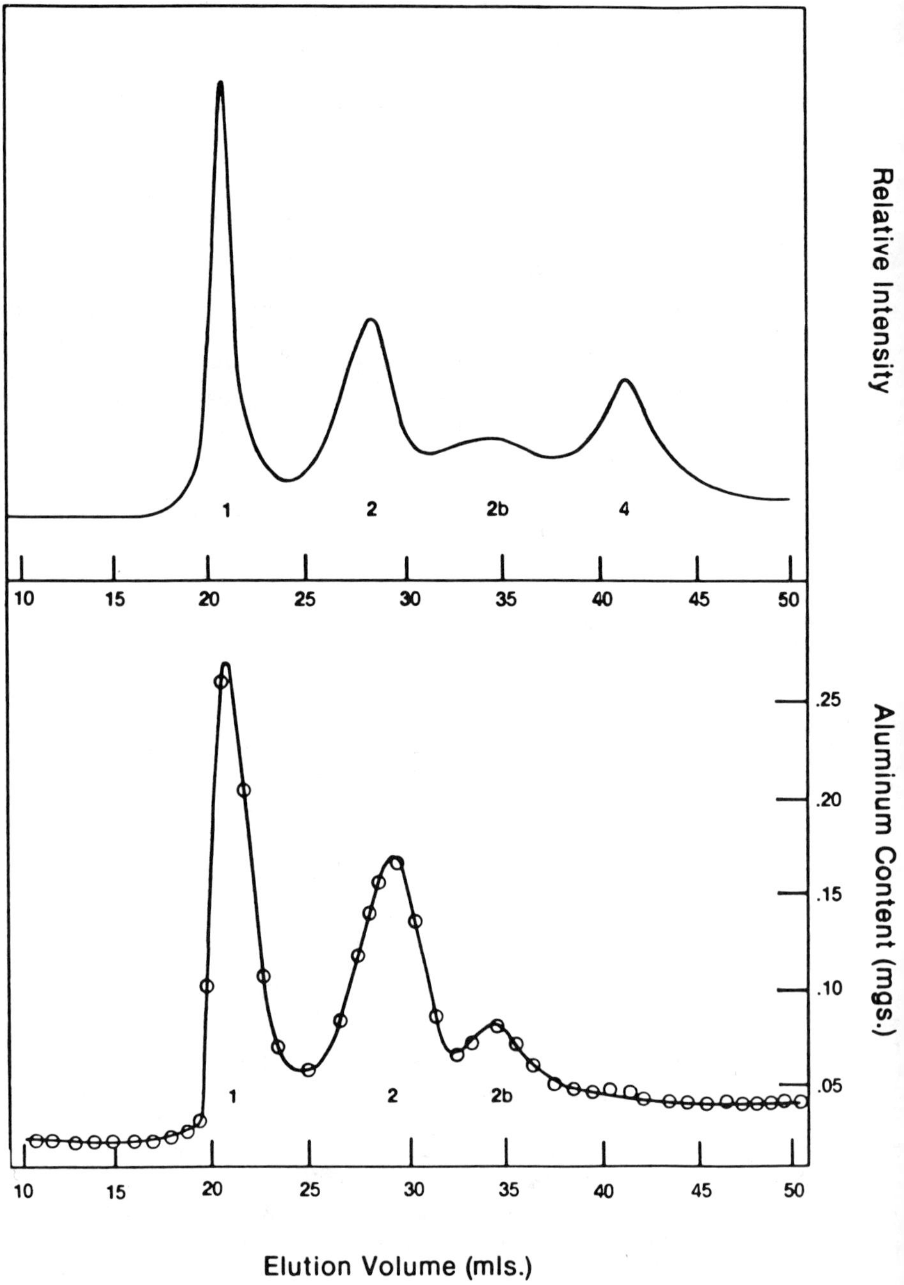

Figure 12 Sephadex G-25 gel filtration chromatogram of 10% aluminum chlorohydrate: (above) refractive index profile and (below) aluminum content profile (Fitzgerald, 1975).

Table 12 GFC Parameters of 10% ACH Solutions on Sephadex G-25 Columns

Peak number	K_d[a] (V_e, ml)	Normalized peak height
1	0.00 (20.6)	57.4
2a	0.25 (25.6)	—
2	0.40 (28.6)	32.2
2b	0.68 (34.1)	10.4
4 (non-Al)	1.00 (40.5)	—[b]

[a]Calculated from the equation $K_d = V_e - V_0/V_i$ where $V_0 = 20.6$ ml, $V_i = 19.9$ ml.
[b]Peaks 2a and 4 are not considered in normalization of relative peak heights.

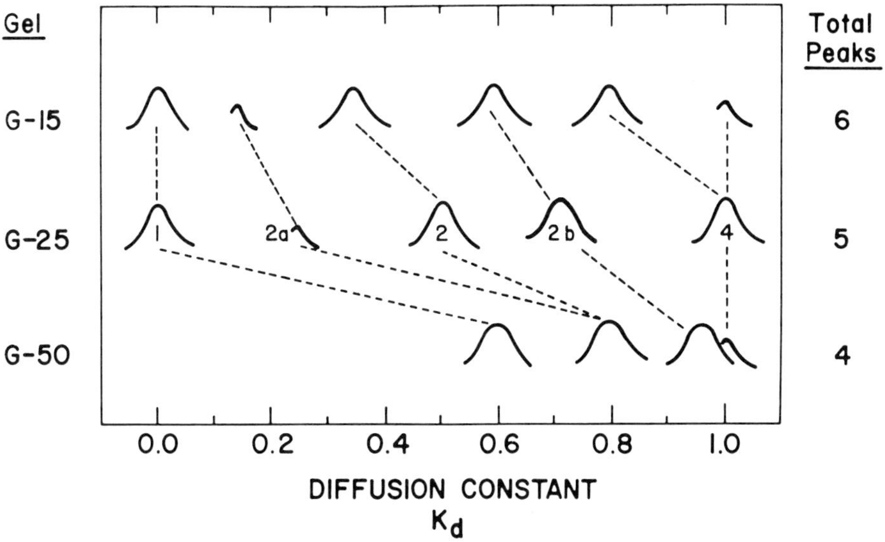

Figure 13 Depiction of peak diffusion constants for ACH chromatographic peaks on various Sephadex columns (Fitzgerald, 1975).

using various Sephadex columns, e.g., Sephadex G-15, G-25, and G-50, having exclusion limits of MW 1500, 5000, and 10,000–30,000, respectively. Figure 13 depicts the variation of peak K_d values for ACH components as recorded on these different Sephadex columns. ACH exhibits six, five, and four chromatographic peaks, including shoulders, on G-15, G-25, and G-50 columns, respectively. All chromatograms exhibit the non-aluminum-containing ghost peak 4. Increases in the gel pore size in going from G-15 to G-50 result in decreased separation of the solute peaks, as expected theoretically. The data shown in Fig. 13 allow comparison of the variable behavior of an ACH solution on different Sephadex gel. The peak intensities are not depicted, but rather the peaks are identified as resolved (full peaks) or unresolved shoulders (partial peaks).

Considering peak 1, the K_d value for the largest species distribution indicates exclusion of these solutes on both G-15 and G-25 gels (K_d = 0.00) but partial inclusion (K_d = 0.61) on G-50 gels. This peak corresponds to a unique species distribution in the range 5000–8000 daltons. Peak distributions 2a and 2 are separated on G-15 and G-25 gels, but coalesce into a single distribution on the larger pore size G-50 gel. A range of 1500–3000 daltons is consistent with this behavior. Peak 2b is resolved on all gels examined and has a G-25 K_d range of 0.65–0.72, corresponding to an intermediate polymeric solute distribution in the 900–1200-dalton range. Peak 3 probably corresponds to monomeric or dimeric species and is observed only for ACH solutions on G-15 gel columns. In summary, ACH solutions are polydisperse and contain at least four polymeric aluminum solute species distributions of 900–8000 daltons, in addition to a minimal concentration of monomeric or dimeric Al^{3+} species.

The chromatographic behavior of ACH on Sephadex gels may be contrasted with chromatograms obtained for $AlCl_3 \cdot 6H_2O$ and other partially neutralized aluminum solutions (Fitzgerald, 1981; Fitzgerald and Johnson, 1980) of $\bar{n}$ = 2.4–2.5. Typical Sephadex G-25 chromatograms of a 1.0 M aluminum chloride solution and a neutralized ($\bar{n}$ = 2.5) aluminum chloride solution are shown in Figs. 14 and 15, respectively. Calculated GFC parameters for these two chromatograms, the ACH chromatogram of Fig. 12, and a number of other neutralized aluminum chloride solutions are given in Table 13. The experimental chromatograms of all four aluminum chloride solutions readily illustrate the difference in the species compositions of these systems. Aqueous solutions of $AlCl_3 \cdot 6H_2O$ from 10 to 20% concentration exhibit chromatograms with a single aluminum solute peak distribution (Fig. 14) with a peak width of 5–7 ml. This monodisperse peak distribution (K_d = 0.81) is assigned to the aquated monomer. The GFC data for the three partially neutralized aluminum solutions prepared by the $AlCl_3/NaHCO_3$ and $AlCl_3/Al(OH)_3$ methods contain principally (>90% Al) one aluminum species distribution (peak 2b,

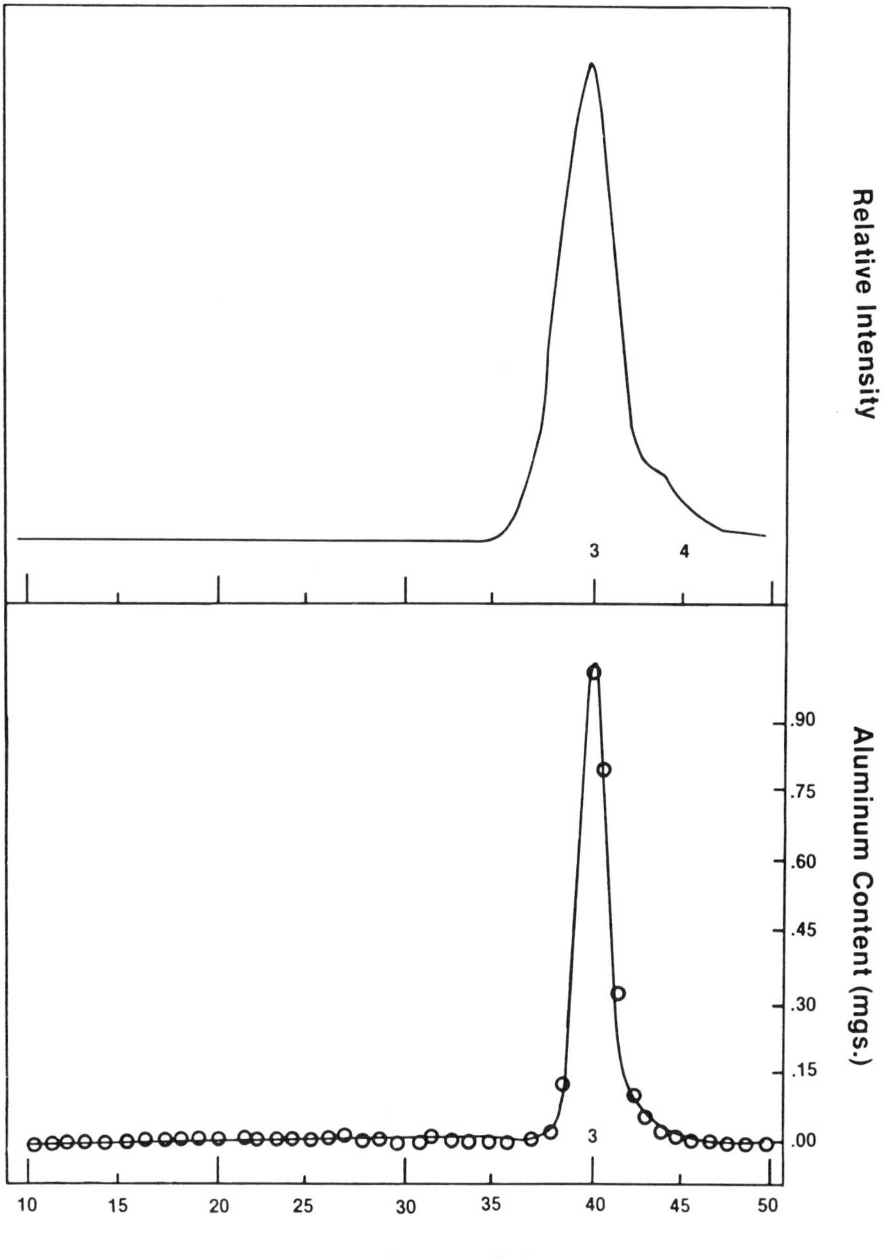

Figure 14 Sephadex G-25 gel filtration chromatogram of 20% $AlCl_3 \cdot 6H_2O$ solution: (above) refractive index profile and (below) aluminum content profile (Fitzgerald, 1981).

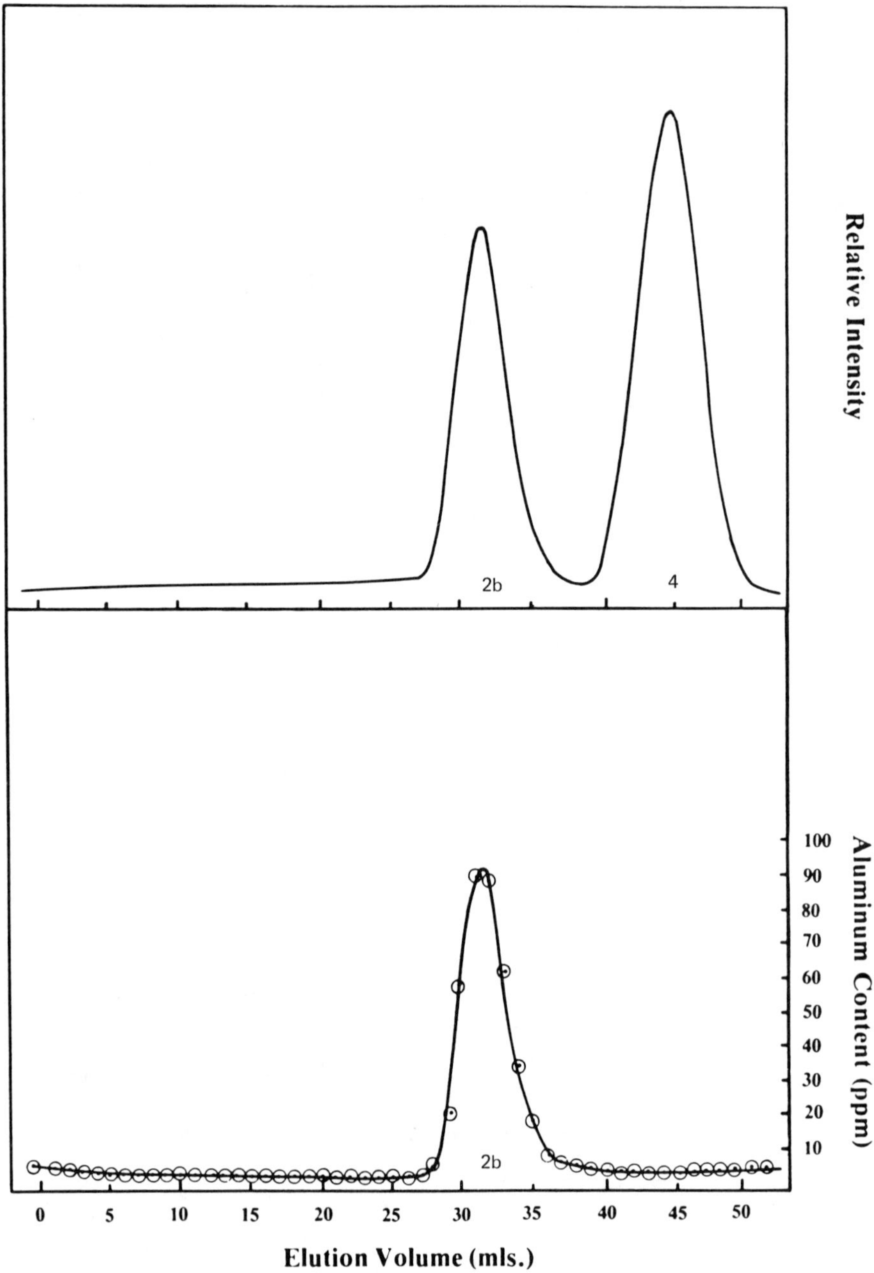

Figure 15 Sephadex G-25 gel filtration chromatogram of $Al_{13\text{-mer}}$ solution (0.1 M Al, $\bar{n}$ = 2.45): (above) refractive index profile and (below) aluminum content profile (Fitzgerald, 1981).

Table 13 GFC Sephadex G-25 Parameters for Aluminum Chlorohydrate and Reference Compounds

Aluminum system	Concentration, wt% (Al, M)	$\bar{n}$	Peak K_d				% normalized peak heights				
			1	2	2b	3	1	2	2b	3	4
ACH	10% (0.97 M)	2.49	0.00	0.40	0.69	—	57.4	32.2	10.4	—	—
$AlCl_3 \cdot 6H_2O$	20% (1.2 M)	0.00	—	—	—	0.81	—	—	—	100	—
Al[b] - 1[a]	— (0.12 M)	2.40	—	—	0.71	0.83	—	—	92.0	8	—
Al[b] - 2[a]	— (0.91 M)	2.45	0.00	—	0.62	0.82	2.0	—	93.0	5	—
Al[b] - 3[b]	— (0.12 M)	2.43	—	—	0.72	0.82	—	—	88.0	12	—

[a]Prepared by neutralization of $Al(OH)_3$ with $AlCl_3$ solutions.
[b]Prepared by $NaHCO_3$ neutralization according to Aveston (1965).

Fig. 15) of K_d from 0.72 to 0.62, with small concentrations of the monomeric species (peak 3, Fig. 15) at $K_d = 0.82$. The GFC data of Table 13 show that all three partially neutralized solutions are nearly monodisperse in nature and that they probably contain an intermediate-sized polymer of 900–1200 daltons, possibly corresponding to the $Al_{13\text{-mer}}$ species or similar complexes. The chromatogram of the sample labeled Al-2 ($\bar{n} = 2.43$) given in Fig. 15 may be contrasted with the other two aluminum systems due to the narrow width observed for peak 2b (5 versus 9–11 ml) and the increased resolution of the polymer and monomer peaks. In summary, all the reference compounds noted in Table 13 contain aluminum species with substantially smaller solute species than ACH samples and of narrower molecular weight range. A summary of the various GFC peaks, their peak numbers, estimated MW ranges, corresponding K_d values on G-25 columns, and the nature of the various sample types which exhibit these peaks is given in Table 14.

The polydisperse nature and variable species contents of ACH solutions (Fitzgerald, 1981) can readily be seen by carrying out computer simulations of Sephadex G-25 chromatograms and obtaining deconvolution information for these multiple-peak chromatograms. Using the experimental Sephadex G-25 chromatogram of ACH shown in Fig. 16, a computer-generated simulated chromatogram was obtained with the aid of a Dupont 310 curve resolver as shown in Fig. 17. Deconvolution analysis of this simulated chromatogram reveals the individual Gaussian peak distributions that constitute it, as shown in Fig. 18. The deconvolutions of experimental and simulated Sephadex G-25 ACH chromatograms show that these solutions may

Table 14 GFC Aluminum Species

Peak number	MW range	K_d values	Species type	Type of samples
1	8000–5000	0.00	Large polymer	BAC (ACH)
2a	4000–3000	0.35–0.25	Medium polymer	All BAC
2	3000–1500	0.50–0.40	Medium polymer	All BAC
2b	1500–500	0.70–0.50	Small polymer	$Al_{13\text{-mer}}$
3	270–135	0.83–0.75	Dimer/monomer	BAC/$AlCl_3$
4	—	1.00	Nonaluminum	—

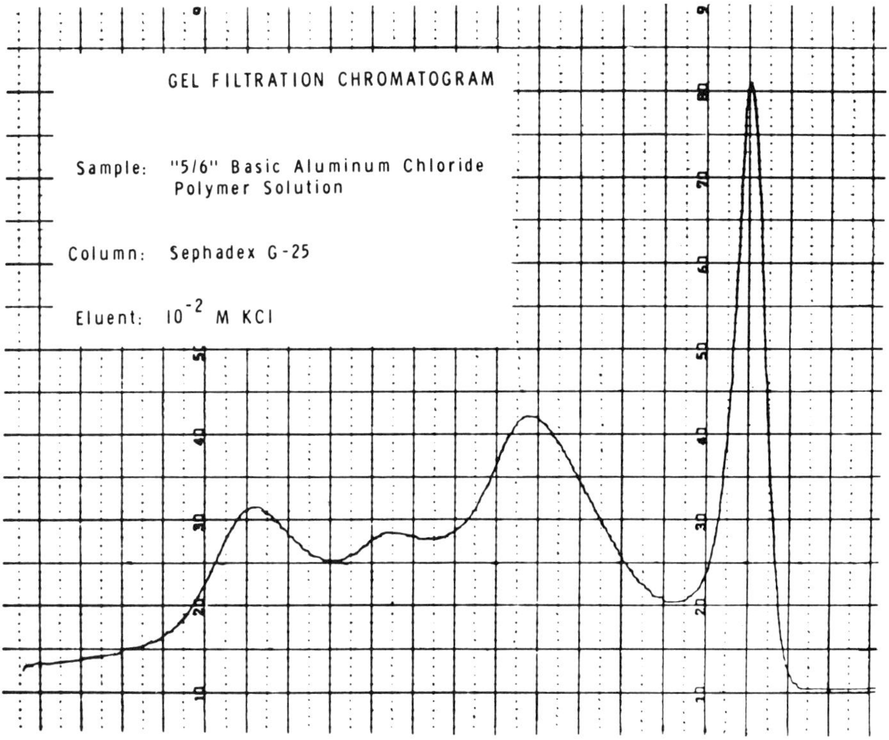

Figure 16 Experimental Sephadex G-25 chromatogram of 10% ACH solution using 0.01 M KCl eluent. (Fitzgerald, unpublished results).

be best fit by five Gaussian peaks as shown in Fig. 18. These peaks may be interpreted as arising from four polymeric peak distributions corresponding to the K_d values and estimates of the molecular size of species in peaks 1, 2a, 2, and 2b summarized in Table 14. Simulation of less basic aluminum chloride systems by this technique has also shown that deconvolution analysis must include a Gaussian peak distribution corresponding to monomeric peak 3 as summarized in Table 14 in order to obtain best-fit simulation chromatograms of ASCH and ADCH solutions.

Computer simulation chromatograms of ACH solutions can also be used to calculate statistical molecular weight data, as reported by Fitzgerald and Johnson (1978). Using the experimentally derived GFC calibration curve, $\log MW = 3.46 - 1.60\ K_d$, for a range of

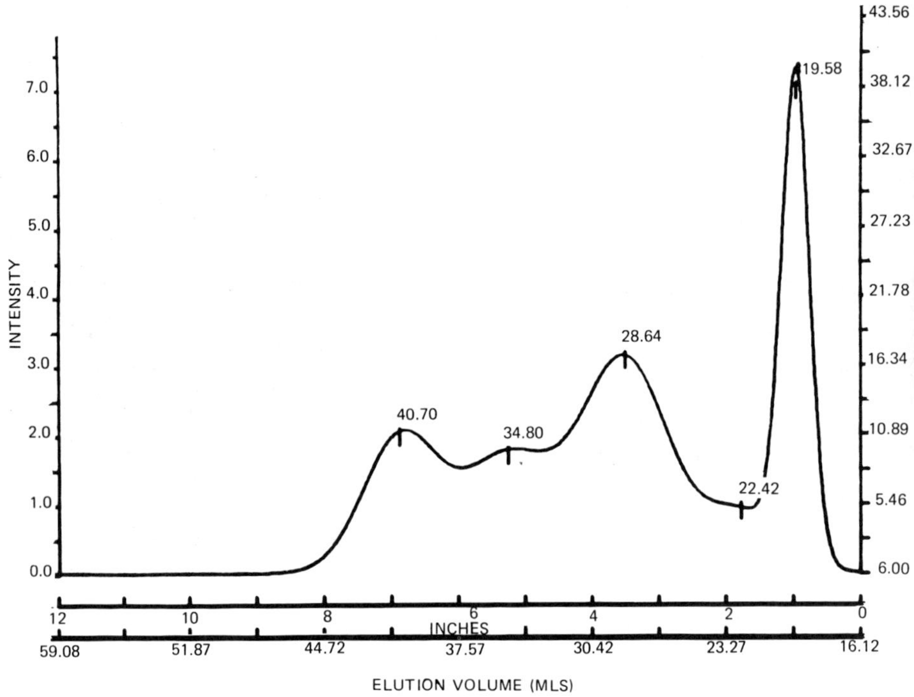

Figure 17 Simulation of Sephadex G-25 chromatogram of 10% ACH solution. (Fitzgerald, unpublished results).

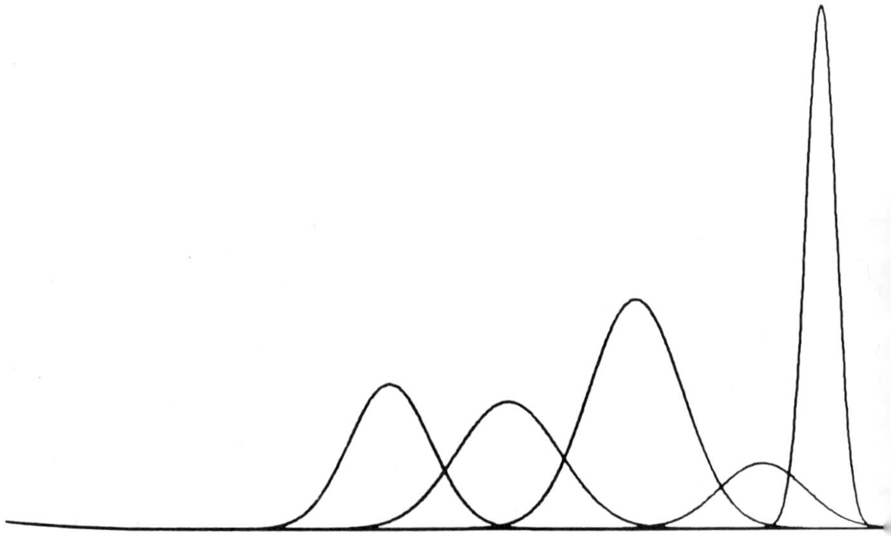

Figure 18 Deconvolution analysis of Sephadex G-25 chromatogram of 10% ACH solution. (Fitzgerald, unpublished results).

linear polypropylene glycol standards on Sephadex G-25 columns, the following molecular weight values were calculated for fresh ACH solutions: MW_w = 5950 and MW_n = 1122. The polydispersity index of ACH was found to be $I = MW_w/MW_n = 5.57$.

Light scattering (*LS*) studies of freshly prepared ACH solutions in the concentration range 0.4–0.1% ACH (0.4–0.1 M Al) have been reported by Riesgraf (1970) in both 1.0 M KCl and 1.0 M KNO_3. Using the techniques of Patterson and Tyree, values of the gross turbidity of this range of solutions prepared fresh from concentrated ACH solutions following filtration through a 0.45-μm filter (to minimize the influence of colloidal particles) were determined at 25°C. A plot of $1/N_{z'}$ versus aluminum concentration derived from corrected net turbidity values is shown in Fig. 19 for ACH solutions in the

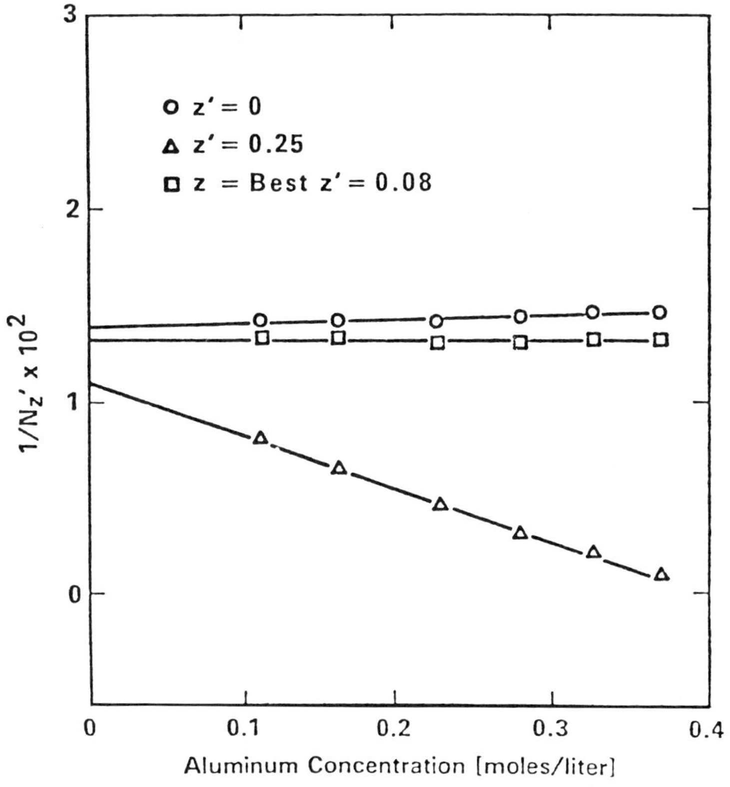

Figure 19 Plot of $1/N_{z'}$ versus aluminum concentration for various ACH solutions in 1.0 M KCl obtained from light scattering. (Reproduced by permission of Riesgraf, 1979.)

Table 15 Average Aggregate Size $N_{z'}$ of ACH Solutions

	$N_{z'}$ average aggregate size			
Electrolyte	z' = 0.00	0.25	Best	Best z'
1 M KCl	67	84	70	0.08
1 M KNO_3	63	78	67	0.08

Source: Riesgraf et al. (1979).

1.0 M KCl solutions. The polymeric aggregate size is determined from these plots at the zero aluminum concentration intercept for a variety of charge values, z', from 0.0 to 0.25. The best-fit relationship of these plots for $z' = 0.08$ is obtained for the $1/N_{z'}$ versus aluminum concentration graphs that have a slope of zero. A summary of the average aggregate size of ACH solutions in the two different supporting electrolytes is given in Table 15. An average degree of aggregation of $N_{z'} = 70$ and charge of $z' = 0.08$ was obtained for 1.0 M KCl solutions of ACH (0.4–0.1%). Based on these light scattering results, pCl measurements of Cl^- ion concentrations, and estimates of OH^- and H_2O contents, these workers proposed that the average polymeric species in ACH solutions corresponded to the species

$$Al_{70}(OH)_{188}Cl_{16}(H_2O)_{144}^{6+}$$

This average polymeric species has an $\bar{n}$ or OH/Al ratio of 2.69, an average molecular weight of 6444, and contains about 45% of the Cl^- ion in the "bound" form. These workers concluded that the average degree of aggregation of aluminum hydroxide species in ACH solutions was readily distinguishable from the $Al_{13\text{-mer}}$-containing solution of $\bar{n} = 2.43$ studied by Patterson and Tyree (1973). In these solutions, $N_{z'}$ values of up to 13 aluminums per aggregate and a corresponding average molecular weight of 1430 were reported.

Ultracentrifugation studies have been carried out by Langerman (1986) for a range of ACH solutions by means of both sedimentation velocity and sedimentation equilibrium experiments. Sedimentation equilibrium measurements of solutions over the concentration ranges 2–0.5% ACH and 0.1–0.03% ACH were carried out in both 0.1 M and 1.0 M KCl supporting electrolytes. The parameter σ_W, defined as the reduced weight-averaged molecular weight by the equation

$$\sigma_w = \frac{M_w(1 - \nu\rho)\omega^2}{RT} \tag{28a}$$

according to the procedure of Roark and Yphantis, was used to evaluate sedimentation equilibrium data as well as to define the other reduced number-averaged and *z*-averaged molecular weight values, σ_n and σ_z, respectively. Figure 20 shows a plot of σ_w versus concentration of solutes in a centrifuge cell at equilibrium for two ultracentrifuge (UC) cell sectors (sectors L and M) containing 0.1 and 0.07% ACH solutions, respectively. Figure 21 illustrates an analysis of the various reduced molecular weight parameters (σ_z, σ_w, and σ_n) for the 0.1% ACH solution (sector L) at sedimentation equilibrium. The reduced molecular weight parameters may be related to the corresponding average molecular weight parameters by

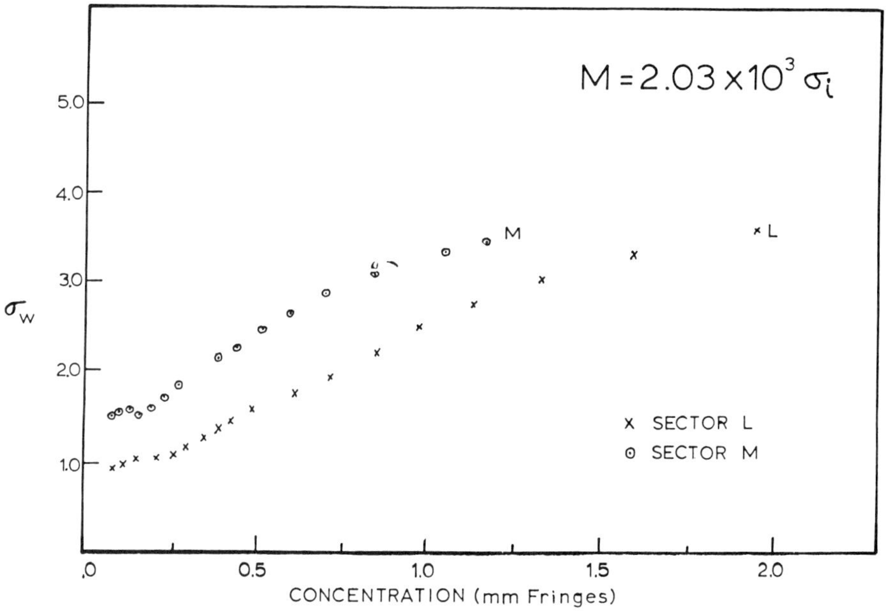

Figure 20 Ultracentrifugation cell distribution of σ_w, reduced weight-averaged molecular weight versus ACH cell concentrations. Sector L, 0.7 mg/ml ACH; sector M, 0.1 mg/ml ACH. (From Langerman, 1986.)

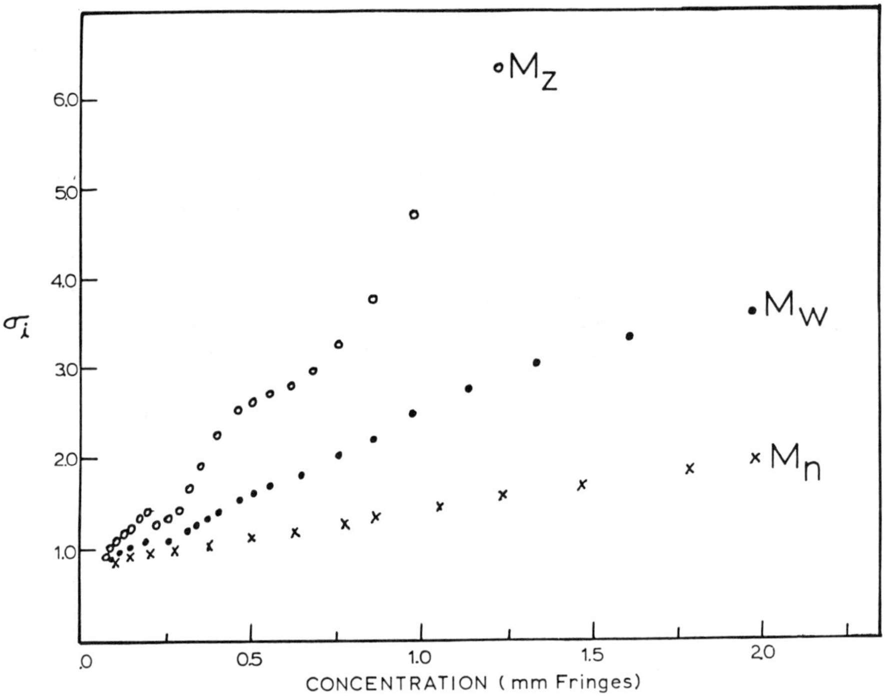

Figure 21 Ultracentrifuge plot of various M_i values for ACH solutions under equilibrium sedimentation conditions of sector L of Fig. 20. (From Langerman, 1986.)

$$M_i = 1.698 \times 10^3 \sigma_i \tag{28b}$$

The results of the equilibrium ultracentrifuge experiments shown in Figs. 20 and 21 provide the following information about ACH solutions:

1. The equilibrium distribution of aluminum solute species in the UC cells labeled L and M shows that the species distributions are very dependent on sample loading concentration, indicative of polydisperse solutions (See Fig. 20).
2. Analysis of the various reduced molecular moments of the molecular weight distribution of the ACH solutions in sector L (0.07% ACH) of Fig. 20, as shown in Fig. 21, reveals that the reduced moment σ_i distribution values behave in a manner consistent with polydisperse systems. Polydispersity

indices σ_z/σ_w and σ_w/σ_n of 1.66 and 1.42, respectively, can be obtained from these data.

3. Estimates of the reduced moment data, in terms of Eqs. 27 and 28 for this experiment, indicate that ACH solutions have polymeric species in the 1500–6800 g/mole range (M_z varies from 1500 to 10,7000, M_w from 1500 to 6000, and M_n from 1400 to 2900).

These sedimentation equilibrium experiments also indicated that the species do not exist in a reversible associative equilibrium and that a model consisting of only two species with molecular weights differing by a factor of 2.5–5.0-fold (average 4.3) was consistent with the species distribution throughout the centrifuge cells. Additional sedimentation velocity experiments at 60,000 rpms yielded a number of other important observations regarding ACH solutions. First, gibbsite particles were easily observed in dilute ACH solutions in ultracentrifuge cells and are also observed in aged ACH solutions of only 2% or less in 0.1 M KCl supporting electrolyte. Second, the concentration of supporting electrolytes significantly altered the sedimentation behavior of polymeric species in ACH solutions.

In addition to the GFC, LS, and UC studies of ACH, experimental examinations of ACH and ADCH solutions have been reported by Bretschneider et al. (1975) using *vapor phase osmometry (VPO)* measurements. These studies have not been fully described in the literature; however, the work at 37°C using VPO data extrapolated to infinite dilution (probably in the absence of supporting electrolyte) provide results in general agreement with the UC studies of Langerman.

A number-averaged molecular weight of 1571 was obtained for ACH solutions at infinite dilution, in agreement with the M_n value range of 1400–2900 from ultracentrifugation experiments. In addition, Bretschneider et al. reported an M_n value of 975 for ADCH solutions under similar experimental conditions.

Molecular weight range and polydispersity of ACH solutions. The macromolecular studies summarized above establish unequivocally that ACH solutions contain a broad range of aluminum hydrolysis species of different molecular weights, undoubtedly larger in size than the $Al_{13\text{-mer}}$ species. The variable species content and polydispersity of ACH solutions are most readily observed from Sephadex G-25 chromatograms or from optical measurement profiles of sample solute distributions observed in ultracentrifugation cells in equilibrium sedimentation experiments at high rotational speeds.

A summary of the molecular weight ranges and various molecular weight parameters derived from studies of ACH solutions by different macromolecular techniques is given in Table 16. Also included are similar data for partially hydrolyzed $\bar{n}$ = 2.4–2.5 aluminum solutions containing principally the $Al_{13\text{-mer}}$ species. The molecular

Table 16 Summary of Molecular Weight Ranges and Molecular Weight Parameters for ACH and $Al_{13\text{-mer}}$ Solutions

		Statistical MW parameters			
Technique	MW range	MW_n	MW_w	MW_z	Polydispersity[a]
ACH solutions ($\bar{n}$ = 2.5)					
GFC	8000–135	1122	5954	—	5.3
LS	—	—	6444	—	—
UC	6800–1500	2900–1500	6000–1500	10,700–1500	1.5
VPO	—	1571	—	—	—
$Al_{13\text{-mer}}$ solutions ($\bar{n}$ = 2.4–2.5)					
GFC	1500–500	940	950	—	1.02
LS	—	—	1430	—	—
UC	—	—	1250	—	—
VPO	—	—	—	—	—

[a]The polydispersity index equals the MW_w/MW_n ratio.

weight range 8000–135 from GFC and the range 6800–1500 from UC studies are in good agreement. The MW_w values of 5950 from GFC and 6444 from LS measurements are also internally consistent, as are the MW_n values of 1122 from GFC and 1571 from VPO studies. While the numerical agreement of various molecular weight parameters may be considered at length, the most important conclusion from these studies using a range of techniques is that ACH solutions contain solute species substantially higher in molecular weight than $Al_{13\text{-mer}}$ solutions. Furthermore, ACH solutions have considerably higher molecular weight ranges and are more polydisperse than $Al_{13\text{-mer}}$-type solutions, as evidenced from the GFC and UC measurements in particular.

B. Ferron Reaction Studies of ACH Solutions

Studies using ferron spectrophotometric analysis for ACH solutions have been reported by Riesgraf et al. (1979), Fitzgerald (1981), Fitzgerald and Johnson (1980), and Waters and Henty (1977). In addition, the ferron studies of Schonherr and Frey (1979) and

Schonherr et al. (1981a, 1983) have involved examination of a range of hydrolyzed aluminum solutions prepared at high concentration (0.2–2.0 M Al) by the reaction of aluminum metal with aqueous aluminum chloride solutions. The latter studies are of significance in understanding the commercial synthesis of ACH and will be discussed later in Sec. VI.I.

The work of Fitzgerald et al. (1977, 1980, 1981) and Riesgraf et al. (1979) represents the most extensive investigations reported using the ferron method to characterize ACH solutions over a wide concentration range. In addition, these workers have carried out comparative studies to determine how the reactivity of species in ACH solutions differs from that of hydrolysis species contained in hydrolyzed aluminum solutions synthesized at lower concentrations. Furthermore, Fitzgerald and co-workers (1981) examined the ferron reactivity of the complexes contained in various Sephadex G-25 GFC peaks to determine the relative reactivity and ferron reaction types for aluminum solutes obtained following chromatographic separation.

Figure 22 shows a typical ferron reaction profile for six aluminum solutions obtained by the ferron method of Smith and Hem (1972). In the reaction profiles shown, the amount of aluminum reacting to form the ferron-aluminum complex is plotted as a function of time over a 2.5-hour period. The concentration of aluminum reacting is calculated from standard curves of aluminum concentration and measurement of the concentration of the aluminum-ferron complex at 368 nm for the samples analyzed over the reaction time period. The six aluminum samples shown in Fig. 22 are a 1.0 M $AlCl_3 \cdot 6H_2O$ solution, three samples of dilute partially neutralized aluminum chloride solutions from $\bar{n}$ = 2.40 to 2.45, and two 10% ACH solutions, a freshly prepared solution and one aged for 1 month. The three partially neutralized aluminum chloride solutions examined were prepared by various neutralization methods at the $\bar{n}$ and solution concentrations shown in Table 17. Also noted in Fig. 22 are the Al_t values for the total aluminum content of each of the six samples analyzed.

The reaction profiles for these six aluminum systems indicate that the $AlCl_3$ solution reacts almost instantaneously. The three solutions labeled Al-1, Al-2, and Al-3 react more slowly but completely over a 4-hour reaction period and can be classified as Al^b-containing solutions. Both fresh and aged ACH solutions react very slowly, remain essentially unreacted after the 4-hour period of the assay, and are classified as Al^c species.

Results of quantitative computer analysis of the reaction profiles for these six solutions are shown in Table 18, where the % Al of each species is given for each solution. The ferron assay results of Table 18 establish that the six solutions examined consist principally of single species according to the ferron technique. The aluminum

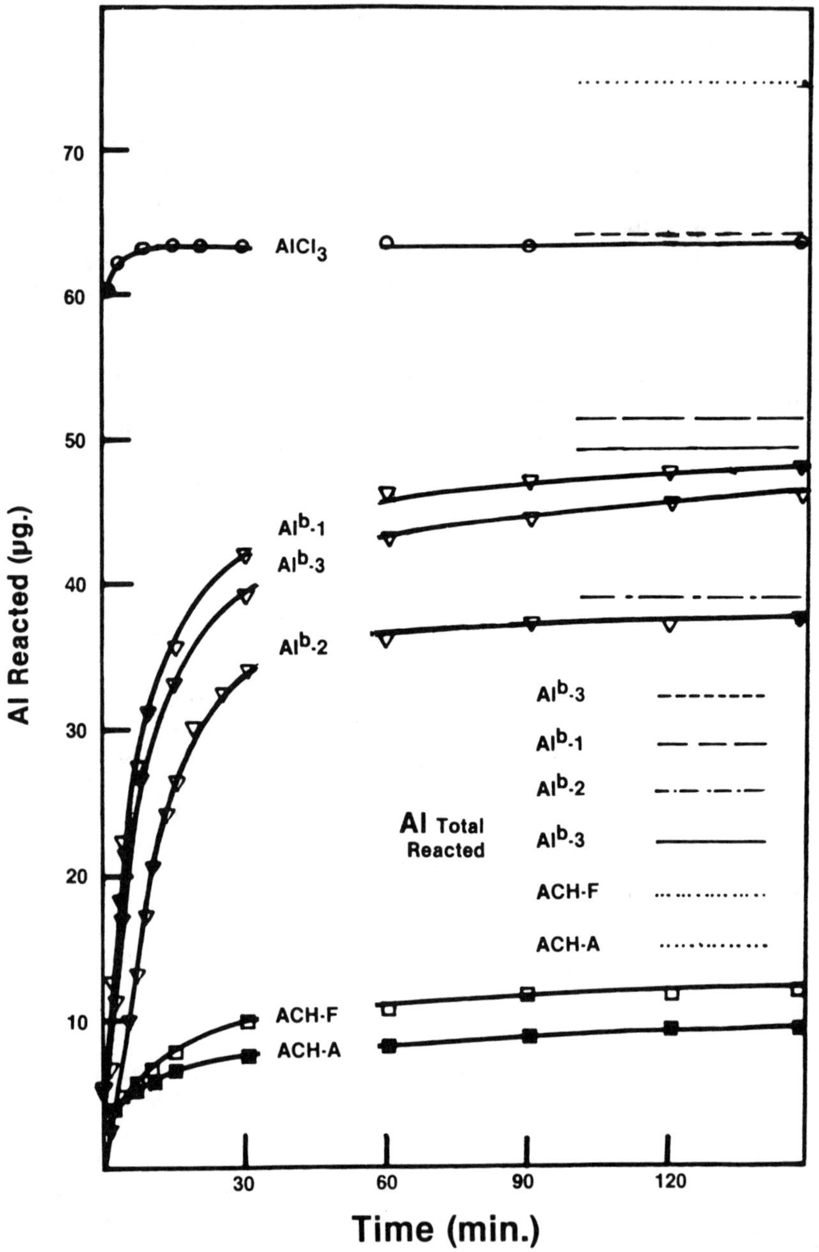

Figure 22 Ferron reaction profile (plot of micrograms Al reacted versus time) for an aqueous aluminum chloride solution, various Al^b solutions, and fresh and aged 10% ACH solutions. (From Fitzgerald, 1977.)

Table 17 Summary of Various Al^b-Type Aluminum Solutions Examined by Ferron Method

Sample	Solution concentration	$\bar{n}$	Synthesis method	Al^b reaction rate $k_1 \times 10^2$ min^{-1}
Al-1	0.12	2.40	Slow neutralization $Al(OH)_3/AlCl_3$	9.8
Al-2	0.91	2.45	Slow neutralization $Al(OH)/AlCl_3$	7.9
Al-3	0.12	2.43	Slow neutralization $NaHCO_3/AlCl_3$	9.6

chloride solution is principally Al^a (>95%), the three Al^b solutions vary from 79 to 87% Al^b, and the ACH solutions (fresh and aged) contain 90 and 87% Al^c, respectively. The three Al^b samples investigated have pseudo-first-order rate constants in the 7.9–9.8 × 10^{-2} min^{-1} range, consistent with literature values.

Similar ferron results have been reported by Riesgraf et al. (1979) for both ACH solutions and various partially neutralized aluminum chloride solutions. These investigators reported that $AlCl_3 \cdot 6H_2O$ solutions consisted of 96% Al^a solutes. They obtained the

Table 18 Summary of Ferron Analysis of Aluminum Chloride and Basic Aluminum Solutions

Sample	Ferron aluminum analysis (% Al)		
	Al^a	Al^b	Al^c
$AlCl_3 \cdot 6H_2O$	96	4	—
Al-1	10	80	10
Al-2	6	87	7
Al-3	11	79	10
ACH (fresh)	2	5	93
ACH (aged)	4	6	90

% Al species compositions noted for the aluminum chloride solutions partially neutralized with $NaHCO_3$ at $\bar{n}$ = 1.5 (29% Al^a, 71% Al^b), $\bar{n}$ = 2.0 (94% Al^b, 6% Al^c), and $\bar{n}$ = 2.4 (94% Al^b, 6% Al^c). These workers showed by the ferron method that the systems contained primarily Al^b solute types. A ferron study of a 0.01 M ACH solution and ACH solutions prepared by the $Al/AlCl_3$ oxidation-reduction method for 48 hours showed that these solutions contained principally Al^c solutes (86–90% Al^c).

Additional reactivity studies using Turner's 8-hydroxyquinoline assay are described by Waters and Henty (1977) for three solutions with $\bar{n}$ = 2.45–2.49: commercial ACH, ACH prepared from Al/HCl in the presence of mercury, and an $\bar{n}$ = 2.45 solution prepared by neutralization of $Al(OH)_3$ with Na_2CO_3. In the Turner nomenclature, this work showed that these aluminum systems contained 90–93% polymeric species, with about 5% Al^a and 3–5% filterable solid species. No distinction was made in this work between polymeric species of the Al^b and Al^c types.

C. Correlation of GFC and Ferron Assay

The monodisperse nature of $AlCl_3$ and Al^b solutions reported from the GFC work of Fitzgerald (1981b) and the single-species ferron reactivity observed for these solutions suggested that a useful species correlation existed between the ferron assay and GFC descriptions of aluminum hydrolysis complexes. Fitzgerald (1977, 1981b) reported experimental studies which established a relationship between GFC and the ferron assay for five of the six solutions previously described. First, a series of chromatographic experiments were conducted in which $AlCl_3$, Al-1, Al-2, Al-3, and ACH solutions were chromatographed to obtain their GFC profiles. Second, the various samples were chromatographed, the eluted 1.0-ml fractions were combined on the basis of the GFC profiles into individual eluted peaks, and finally these combined fractions were assayed by the ferron technique. The ferron assay of these five samples established a correlation between the GFC behavior and the ferron aluminum types. For example, peak 3 was shown to contain fast-reacting monomeric type Al^a species, peak 2b contained principally type Al^b species, and peaks 1, 2a, and 2 (found principally in ACH and other BAC solutions) contained various Al^c solutes of very slow ferron reactivity. The correlation between G-25 chromatographic behavior and ferron species types is summarized in Table 19. This relationship suggests that polymeric species reactivity as defined by the ferron assay is related to molecular size of the species from Sephadex G-25 GFC experiments. Increases in species molecular size lead to increased rate of reaction in the ferron assay for the various aluminum hydrolysis species studied.

Table 19 Ferron-GFC Relationship on Sephadex G-25 Columns

Ferron notation	GFC G-25 peak number	K_d	Molecular weight range
Al^a	3	0.91–0.80	135–270
Al^b	2b	0.70–0.58	800–1200
Al^c	2, 2a, and 1	0.55–0.00	1200–8000

D. Role of Chloride Ion in ACH

The role of the chloride ion in solutions of aluminum chlorohydrate has been the subject of conflicting interpretation regarding the degree of ionization following dilution of ACH solutions. Measurements of pCl by Teagarden et al. (1981b), Riesgraf et al. (1979), and Fitzgerald et al. (1981) exemplify the problems in measuring free chloride in ACH solutions. Measurements by Teagarden et al. (1981b) of the free Cl^- ion contents using the Volhard method in dilute ACH solutions and a chloride specific electrode in ACH solutions containing KNO_3 or HNO_3 indicate that Cl^- is completely ionized. By contrast, pCl measurements of free Cl^- ion in dilute ACH solutions containing no other anions (Riesgraf et al., 1979) indicated that about 45% of the chloride ion is bound to polymeric aluminum solutes. The measurements led these workers to conclude that Cl^- ion is associated with or bound to aluminum hydroxide polymeric species; thus, they proposed that the average polymeric species in the ACH solutions corresponded to

$$Al_{70}(OH)_{188}Cl_{16}(H_2O)_{144}^{6+}$$

Measurements of pCl of Fitzgerald et al. (1981) support the conclusions of both workers. Using a chloride specific electrode the free $[Cl^-]$ of solutions prepared at 2, 5, and 10% ACH was examined as a function of time following dissolution of ACH over a 3-month period. The results of these measurements are shown in Fig. 23 as % free Cl^- ion in comparison with the total chloride of each solution. At time zero following dissolution of the powdered ACH, it is apparent that only a fraction of the Cl^- is in the ionized form. For example, the % Cl^- free varies from 8 to 16 to 26% for 10, 5, and 2% ACH, respectively. Initially, on dissolution, from 74 to 92% of the Cl^- ion is associated with polymeric solutes. Dilution of ACH increases the % Cl^- in the free ionized form.

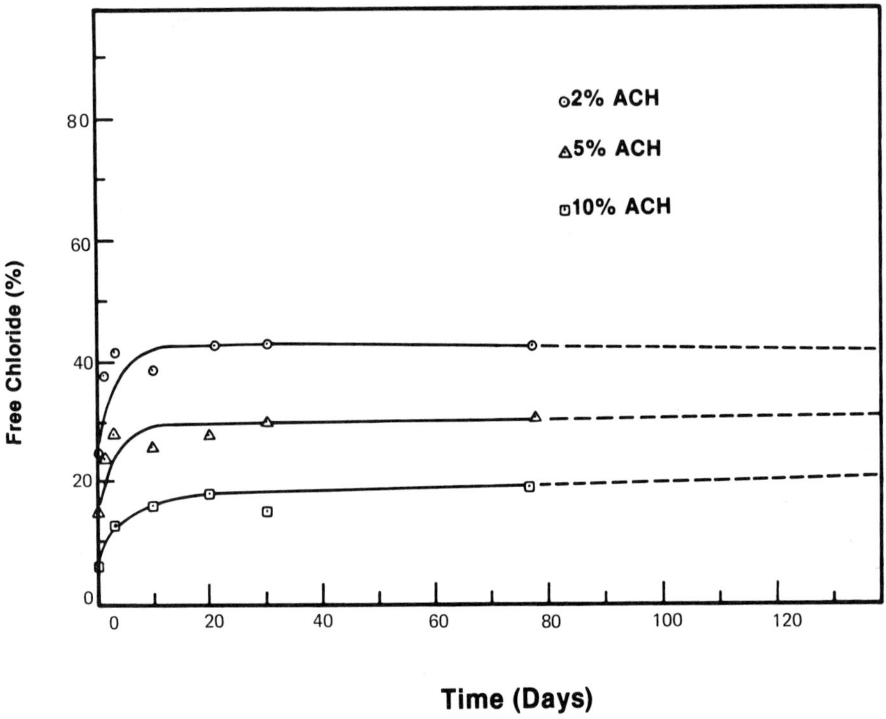

Figure 23 Dependence of % free chloride ion on time of aging for 10, 5, and 2% ACH solutions as determined by specific chloride ion electrode measurements. (From Fitzgerald, 1981.)

The dependence of the % free Cl^- ion on time of aging is also shown in Fig. 23. Aging results in a rapid release of Cl^- ion for all ACH concentrations, reaching about 45, 30, and 18% free Cl^- ion for 2, 5, and 10% ACH solutions over 3 months. The Cl^- ion release is found to be substantially faster for dilute solutions; e.g., the $t_{1/2}$ for Cl^- release is about 1.5 days for 2% solutions and 5–6 days for 10% solutions. These workers showed that the slow rate of Cl^- ion release is comparable in order of magnitude to the depolymerization rates from GFC measurements of similar aged ACH solutions.

While these measurements support the conclusions based on similar measurements by Riesgraf, additional studies of free Cl^- levels in similar ACH solutions in the presence of NO_3^- ion by Fitzgerald also agree with the work of Teagarden. The pCl measurements showed that while 50% of the chloride was bound in 2% aqueous ACH solutions, the addition of nitrate ion at 0.1 and 1.0 M NO_3^- decreased the % bound Cl^- ion to 36 and 20%, respectively, of the total solution chloride ion concentration. These results indicate that the addition of NO_3^- ion results in liberation of bound Cl^- ion in dilute ACH solutions. It was suggested that the role of the bound Cl^- ion in ACH solutions involves stabilization of the charge of these higher molecular weight polymeric species. Addition of supporting electrolytes like NO_3^- ion caused a release of "polymer-bound" Cl^- ion with added supporting electrolytes. The exact details of this anion competition and the overall role of Cl^- ion in stabilizing ACH polymeric species warrant further investigation.

E. Characterization of ACH Using Spectroscopic Methods

The emergence of techniques such as Raman spectroscopy, Fourier transform infrared spectroscopy, and ^{27}Al nuclear magnetic resonance spectoscopy has resulted in new structural information about aluminum hydrolysis species in ACH solutions and the solid state. These three spectroscopic techniques have been applied with varying degrees of success to examine the structure of species in ACH solutions. In addition, x-ray diffraction and solid-state ^{27}Al NMR spectroscopy have been applied to study ACH powders.

1. *Infrared and Raman Studies of ACH*

Raman spectroscopic investigations by Waters and Henty (1977) included studies of the dimer and the $Al_{13\text{-mer}}$ in solution and the solid state, in addition to studies of commercial ACH and ACH prepared from the redox reaction involving Al/HCl in the presence of mercury. Solution Raman spectra of ACH were obtained in the concentration range 2–7 M Al. Assignments of a Raman band at 530 cm^{-1} to the dimer species and three bands at 525, 440, and 330 cm^{-1} to the $Al(H_2O)_6^{3+}$ ion were made by these workers. Solid-state Raman spectra of the $Al_{13\text{-mer}}$ complex and ACH solutions exhibited similar bands. These workers concluded that the $Al_{13\text{-mer}}$ was the major species in ACH solutions. They assigned the Raman band at 542 cm^{-1} to a vibration of the $Al{-}H_2O$ group and the band at 380 cm^{-1} to the bridging Al–OH–Al group vibration from deuterium substitution experiments. The band at 625 cm^{-1} was assigned to the central tetrahedral AlO_4 moiety in the $Al_{13\text{-mer}}$ structure.

Infrared studies of ACH solutions and ACH solids have been reported by Waters and Henty (1977), Teagarden et al. (1981a, 1981b, 1981c, 1982), and Riesgraf and May (1978). The work of Waters and Henty was associated with Raman measurements of ACH and other aluminum hydrolysis systems discussed previously. Riesgraf and May

(1978) reported FTIR measurements of ACH and related aluminum hydrolysis systems in an effort to identify differences in the IR spectra of Al^c- and Al^b-containing solutions. Their work, together with additional assignments of the IR spectrum of solid ACH by Teagarden et al. (1981a), provides the most definitive evidence for the presence of various structural moieties in polymeric species in ACH solutions. The FTIR spectrum of ACH solutions (following subtraction of H_2O) in the 4000–500 cm^{-1} region is shown in Fig. 24a from the work of Riesgraf and May. The FTIR spectrum of an ACH solution in the 1200-500 cm^{-1} region is also given in Fig. 24b from this report. Table 20 summarizes the FTIR spectral bands for ACH solutions and films, together with the assignment of the bands from the work

Table 20 FTIR Spectral Bands and Assignments for ACH Solutions

IR spectral band (cm^{-1})	Assignment (author)
3600 (m)	OH stretch of Al-OH-Al moiety in polymeric species (Riesgraf)
3300 (s)	OH stretch of Al-OH-Al moiety in polymeric species (Riesgraf)
3150 (s)	OH stretch of Al-OH_2 group, antisymmetric (Riesgraf)
2500 (w)	OH stretch of Al-OH_2 group, symmetric (Riesgraf)
1920 (w)	Unassigned
1080 (s)	Al-O-H bend of bridging Al-OH-Al moiety (Riesgraf)
970 (s)	Al-O-H deformation bend of Al-OH and Al-OH_2 moiety (Riesgraf)
790 (s)	AlO_4 antisymmetric stretch (Teagarden)
630 (s)	AlO_4 symmetric stretch (Teagarden)
345[a]	AlO_4 antisymmetric bend (Teagarden)

[a]Work of Teagarden et al. (1981a) using KBr pellet of air-dried ACH solutions. Not reported in FTIR spectrum by Riesgraf and May.
Source: Riesgraf and May (1978).

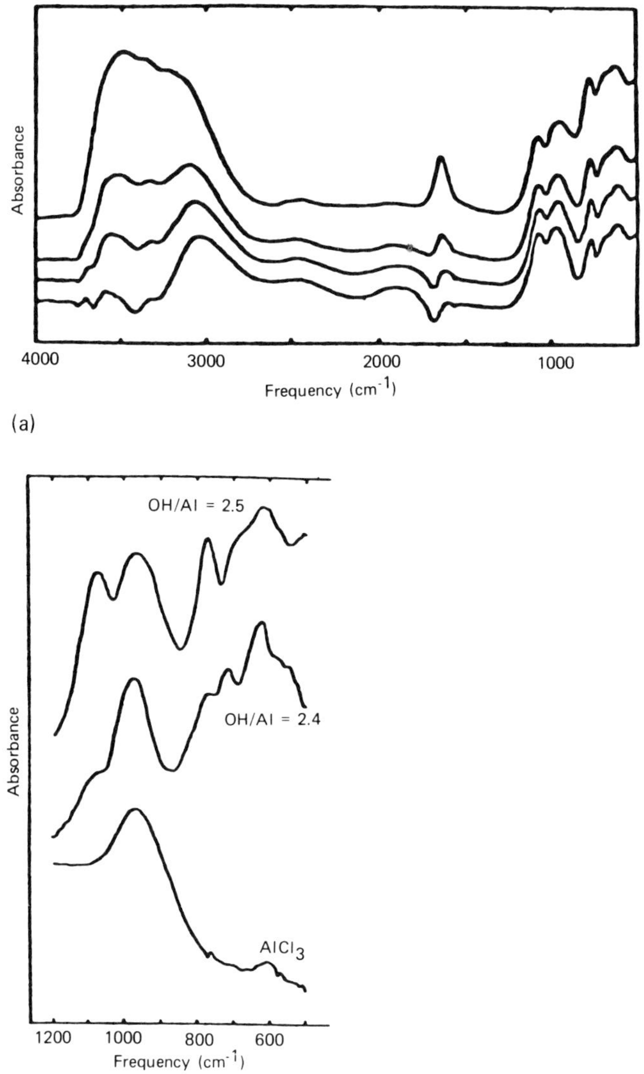

(b)

Figure 24 Infrared spectra of various aluminum hydrolysis solutions: (a) FTIR spectra of $\bar{n}$ = 2.5 ACH solution over the 4000–500 cm^{-1} region; (b) FTIR spectra of ACH solution (top), $\bar{n}$ = 2.4 Al^b-containing solution (middle), and $AlCl_3 \cdot 6H_2O$ solution (bottom) over the 1200–50 cm^{-1} region. (From Riesgraf and May, 1978.)

of both Riesgraf and May and Teagarden et al. The FTIR spectra of ACH solutions and solid films exhibit four specific regions: 3700–3300 cm^{-1}, consisting of bands assigned to OH stretches due to surface and bridging OH groups coordinated to octahedral aluminum; 3200–2500 cm^{-1}, consisting of bands assigned to OH stretching modes of water coordinated to aluminum; 1100–900 cm^{-1}, consisting of Al–O–H bending modes due to bridging hydroxide (1080 cm^{-1}) and hydroxide and water coordinated to octahedral aluminum; and three spectral bands at 790, 630, and 345 cm^{-1} attributed to Al–O stretches in AlO_4 tetrahedra. The FTIR spectra of ACH solutions (containing primarily Al^c species by the ferron method) reported by Reisgraf and May were compared to FTIR spectra of hydrolyzed aluminum solutions containing Al^b species. The Al^b-containing solutions exhibited different absorption bands in the 3700–3300 and 790–630 cm^{-1} regions in comparison with ACH solutions. These workers concluded that the IR spectral differences for Al^b- and Al^c-containing solutions supported the conclusion that these systems contained different polycationic aluminum species.

2. *^{27}Al NMR Spectroscopy of ACH*

The chemical nature and structural characteristics of Al^{3+} ion hydrolysis species have been extensively studied by ^{27}Al NMR spectroscopy (Akitt, 1972; Delpuech, 1983) over the past 15 years because of the relatively high NMR sensitivity of the ^{27}Al nucleus (100% natural abundance). The work of Akitt and co-workers (1969a, 1969b, 1972, 1978, 1981a–1981c, 1984, 1985a–1985b, 1986) and Schonherr et al. (1981a, 1981b, 1983) is most notable, but investigations by Bottero et al. (1980), Fedotov et al. (1978), Teagarden et al. (1981a–1981c, 1982), and Fitzgerald and co-workers (1981, 1983, 1984b, 1986) must also be cited. Early investigations by Akitt et al. (1969a, 1969b, 1972) showed the first successful use of ^{27}Al NMR spectroscopy to obtain direct spectroscopic evidence for the presence of the aquated monomer, the hydroxy-bridged dimer $Al_2(OH)_2(H_2O)_8{}^{4+}$, and the $Al_{13\text{-mer}}$ species in hydrolyzed aluminum(III) ion solutions. The ^{27}Al NMR spectra of the monomeric ion, the dimer, and the $Al_{13\text{-mer}}$ (at room temperature and elevated temperature) are shown in Figures 25a, 25b, and 25c, respectively (Fitzgerald, 1984, 1983; Fitzgerald et al., 1981). A summary of ^{27}Al NMR spectral parameters is given in Table 21 for these complexes, together with relevant data for the basic aluminate and dimeric aluminate species. The known chemical shifts $\delta(^{27}Al)$, in parts per million, are reported relative to the reference hexaaquo species. The ^{27}Al NMR chemical shift region from 0 to 30 ppm corresponds to octahedral AlO_6 aluminum sites in the various hydrolysis species, whereas

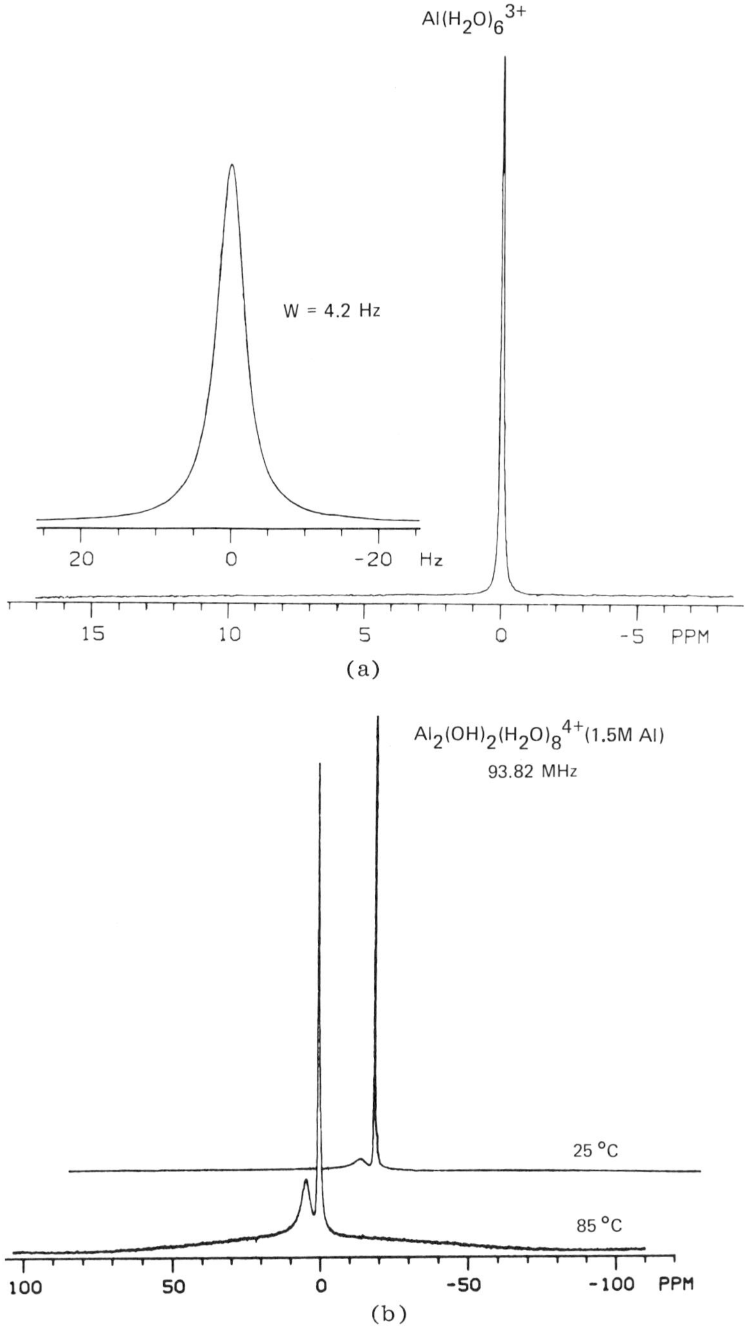

Figure 25 Room temperature ^{27}Al NMR spectra at 93.82 MHz: (a) 1.0 M $AlCl_3 \cdot 6H_2O$ solution, (b) $Al_2(OH)_2(H_2O)_8^{4+}$ ion in equilibrium with hexaaquo complex, and (c) $Al_{13\text{-mer}}$ (at room temperature and 65°C). (From Fitzgerald, 1981.)

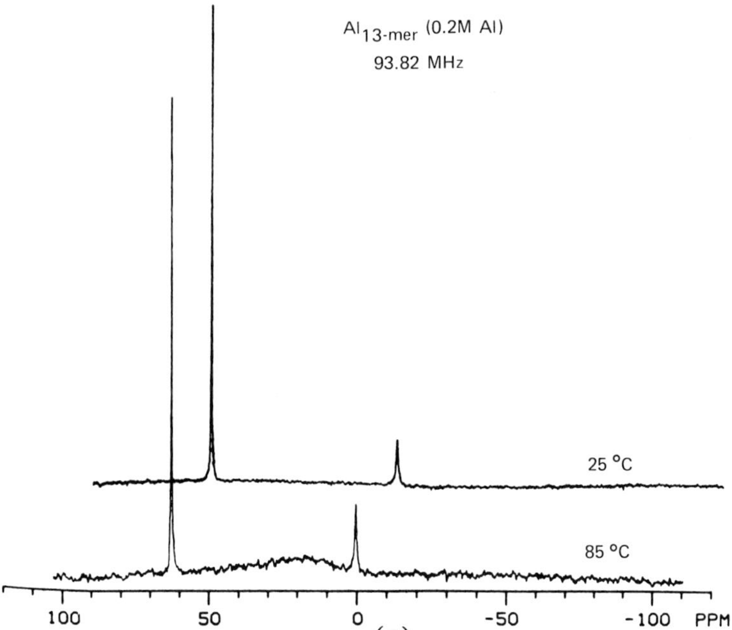

(c)

Figure 25 (continued)

Table 21 Summary of ^{27}Al NMR Parameters for Various Aluminum(III) Ion Hydrolysis Species

Complex ion (Al site)[a]	Solution NMR parameters	
	δ(^{27}Al) (ppm)[b]	(Hz)[c]
$Al(H_2O)_6^{3+}$	0.0 (Oh)	4
$Al_2(OH)_2(H_2O)_8^{4+}$	4.2 (Oh)	40
$Al_{13}O_4(OH)_{24}(H_2O)_{12}^{7+}$	17.3 (Oh)	400
	62.8 (Td)	15
$Al_2O(OH)_6^{2-}$	76.3 (Td)	40
$Al(OH)_4^-$	80.0 (Td)	25

[a]Aluminum sites denoted as tetrahedral or octahedral geometry.
[b]Relative to the reference standard, 1 M $AlCl_3 \cdot 6H_2O$ in D_2O.
[c]Peak widths at half-height in hertz (Hz).
Source: Fitzgerald and Maciel (1984b).

signals due to tetrahedral AlO_4 sites are observed in the 50–80 ppm chemical shift region (Akitt, 1972; Delpuech, 1983). While the ^{27}Al NMR resonances for the octahedral monomeric and dimeric complexes are readily observable at room temperature (Figs. 25a and 25b), the ^{27}Al NMR spectrum of the $Al_{13\text{-mer}}$ complex is highly temperature-dependent (Fig. 25c). At room temperature, only a single sharp resonance at 62.8 ppm has been observed and is assigned to the highly symmetrical central tetrahedral AlO_4 moiety in the $Al_{13\text{-mer}}$ ion. At elevated temperature, however, a broad octahedral region resonance (17.3 ppm) and the tetrahedral peak at 62.8 ppm have been reported in a 12:1 ratio at 104.2 MHz (Akitt and Mann, 1981e). The octahedral resonance corresponds to the 12 highly distorted AlO_6 units in the $Al_{13\text{-mer}}$ species and is observable only at elevated temperature.

The use of these ^{27}Al NMR spectral assignments to study aluminum hydrolysis systems containing multiple species in aluminum(III) solutions partially neutralized with Na_2CO_3 or NaOH has been reported. Akitt and co-workers (1972, 1978, 1981a–1981e), Bottero et al. (1980), Fedotov et al. (1978), and Schonherr et al. (1981a, 1981b, 1983) used ^{27}Al NMR spectroscopy to determine the relative amounts of the monomer, dimer, and $Al_{13\text{-mer}}$ as a function of degree of hydrolysis from $\bar{n}$ = 0.0 to 2.4. In addition, Bottero et al. (1980) interpreted the increased peak width of the 0.0 ppm octahedral resonance in the $\bar{n}$ = 0.5–2.2 solution range as due to the presence of three different monomeric octahedral species, including the Al^{3+}, $Al(OH)^{2+}$, and $Al(OH)_2^+$ ions. These investigators also examined aluminum(III) solutions partially neutralized with either NaOH, or Na_2CO_3.

Akitt and Farthing (1978, 1981a–1981e) in particular has carried out extensive ^{27}Al NMR investigations to clarify differences in the polymeric species in aluminum(III) solutions rapidly hydrolyzed with Na_2CO_3 in comparison with those prepared by slow hydrolysis using $AlCl_3$/Al reactions. These investigators studied the ^{27}Al NMR spectral characteristics of the aluminum hydrolysis species in two distinct $\bar{n}$ = 2.5 solutions, one prepared by rapid neutralization of 1 M $AlCl_3$ with Na_2CO_3 at 100°C (type 1) and one prepared by slow hydrolysis involving the dissolution of aluminum metal in boiling $AlCl_3$ solutions for 48 hours with mercury as a catalyst (type 2). Both $\bar{n}$ = 2.5 solutions were examined by ^{27}Al NMR spectroscopy (Akitt and Farthing, 1981b–1981e) and gel filtration chromatography (Akitt and Farthing, 1981a). In some ^{27}Al NMR measurements, stopped-flow kinetic techniques were used to aid the study of the acid decomposition products formed following neutralization of the aluminum hydrolysis complexes with acid to obtain final reaction solutions of $\bar{n}$ = 2.0 or 1.3.

In their GFC work, Akitt and co-workers chromatographed these two solution types and monitored the eluted components using Al content analysis, pH, pCl, conductivity, and ^{27}Al NMR measurements. Solutions of type 1 (sodium carbonate synthesis) contained only a high-molecular-weight peak at the column void volume ($K_d = 0.0$) and the $Al_{13\text{-mer}}$ (approximately 75% of sample). By contrast, GFC separation and eluent analysis of the type 2 ($Al/AlCl_3$ synthesis) sample showed that these solutions contained at least four components: the monomer, the $Al_{13\text{-mer}}$, and two higher-molecular-weight polymers which are distinctly different from the $Al_{13\text{-mer}}$ and the higher-molecular-weight peak in the type 1 sample.

The ^{27}Al NMR spectra of these two different samples obtained by stopped-flow kinetic analysis of their acid decomposition products were also reported. This work showed distinct differences in the solution Al^{3+} hydrolysis species from NMR data for the two sample types. The type 1 sample, which contained about 75% $Al_{13\text{-mer}}$ and a high-molecular-weight GFC component, reacted to form the Al^{3+} monomeric ion with a pseudo-first-order rate constant of $k = 0.20$–$0.11\ sec^{-1}$ depending on temperature. A mechanism of acid-catalyzed decomposition of the $Al_{13\text{-mer}}$ involving multiple protonation prior to depolymerization to the monomer was suggested. The ^{27}Al NMR measurements of the acid decomposition products of the two aluminum hydrolysis components in the type 2 solutions (both of higher molecular weight than the $Al_{13\text{-mer}}$) showed that these complexes exhibited different rates of acid decomposition, suggesting that the major intermediate-MW component was structurally related to but larger than the $Al_{13\text{-mer}}$ and that the highest-MW polymer species had a different structure than the $Al_{13\text{-mer}}$ complex. This work showed conclusively that 1. different aluminum hydrolysis species were present in type 1 and type 2 solutions at $\bar{n} = 2.5$, 2. the acid decomposition rates were much lower for hydrolysis species in the $AlCl_3/Al$ metal reaction solution products than solutions containing primarily the $Al_{13\text{-mer}}$ (type 1), and 3. aging of hydrolyzed solutions of the $Al_{13\text{-mer}}$ leads to the production of larger polymeric species, similar to those in solutions prepared by the $AlCl_3/Al$ metal redox reaction. This work supports many of the fundamental conclusions of a number of other investigations, where Al^c- and Al^b-containing aluminum hydrolysis systems can be differentiated on the basis of various other physical-chemical approaches such as the ferron method and gel filtration chromatgraphy. Similar conclusions were reached by Schonherr and co-workers (1981a, 1981b, 1983) in their ^{27}Al NMR, ferron, and gel filtration chromatographic studies of aluminum hydrolysis complexes obtained from the dissolution of aluminum metal in HCl over the concentration range 0.1–2 M Al.

The results of detailed ^{27}Al NMR spectroscopic studies of aluminum hydrolysis systems prepared by dissolution of aluminum metal

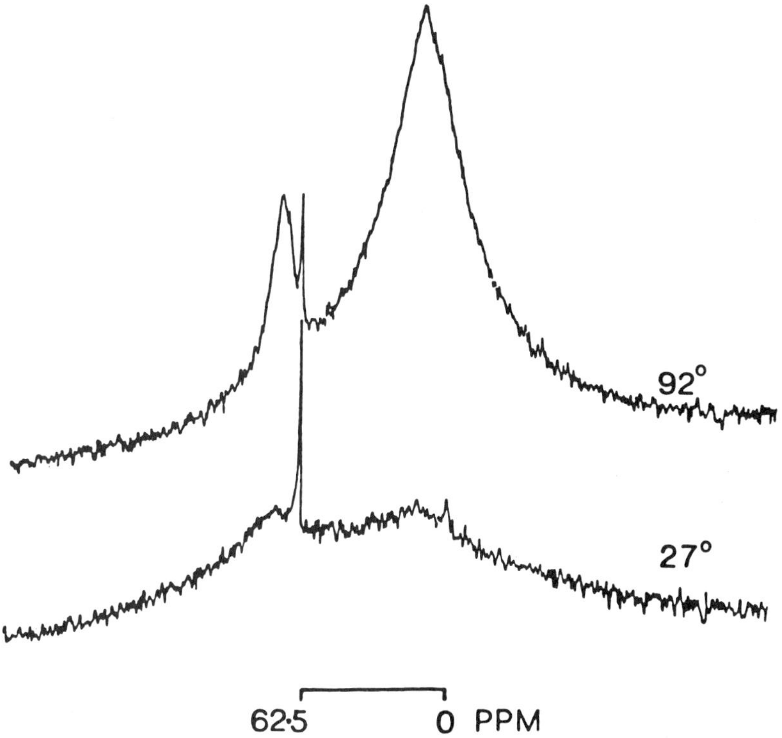

Figure 26 ^{27}Al NMR spectra at 27 and 92° for 1.0 M Al solution at $\bar{n}$ = 2.5 prepared from Al/$AlCl_3$ reaction. (From Akitt and Farthing, 1978.)

in the presence of HCl or $AlCl_3$ solutions provide the most relevant information in terms of the chemistry of ACH solutions. Studies by Akitt and Farthing (1978, 1981d) and Akitt and Mann (1981e) on basic aluminum chloride solutions prepared by the Al metal/$AlCl_3$ solution reaction have been reported. Akitt and Farthing (1978) initially reported ^{27}Al NMR spectra at 23.45 MHz for concentrated (1 M Al) hydrolyzed $\bar{n}$ = 2.5 solutions prepared by the "commercial" ACH method involving the reaction of $AlCl_3$ solutions with Al metal at elevated temperatures. The ^{27}Al NMR spectrum of this system was obtained at 27 and 92°C as shown in Fig. 26. The spectra were quite different from those of "$Al_{13\text{-mer}}$" solutions and exhibited two broad resonances, including an octahedral signal at 8.5 ppm and a broad tetrahedral resonance at 72.6 ppm. The ^{27}Al NMR spectra of these Al metal-hydrolyzed $AlCl_3$ solutions were observed to be both temperature-dependent (see Fig. 26) and concentration-dependent. These workers interpreted the new ^{27}Al NMR resonance signals as being due to higher-molecular-weight polymeric species different from the $Al_{13\text{-mer}}$ complex. Later investigations allowed Akitt and Farthing (1981d) to more fully described the temperature and concentration dependence of the ^{27}Al NMR spectra of these complexes as a function of different magnetic field strengths. Akitt and Farthing's work provided the following information: at 62.86 MHz the ^{27}Al NMR spectra of $\bar{n}$ = 2.5 ACH solutions are markedly temperature-dependent, and four visible resonances are observed: 1. a narrow peak at 0.0 ppm due to monomer, which broadens due to exchange with the dimer at elevated temperature, 2. a narrow tetrahedral (Td) AlO_4 peak in the $Al_{13\text{-mer}}$ at 62.5 ppm, 3. a broad and at least two-component peak due to two octahedral (Oh) AlO_6 moieties at 8.5 ppm, and 4. a broad Td AlO_4 peak at 71 ppm. These workers also noted that approximately 90% of the aluminum is seen at 27°C and virtually all at 80°C. MNR spectra of solutions from 1 to 5 M Al showed changes in the shapes of the octahedral region.

Although a detailed interpretation of the NMR spectra was not made, Akitt (1981d) proposed that an $Al_{13\text{-mer}}$-like species such as an $Al_{20\text{-mer}}$ complex of the general framework $(AlO_4)_2Al_{18}O_{48}$ would be consistent with the NMR signals of the species in these solutions. Although the exact nature of these species was not established, this work suggested the advantages of using elevated-temperature NMR measurements to narrow the complex NMR resonances observed for these systems.

Higher-resolution ^{27}Al NMR studies at 104.2 MHz (Akitt, 1981d) confirmed that these $\bar{n}$ = 2.5 ACH-like solutions exhibited ^{27}Al NMR spectra showing a broad octahedral region resonance near 0.0 ppm, a broad Oh resonance consisting of two or more overlapping signals, and two tetrahedral signals at 62.6 and 71.7 ppm. They interpreted the spectra as due to some species different from but related to the $Al_{13\text{-mer}}$ complex containing several distinct octahedral AlO_6 and tetrahedral AlO_4 sites.

More detailed ^{27}Al NMR studies of hydrolyzed $\bar{n}$ = 2.5 solutions prepared from the Al metal/$AlCl_3$ solution, as well as detailed NMR studies of commercial ACH, have been carried out by Fitzgerald and co-workers (1981a, 1983, 1984b, 1986). These included detailed studies of freshly diluted and aged ACH solutions at various concentrations and temperatures and a detailed investigation of ACH solutions during the commercial redox synthesis. Simultaneous studies of these aluminum solutions were carried out using both Sephadex G-25 GFC and ferron assay measurements. The results of the investigations of the redox synthesis will be summarized in Sec. VI.I.

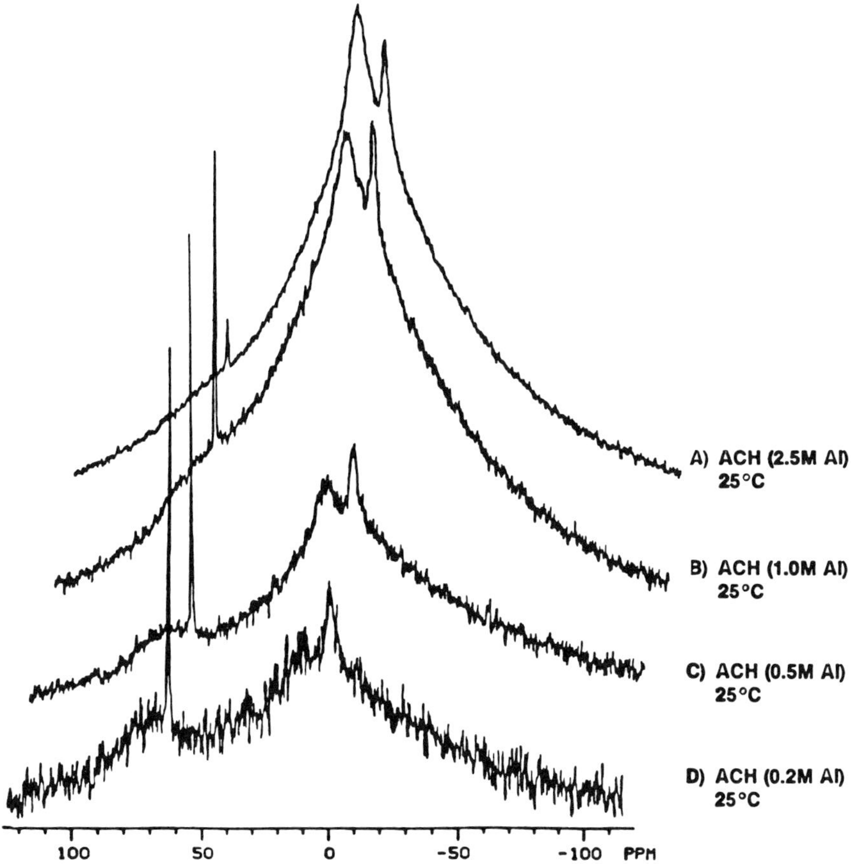

Figure 27 Aluminum-27 NMR spectra of ACH at 25, 10, 5, and 2% concentrations. (From Fitzgerald, 1981.)

The study of various ACH solutions of $\bar{n}$ = 2.5 by Fitzgerald and co-workers, however, provided new insights into the interpretation of the ^{27}Al NMR spectra of ACH solutions and a number of important conclusions regarding the species contained in ACH solutions. First, ACH solutions from 25 to 2% (2.5 to 0.2 M Al) exhibited dramatic changes in spectral resonance shapes and intensities as shown in Fig. 27. The four resonances reported by Akitt are most readily observed at lower concentration, i.e., two broad Oh signals at 0.0 and 10.4 ppm and one sharp and one broad Td signal

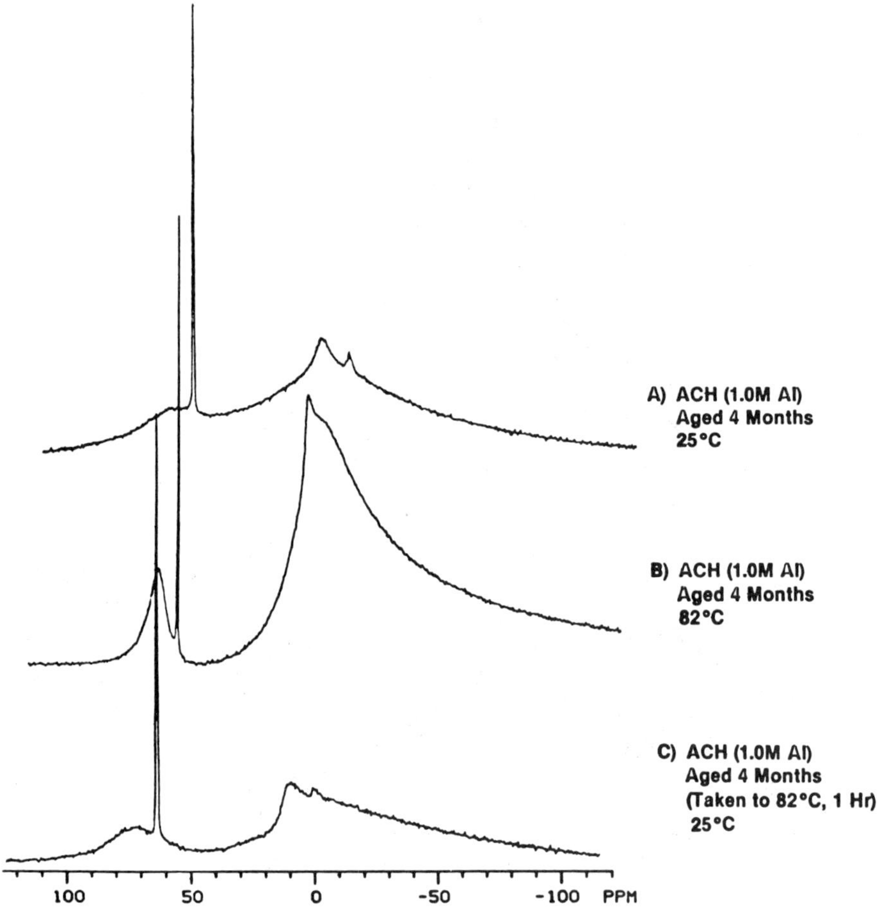

Figure 28 Temperature dependence of 10% ACH solution aged for 4 months. (From Fitzgerald, 1981.)

at 63.0 and 70.4 ppm. Assignment of these complex resonances observed for ACH solutions to specific species is not readily apparent from these room temperature (RT) measurements at 93.8 MHz. Second, the ^{27}Al NMR spectra of more dilute and aged 10% ACH solutions exhibited simpler NMR spectra than those of concentrated or freshly prepared ACH solutions. Aged (4 months) 10% ACH solutions exhibited RT NMR spectra shown in Fig. 28, consiting of two octahedral resonances at 0.1 and 10.4 ppm and probably the 17 ppm Oh signal

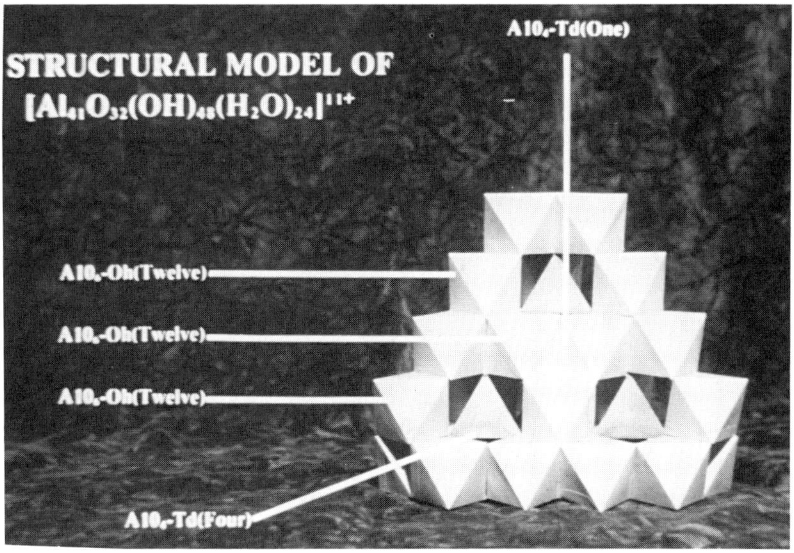

Figure 29 Photograph of structure of $Al_{41\text{-mer}}$. (From Fitzgerald et al., 1981.)

due to an $Al_{13\text{-mer}}$-type site, which was obscured by the equally broad and intense other two Oh resonances. In addition, the two Td signals reported by Akitt at 63.3 and 71.8 ppm were observed at RT and at 82°C, as seen in Figs. 28A and 28B. Fitzgerald proposed that dilute and aged ACH solutions contained a single principal component possibly corresponding to a polymeric species containing 41 aluminums per aggregate (MW about 3000), in agreement with the molecular weight range of 2500–3500 observed for the principal polymeric peak observed in Sephadex G-25 GFC studies of aged ACH solutions. This complex, as shown in Fig. 29, consists of three different octahedral aluminum sites and two different tetrahedral aluminum sites, giving five distinct ^{27}Al NMR resonances. The proposed complex is interestingly built up from an $Al_{13\text{-mer}}$ core and consists of alternating layers of tetrahedral and octahedral aluminum sites. The five tetrahedral aluminum sites include one central AlO_4 site and four others (71 ppm) interspaced between two octahedral layers of AlO_6 units. The three AlO_6 sites indicated by the NMR spectra included an inner layer of 12 AlO_6 sites (at 17 ppm) related to simi-

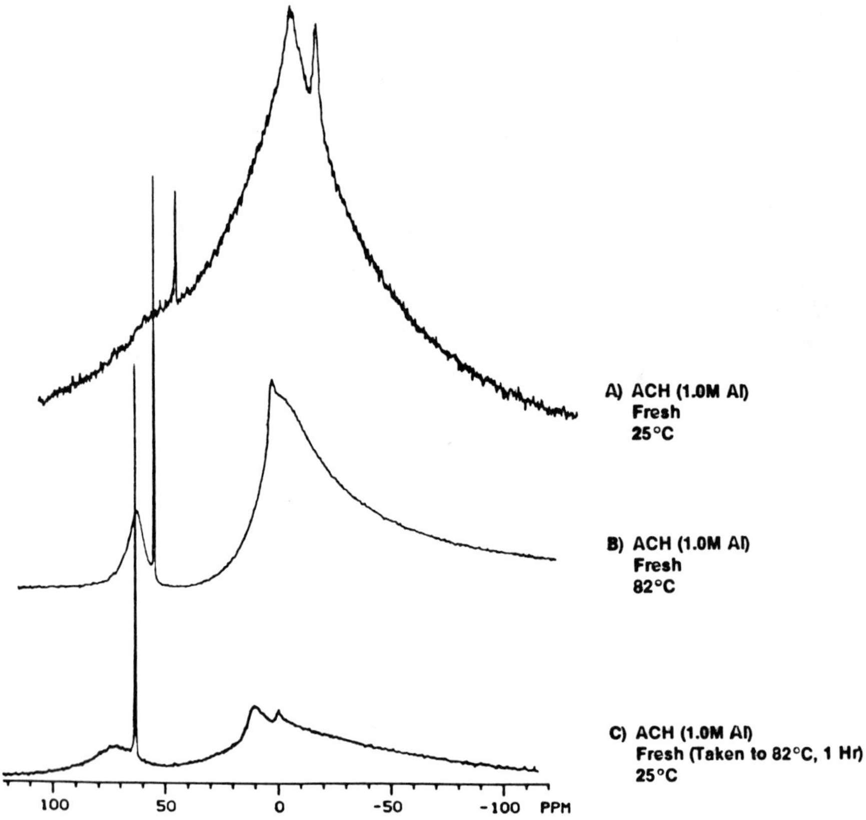

Figure 30 Temperature dependence of freshly prepared 10% ACH solution. (From Fitzgerald, 1981.)

lar sites in the $Al_{13\text{-mer}}$ complex and two sets of 12 chemically distinct AlO_6 outer octahedral sites. The overall Td:Oh ratio of 5:36 was consistent with integration of the tetrahedral and octahedral regions of the NMR spectra of dilute and aged ACH solutions.

Freshly diluted 10% ACH solutions, by contrast, exhibit broader ^{27}Al spectral resonances as shown in Fig. 30A. When the temperature of this sample was elevated, the ^{27}Al NMR spectrum of a 1.0 M (10%) ACH solution was converted to a spectrum similar to that of aged ACH at 82°C (compare Fig. 30B to Fig. 28B). However, the most significant effect of elevated temperature on the ^{27}Al NMR spectra of this 10% ACH solution is seen when comparing the NMR spectra of Figs. 30A and 30C. Thermal treatment of a fresh 10% ACH solution at 82°C for 1 hour converted the RT NMR spectrum to one resembling the ^{27}Al NMR spectra of dilute or aged ACH solutions at RT (compare Fig. 28A or 28C to Fig. 30C). The results of these NMR experiments suggest that thermal treatment and/or dilution of a 10% ACH solution result in major changes in the species composition of fresh concentrated ACH solutions. Fitzgerald interpreted these changes as due to the depolymerization of higher-MW species in fresh ACH solutions to primarily $Al_{41\text{-mer}}$-type species in aged ACH solutions. The interpretation of these results is supported by GFC Sephadex G-25 measurements on dilute and aged ACH solutions as will be discussed subsequently.

F. X-Ray Diffraction and Solid-State ^{27}Al NMR of ACH Powders

X-ray diffraction studies of ACH in the solid state have been reported by Teagarden et al. (1981a, 1982). The characteristic x-ray diffratogram of ACH shown in Fig. 31 indicates that ACH powders are

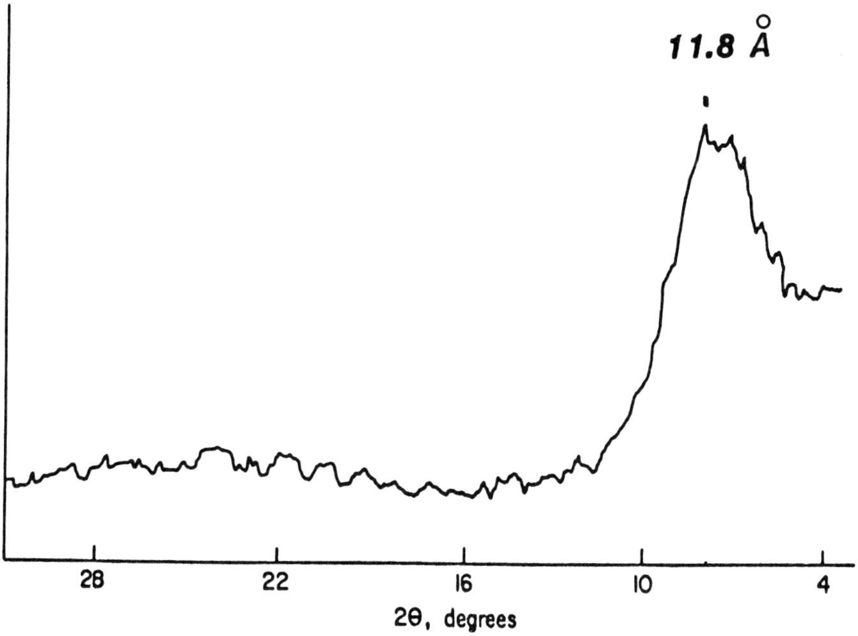

Figure 31 X-ray diffractogram of aluminum chlorohydrate. (From Teagarden et al., 1981a-1981c. Reproduced with permission of the copyright owner, the American Pharmaceutical Association.)

poorly ordered and exhibit a single broad peak in the 2θ range 4–10°, with a corresponding *d* spacing of 11.8 Å. Line broadening suggests the occurrence of crystallites of about 50 Å. Teagarden proposed that the 11.8 Å molecular dimension of hydrolyzed aluminum species in solid ACH was consistent with the size of the $Al_{13\text{-mer}}$ (8.9 Å) from these measurements and the fact that intercalation of aluminum complexes contained in a 25% ACH solution into montmorillonite clays increased the clay layer interspacing by 9 Å. While such intercalation may involve $Al_{13\text{-mer}}$ species, known to exist in small amounts in 25% ACH solutions, the 2.0 Å discrepancy between the x-ray diffraction peak and the molecular diameter of the $Al_{13\text{-mer}}$ complex suggests that a somewhat larger complex would likewise intercalate montmorillonite. The $Al_{20\text{-mer}}$ proposed by Akitt and the $Al_{41\text{-mer}}$ species proposed by Fitzgerald would be of sufficient molecular size (11–13 Å) to be considered as plausible models of the principal components in 25% ACH solution on the basis of this clay intercalation study. Teagarden also suggested that the average crystalline size of 50 Å indicates that the basic crystallite unit would contain up to five layers of $Al_{13\text{-mer}}$ species. Assuming a basic molecular units of an $Al_{20\text{-mer}}$ or $Al_{41\text{-mer}}$, approximately four such units of 80–160 aluminum atoms per structural unit would be contained in the average ACH crystallite. This molecular size is consistent with the average number of 70 aluminum atoms per aggregate determined for ACH solutions by light scattering. In addition, 80–160 aluminum atoms per aggregate is consistent with the molecular weight range of 5000–10,000 proposed for the higher-molecular-weight species in ACH solutions on the basis of gel filtration chromatography.

Solid-state ^{27}Al NMR studies of ACH under magic-angle spinning (MAS) and high-power proton decoupling (HPD) conditions have been described by Fitzgerald and Maciel (1984b). The solid-state MAS/HPD ^{27}Al NMR spectrum of ACH at 93.8 MHz is shown in Fig. 32 together with spectra of solid sampes of $AlCl_3 \cdot 6H_2O$, the hydroxy dimer solid, the sulfate salt of the $Al_{13\text{-mer}}$ complex, and a solid sulfate salt of a sample identified as the "$Al_{41\text{-mer}}$ complex." Resonance peaks due to spinning sidebands are quite easily recognized in the various spectra presented. The ^{27}Al NMR chemical shifts for the resonance peaks of each sample are summarized in Table 22, together with the assignment of their respective Oh or Td site geometries. The solid-state NMR spectrum of the central transition is noted in this table together with corresponding data from solution ^{27}Al NMR spectra of these solids upon dissolution in water. Although the detailed interpretation of these solid-state spectra is reported elsewhere, these data provide very important information on the solid state-solution relationships of these reference aluminum hydrolysis species and lead to a number of significant observations about ACH

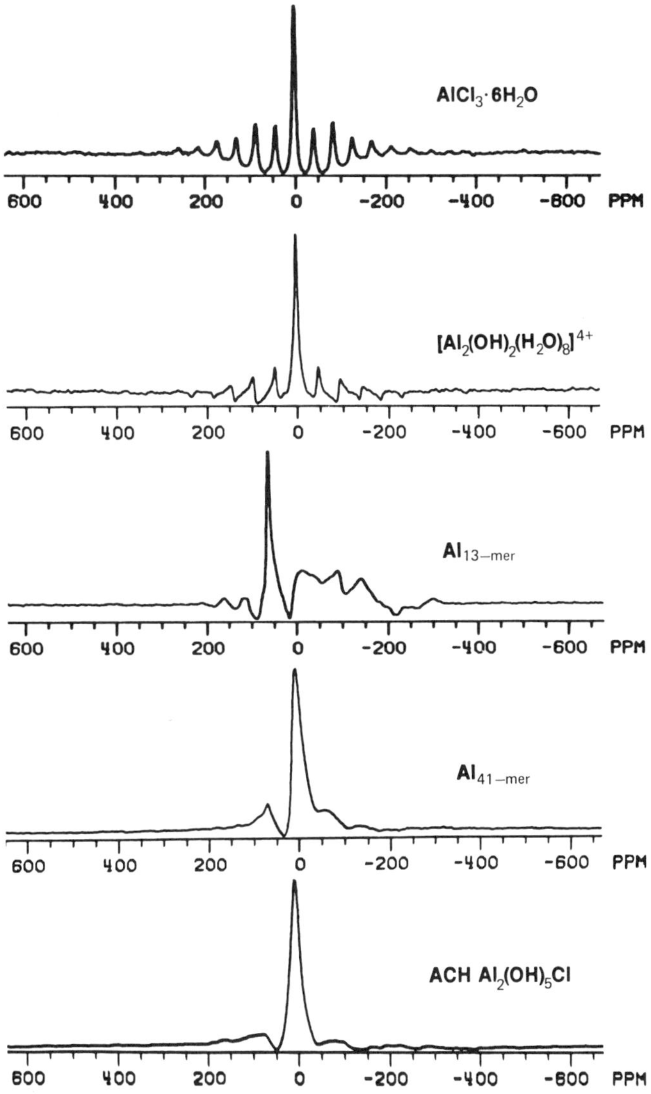

Figure 32 Solid-state magic-angle spinning ^{27}Al NMR spectra of various aluminum complexes. From top to bottom: $AlCl_3{\cdot}6H_2O$, dimer, $Al_{13\text{-mer}}$, $Al_{41\text{-mer}}$, and ACH. (From Fitzgerald and Maciel, 1984b.)

Table 22 ^{27}Al Solution and Solid-State NMR Data for Various Aluminum Hydrolysis Species

Complex ion[a]	Solution parameters		Solid-state parameters[c]	
	δ (ppm)	Δν (Hz)[b]	δ (ppm)	Δν (Hz)[b]
$Al(H_2O)_6^{3+}$	0.0 (Oh)	4	0.7 (Oh)	138
$Al_2(OH)_2(H_2O)_8^{4+}$	4.2 (Oh)	40	4.0 (Oh)	492
$Al_{13}O_4(OH)_{24}(H_2O)_{12}^{7+}$	17.3 (Oh)	400	7.7 (Oh)	856
	62.8 (Td)	15	62.4 (Td)	—
$Al_{41}O_{32}(OH)_{48}(H_2O)_{24}^{11+}$	4.3 (Oh)	25	3.4 (Oh)	1765
	11.7 (Oh)	650		
	15.0 (Oh)			
	63.0 (Td)	150	67.8 (Td)	2337
	71.1 (Td)			
$Al(OH)_4^-$	80.0 (Td)	25	79.9 (Td)	389
$Al_2O(OH)_6^{2-}$	76.3 (Td)	40	75.2 (Td)	393

[a]Complexes in the solid state are either potassium or sulfate salts.
[b]Δν refers to line widths at peak half-height.
[c]Solid-state spectra taken under MAS conditions with high-powder proton decoupling at 93.82 MHz.
Source: Fitzgerald and Maciel (1984b).

in the solid state. Most notably, the solid-state NMR chemical shifts for resonances due to the monomer, dimer, and $Al_{13\text{-mer}}$ are consistent with the chemical shifts of the same resonances for these species in solution. The monomer and dimer spectra exhibit an intense central transition at 0.7 and 4.0 ppm, respectively. The solid-state NMR spectrum of the $Al_{13\text{-mer}}$ consists of a broad, distorted octahedral region resonance centered at 8 ppm and a sharp Td signal at 62.4 ppm. The spectra of the ACH powder and the $Al_{41\text{-mer}}$ are very similar, showing a broad Oh signal at 4 ppm and a Td signal at 68 ppm in a similar Td/Oh relative signal intensity ratio. The NMR spectral peaks for the $Al_{41\text{-mer}}$ complex are approximately 50% more narrow than the ACH spectra, suggesting a decrease in the number of aluminum site local geometries, expected for a narrowing in the molecular weight range of these aluminum complexes. The $Al_{41\text{-mer}}$

complex was isolated by precipitation from an aged ACH solution and consisted of a narrower MW range of species than observed for normal ACH solutions or ACH powders. The lack of further resolution of the broad Oh resonance precludes more detailed interpretation of the nature of the solid-state complexes in ACH powders or the complexes identified as "$Al_{41\text{-mer}}$" complexes as isolated from aged ACH solutions.

G. Effects of Dilution and Aging of ACH

The chemical stability of ACH solutions has been extensively discussed in the literature in terms of 1. the chemical changes in the hydrolysis species on dilution, and 2. the chemical changes which accompany aging of dilute or concentrated solutions. Concentrated ACH solutions of about 6 M Al remain stable indefinitely up to 4–5 years with little or no change in their clarity or pH values. The chemical parameters most frequently used to monitor equilibrium changes of dilute or aged ACH solutions are pH and free Cl^- ion concentrations, in addition to the time of onset of precipitation. Studies of the effects of dilution and aging of ACH have been reported by Riesgraf et al. (1979 and unpublished reports), Teagarden et al. (1981c, 1982), Bretschneider et al. (1975), and Reheis Chemical Company (1977). The general chemical features of ACH on dilution and aging include 1. increases in free chloride ion and conductance, 2. increases in H^+ ion or pH decreases, and 3. accompanying turbidity increases due to the formation of $Al(OH)_3$ (Bretschneider et al. 1975; Reheis Chemical Co., 1977). Reheis proposed the following general reaction scheme:

$$[\text{inorganic polymer}]^{+}Cl^{-} \rightarrow [\text{inorganic polymer}]^{+} + Cl^{-} \quad (29)$$

$$[\text{inorganic polymer}]^{+} + H_2O \rightarrow [\text{inorganic polymer-OH}] + H^{+} \quad (30)$$

$$n[\text{inorganic polymer-OH}] \rightarrow [\text{larger polymer}] + Al(OH)_3\ (s) \quad (31)$$

This general behavior has been confirmed by various workers, depending on the conditions of dilution and aging. The dilution and aging of ACH have been examined in detail by Reisgraf et al. (1979) and Teagarden et al. (1981c, 1982) using pCl, pH, conductance, x-ray diffraction of filtered solids, and the ferron assay method. The ACH solution changes are characterized by two important steps:

1. Initial changes in concentration of H^+ ion, Cl^- ion, and ferron hydrolysis species types
2. Longer-term onset of solution turbidity accompanying nucleation and precipitation of $Al(OH)_3$ solids

The overall state of ACH solutions on dilution and aging has been shown to involve reequilibration to form 14% $Al(H_2O)_6^{3+}$ ion and 86% $Al(OH)_3(s)$ according to the following disproportionation reaction at infinite time and dilution in the absence of other chemical species:

$$Al_2(OH)_5Cl \cdot 2H_2O(aq) \rightarrow Al(H_2O)_6^{3+}(aq) + 3Cl^-(aq) + 5Al(OH)_3(s) \quad (32)$$

For 0.01–0.1 M Al ACH solutions at 30°C, Riesgraf has summarized the descriptions of the system aging in terms of two specific steps. Initially, ACH solutions show increases in Al^b contents to 15–20% accompanied by rapid increases in free Cl^- ion and pH increases or H^+ ion decreases in a 15–20-day time period at 0.1 M Al. At 0.0 M Al, the ferron species types at 1.2 hours after dilution were 5% Al^a, 5% Al^b, and 90% Al^c, which changed to 1% Al^a, 12% Al^b, and 87% Al^c after 4 days aging. Riesgraf interpreted these changes in terms of the following general reaction sequence:

$$H^+ + Al^c \rightarrow Al^a + Cl^- \quad (33)$$

$$Al^a \rightarrow Al^b \quad (34)$$

where H^+ ion consumption leads to release of aluminum Al^a species and Cl^- ion release, followed by polymerization or hydrolysis of Al^a to Al^b species.

In the second step, the Al^b solution content decreases and pH decreases, followed by a slower release of Cl^- ion with simultaneous increases in Al^a and increases in turbidity due to nucleation of $Al(OH)_3$. For a 0.1 M Al ACH solution, the first step occurs in the first 15–20 days, while the second step initially occurs at 36–48 days. At 36 days, for example, approximately 6% of filterable $Al(OH)_3$ has formed. It was noted that either Al^b species or Al^c species are the precursors of aluminum hydroxide. Rapid conductance increases concomitant with the release of H^+ or Cl^- ions as well as reequilibration of polymeric solutes to the hexaaquo aluminum(III) cation occurs:

$$Al^c \text{ or } Al^b \rightarrow Al(OH)_3(s) + H^+ + Cl^- + Al(H_2O)_6^{3+}(aq) \quad (35)$$

Concrete evidence that the $Al(OH)_3$ formed in diluted, aged ACH solutions is the aluminum hydroxide polymorph gibbsite has been obtained by Teagarden et al. (1981c, 1982) using both IR and x-ray diffraction. The IR spectrum of filtered gibbsite obtained from ACH solutions aged for 3 days at 0.1 M Al shown in Fig. 33, together

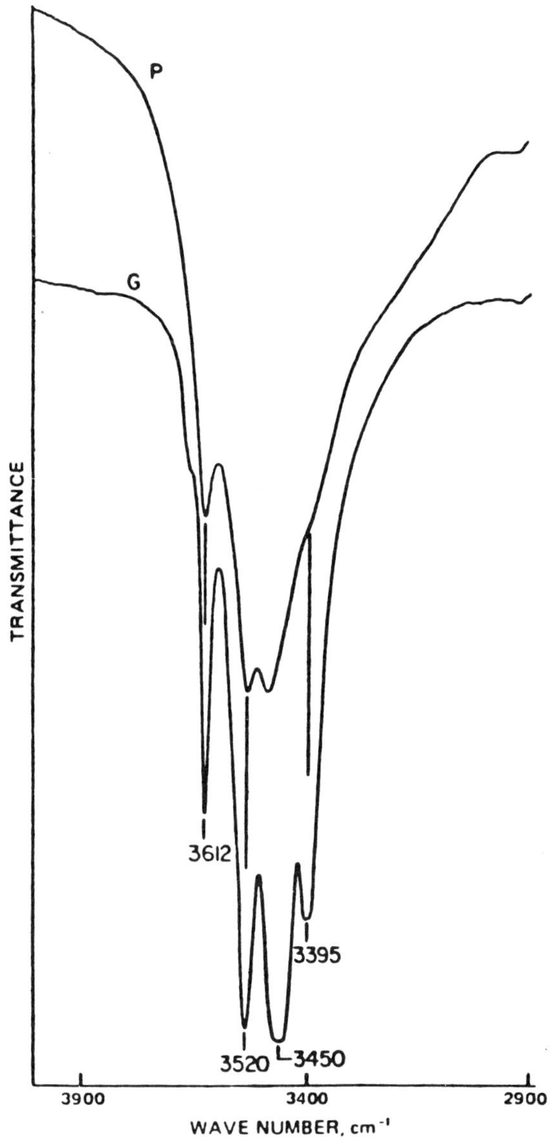

Figure 33 Infrared spectra for gibbsite (G) and precipitated (P) gibbsite obtained from aged ACH solutions. (From Teagarden et. al., 1981a-1981c. Reproduced with permission of the copyright owner, the American Pharmaceutical Association.)

with the corresponding IR pattern for an authentic gibbsite sample. The formation of gibbsite from aged ACH solutions may be contrasted with the fact that bayerite, another aluminum hydroxide solid, is precipitated from ACH solutions which have been neutralized with OH^- ion under neutral or alkaline conditions (Teagarden et al., 1982).

Hydrolysis of diluted, aged ACH solutions to form their ultimate equilibrium hydrolysis species, $Al(H_2O)_6^{3+}$ ion and $Al(OH)_3$ solid, has been found to be dependent on three factors: [ACH], temperature, and solution $[Cl^-]$. Riesgraf and co-workers (unpublished results), for example, have shown that 0.1 M Al ACH solutions at 50°C undergo rapid reequilibration, bypassing the first step involving a Al^b species buildup and pH increase. Thus, the Al^b content decreases and pH decreases with the formation of 20–30% $Al(OH)_3$ in 5 months. At 1 year of aging of a 0.1 M Al ACH solution, 11% Al^a, 42% Al^c, and 47% $Al(OH)_3$ were present, suggesting that the system is still far removed from the final theoretical species distribution of 16% monomer and 84% aluminum hydroxide. These workers noted that the rate of formation of gibbsite was higher at lower concentrations (0.01–0.001 M Al) and that addition of NaCl decreased the rate of $Al(OH)_3$ formation. This work supports the conclusion that increases in solution Cl^- ion retard the rate of formation of gibbsite, consistent with the scheme proposed by Reheis Chemical Co.

Important studies which have clarified the polymeric, oligomeric, and monomeric species interconversions occurring during the early phases of hydrolysis of diluted ACH solutions have been carried out by Fitzgerald (1981) using GFC, ^{27}Al solution NMR, and ferron analysis. This work analyzed 25–2% ACH solutions as a function of time of aging. Gel filtration chromatograms of 25 and 5% ACH solutions aged for a 3-month period at RT are shown in Figs. 34 and 35, respectively. Experimental GFC parameters taken from the chromatograms for the 25% ACH solutions are summarized in Table 23. The GFC data for aged 25 and 5% ACH solutions have been used to construct a plot of the relative intensity (% normalized peak height) for each GFC peak versus time of aging as shown in Figs. 36 and 37, respectively. These aging studies reveal significant new information about the species distributions in dilute, aged ACH solutions. First, the chromatograms for 25 and 5% ACH solutions show that the solutions are metastable polymeric/oligomeric/monomeric systems that undergo extensive species interconversions. For 25% ACH solutions, the chromatograms of Fig. 34 and the GFC data of Table 23 show major species interconversions involving a shift in the relative height of the K_d = 0.41 peak in the first 3 months from about 79% peak 1 versus 21% peak 2a/2 to 39% peak 1 versus 61% peak 2a/2. This interconversion was interpreted as due to depolymerization of peak

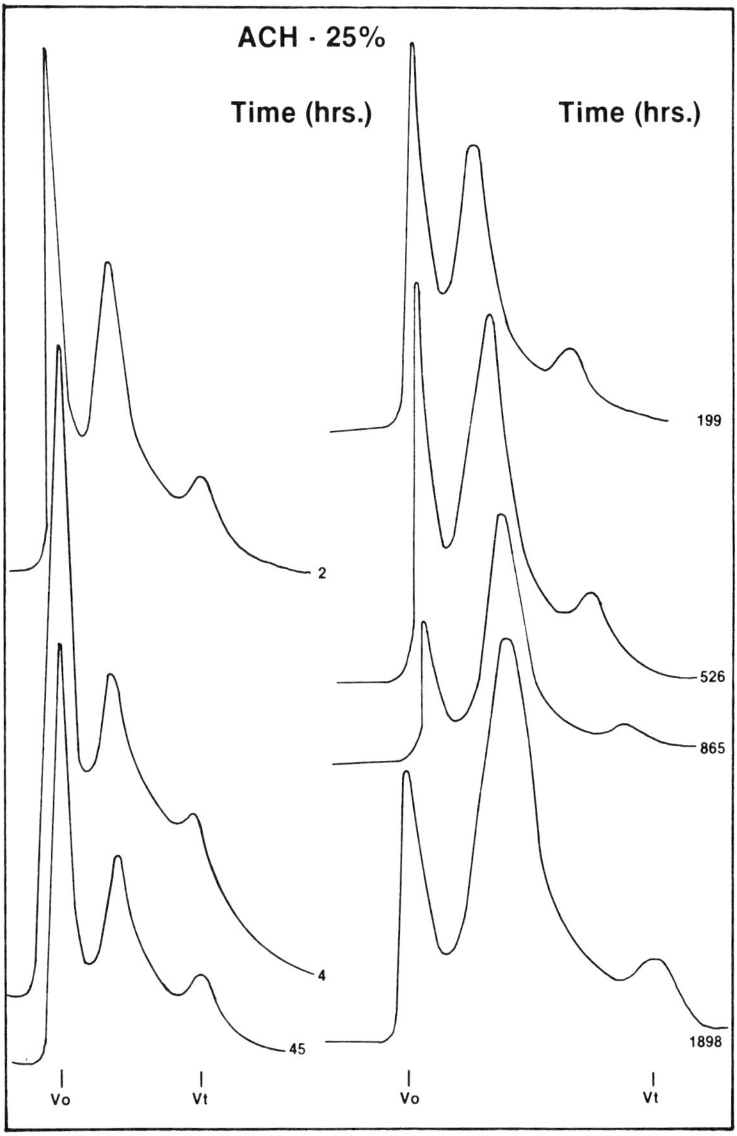

Figure 34 Sephadex G-25 GFC profiles of 25% ACH solutions aged over 1898 hours. (From Fitzgerald, 1981.)

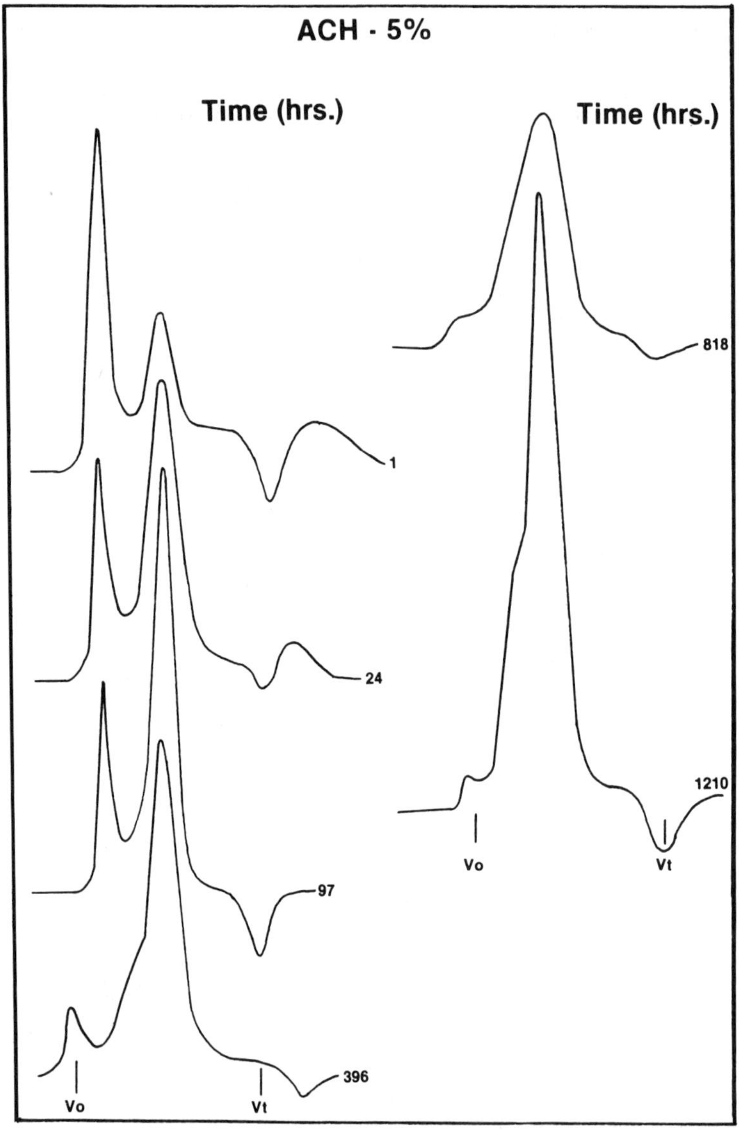

Figure 35 Sephadex G-25 profiles of 5% ACH solutions aged over 1210 hours. (From Fitzgerald, 1981.)

Table 23 GFC Parameters for Aluminum Chlorohydrate, Sephadex G-25 Column, 25% ACH Solution

Time		Normalized height (%)		
Days	Hours	K_d = 0.0	K_d = 0.41 ± 0.01	K_d = 0.65 ± 0.03
0.08	2	78.7	21.8	—
0.17	4	67.7	32.3	—
1.17	28	65.0	35.0	—
1.88	45	67.6	32.4	—
1.96	47	66.7	33.3	—
5.13	123	62.6	37.4	—
5.17	124	63.3	36.7	—
8.29	199	60.2	39.8	—
8.38	201	57.5	42.5	—
21.87	525	53.5	46.5	—
21.92	526	52.5	47.5	—
36.00	864	37.7	62.3	—
79.00	1896	40.0	60.0	—
79.08	1898	39.00	61.0	—

1 higher-molecular-weight species to intermediate-MW species contained in the broad K_d = 0.41 peak containing unresolved peak 2a/2 species. Kinetic treatment of this two-component species interconversion, shown in the plot in Fig. 36, shows that this depolymerization has a half-life of approximately 18–20 days for 25% ACH solutions. Ferron analysis of the same solutions over the first 2 months indicates that the Al^c content decreases slowly from 95 to 82%, with some rapid increases in Al^a content and slower increases in Al^b content. The small changes in the Al^c content suggest that the peak 1 to peak 2a/2 species interconversions involve primarily an Al^c to $Al^{c'}$ species change, although release of Al^a and subsequent hydrolysis of these complexes to Al^b species occur.

Second, the ACH species changes on aging are concentration-dependent, as noted in the gel filtration chromatograms of Figs. 34

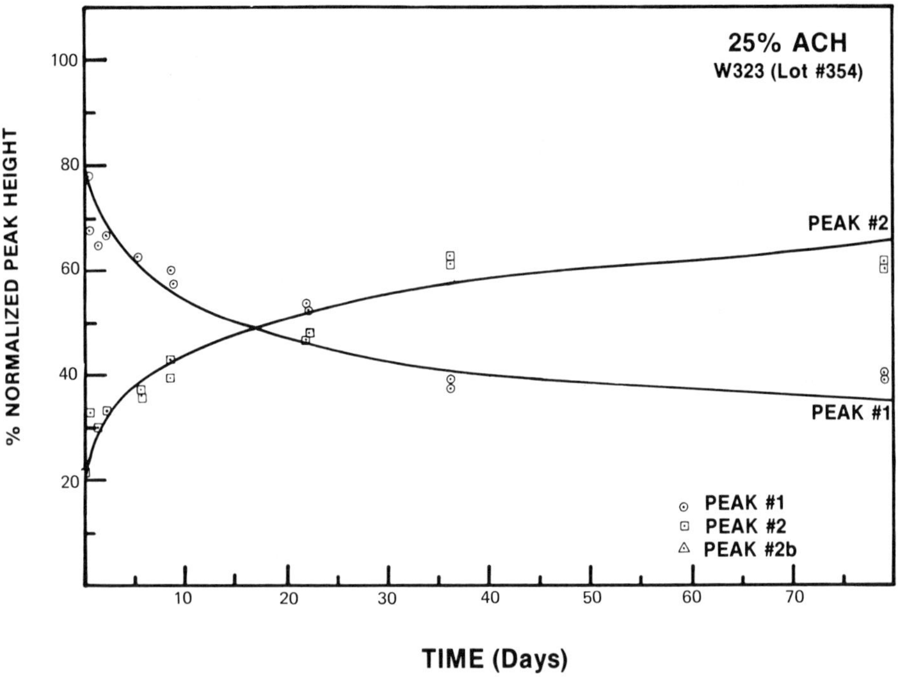

Figure 36 Aging profile for 25% ACH solution depicting relative peak height for Sephadex G-25 peaks 1 and 2 over an 80-day time period. (From Fitzgerald, 1981.)

and 35, the GFC data of Table 23, and the profile of peak 1/peak 2a/2 relative intensity versus time of aging shown in Fig. 35 for 5% ACH solutions. Decreases in [ACH] lead to more rapid depolymerization of peak 1 Al^c solutes to peak 2a/2 $Al^{c'}$ species. Values for the half-life of this depolymerization process are 20 days, 3 days, 1 day, and 2 hours for 25, 10, 5, and 2% ACH solutes, respectively. A 2% ACH solution displays a monodisperse GFC profile consisting of a single solute species corresponding to peak 2 within 6–8 days with a ferron species content of 2% Al^a, 15% Al^b, and 83% Al^c. The ferron changes are most dramatic for the more dilute solutions such as the 2% ACH solution, as shown in more detail in Table 24, over the aging period studied. For this solution, the rapid depolymerization of peak 1 Al^c solutes is also seen to produce rapid increases in the Al^b content to 15% Al^b. More concentrated 25 to 10% ACH solutions exhibit less drastic ferron species type changes.

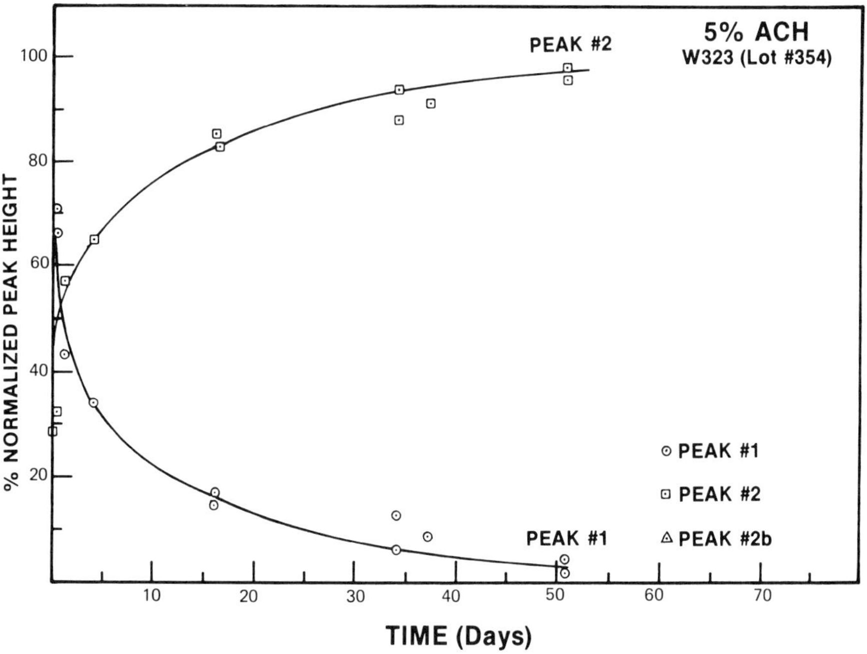

Figure 37 Aging profiles for 5% ACH solution depicting relative peak heights for Sephadex G-25 peaks 1 and 2 over a 50-day time period. (From Fitzgerald, 1981.)

Table 24 Ferron Analysis Results for Aluminum Chlorohydrate, 2% ACH at Various Times

Time		% of total aluminum		
Hours	Days	Al^a	Al^b	Al^c
0	0.00	3	5	92
2	0.08	1	10	89
4	0.17	2	13	85
24	1.00	1	15	84
96	4.00	1	18	81
192	8.00	2	15	83
360	15.00	2	21	77
696	29.00	4	20	76
1320	55.00	0	17	83

The ^{27}Al NMR of diluted and aged ACH solutions also show species interconversions as discussed previously. For example, the ^{27}Al NMR spectra of Fig. 27 are very dependent on ACH concentration for freshly diluted solutions, showing an increase in the $Al_{13\text{-mer}}$ (Al^b-type species) as evidenced by the increase in the tetrahedral 63 ppm signal. Also, the octahedral region of the NMR spectrum for 10% ACH solutions aged 4 months at both 25 and 82°C (Fig. 28) shows the characteristic features which were previously assigned to an intermediate-MW species such as the $Al_{20\text{-mer}}$ (Akitt) or the $Al_{41\text{-mer}}$ (Fitzgerald). These results are consistent with a major depolymerization process involving dissociation of aggregated high-MW species (peak 1 Al^c from GFC data) to intermediate-MW solutes (peak 2a/2 $Al^{c'}$-type complexes).

Additional GFC experiments and ^{27}Al NMR spectral studies have shown that this depolymerization process not only is accelerated by increased dilution of ACH solutions but also is dependent on the solution concentration of Cl^- ion and different anions. For example, GFC data for 2% ACH solutions containing added NaCl from 0.0 to 0.9 M Cl^- ion show dramatic changes in the rate of conversion of peak 1 to peak 2a/2 solutes, as summarized in Table 25, at 40 and 216 days of aging. At 0.9 M added Cl^- ion, 2% ACH solutions still give gel filtration chromatograms which have 62% peak 1 components at 40 or 216 days; in comparison, 2% ACH solutions with no added Cl^- ion gave 17% and 4% peak 1 solutes at 40 and 216 days, respectively. These results indicate that increased Cl^- ion concentrations

Table 25 GFC Peak 1 Relative Intensity (Sephadex G-25), Aging of 2% ACH Solutions at Various Times

Added [Cl^- ion] in 2% ACH	% normalized height (peak 1)	
	40 days aging	216 days aging
0.0	17	4
0.1	22	10
0.3	39	25
0.5	—	44
0.7	—	53
0.9	61	62

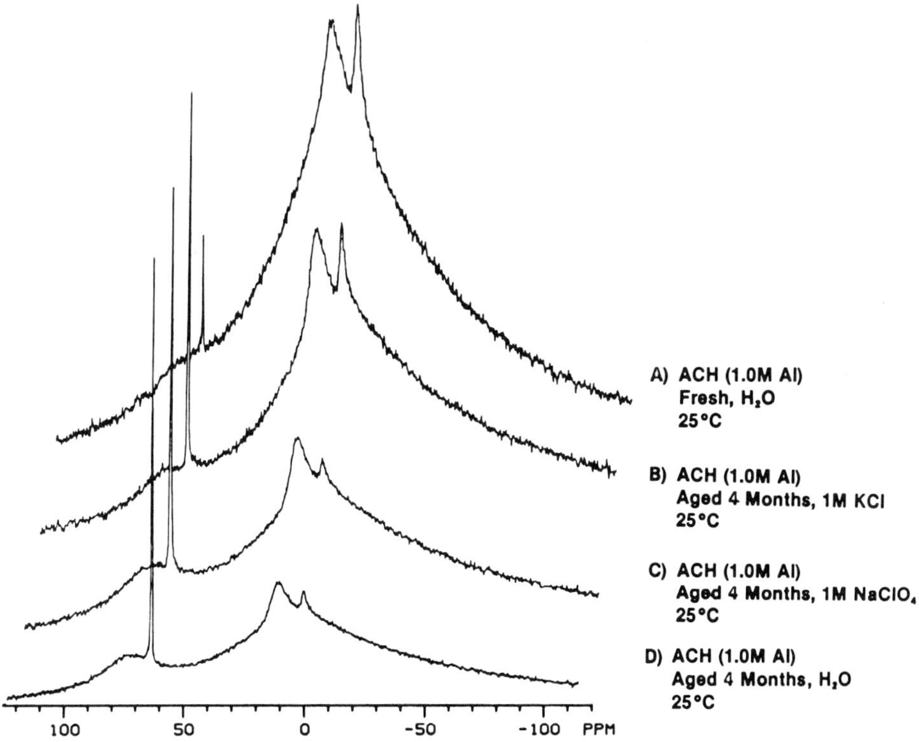

Figure 38 Aluminum-27 NMR spectra of 10% ACH solutions at different aging times in various supporting electrolytes. (From Fitzgerald, 1981.)

retard the depolymerization process, thus stabilizing the higher-MW species observed at $K_d = 0.00$ in Sephadex G-25 analyses.

The ^{27}Al NMR spectra of 10% ACH solutions support the fact that Cl^- ion inhibits the depolymerization reaction, as shown in Fig. 38. The 10% ACH solutions prepared fresh in water and aged for 4 months in 1 M KCl exhibit nearly identical ^{27}Al NMR spectra as shown in Figs. 38A and 38B, in contrast to the "depolymerized" spectrum observed for 10% ACH solutions aged 4 months in water (Fig. 38D). Addition of chloride ion retards the formation of these intermediate-MW species, which give the simpler ^{27}Al NMR spectra shown in Fig. 38D. The dependence of this depolymerization process on anion type was also observed in both GFC and ^{27}Al NMR experiments. For 10% ACH solutions aged in 1 M KCl, 1 M KNO_3, and

1 M $NaClO_4$ (see NMR spectrum in Fig. 38C), it is apparent that different anions inhibit the depolymerization process of 10% ACH solutions. The retarding effect on the depolymerization process follows the sequence $NaClO_4 < KNO_3 < KCl$ at comparable concentrations. Thus, Cl^- ion has the greatest inhibiting effect on the Al^c (peak 1) to $Al^{c'}$ (peak 2a/2) conversion process, and perchlorate ion has the least effect. Since the effect of added Cl^- ion is concentration-dependent, it has been proposed that the outer-sphere complexation of anions, in particular Cl^- ion, leads to charge stabilization of the high-MW aluminum hydrolysis complexes in ACH solutions. Clearly, the complex species interconversions occurring in dilute ACH solutions can readily be monitored by GFC, ferron, and ^{27}Al NMR measurements. This depolymerization is a most important reaction process for dilute and aged ACH solutions. The rate of depolymerization is accelerated at lower [ACH] in the absence of supporting electrolytes, but is inhibited at high concentrations of added electrolytes, particularly Cl^- ion. The polymeric species rearrangements of this depolymerization process occur prior to the more dramatic reequilibration of the aluminum hydrolysis complexes in ACH solution to produce the final equilibrium status of this system, which involves the formation of the aluminum(III) hexaaquo cation and $Al(OH)_3$ solid precipitates.

H. Activated ACH (ACH') Systems

One of the most important aspects of the current manufacturing and processing of ACH has been the recent availability of aluminum chlorohydrate solutions and solids termed activated ACH or ACH'. Activated ACH refers to aluminum chlorohydrates (Al/Cl ratio = 2.00, $\bar{n}$ = 2.50) which have been modified by processing ACH solutions using various proprietary or patented thermal treatments. Such treatments produce ACH' solutions which display increased (or different) biological A/P activity in comparison with normal ACH solutions.

The history of the patent literature pertaining to the preparation of thermally treated ACH solutions as well as other basic aluminum chlorides, nitrates, and bromides includes three series of patents: 1. an early nondescriptive Australian patent (1953), 2. a series of international patents by Gosling et al. (1978)a, 1978b), and 3. two families of patents by Gosling et al. (1980a, 1980b, 1981, 1982a, 1982b) and Fitzgerald et al. (1980). The Australian patent first described the preparation of ACH solutions by thermally treating ACH solutions or similar basic aluminum hydrolysis systems following dilution to a density of 1.25 g/ml (or 25–30% wt/wt ACH). Although these systems were not identified as activated ACHs, they were prepared by processes similar to those described in later patents.

A series of international patents by Gosling et al. (1978a, 1978b) described in more detail a broader range of treatment conditions of ACH solutions as well as HPLC evidence for how these solutions could be distinguished analytically from normal untreated ACH solutions. The method of preparation included heating of BAC solutions at 50–140°C. Using LiChrosorb RP-2 columns in their HPLC analysis with a 0.01 M HNO_3 eluent, these workers showed that thermally treated ACH solutions exhibited four chromatographic bands at K_d = 0.67–0.70 (I), 0.71–0.75 (II), 0.76–0.82 (III), and 0.83–0.97 (IV). Initially, they proposed that the chromatographic evidence showed that normal ACH contained at least 2% polymers greater than 100 Å whereas heat-treated ACH solutions contained no polymers >100 Å, with at least 20% of the polymeric species in band II of the HPLC analysis profiles. Subsequent patents by Gosling et al. (1978c, 1980a, 1980b, 1981, 1982a, 1982b), using HPLC Porasil AX columns with a 0.01 M HNO_3 eluent, showed that heating 7–35% ACH solutions for 0.5–30 days at 80–140°C produced ACH' solutions which contained at least 2% polymers greater than 100 Å (K_d = 0.00). Some of the activated ACH' solutions were found from the HPLC analysis to contain 5–80% larger polymers for the 5/6 basic aluminum chloride, bromide, and nitrate systems, whereas normal ACH solutions did not exhibit the K_d = 0.00 (greater than 100 Å) polymeric peaks on HPLC analysis.

The preparation and description of activated ACH' solutions by Fitzgerald et al. in a UK Patent Application (1980) may be contrasted with the patent descriptions by Gosling et al. primarily on the basis of the chemical methods of analysis used. Fitzgerald and co-workers reported that ACH' solutions may be prepared by thermally aging ACH solutions over a range of temperatures, concentrations, and time periods. They showed that the ACH' solutions consisted of a group of polyhydroxy aluminum species which had not been recognized or identified previously, and referred to them as $Al^{c'}$ complexes on the basis of ferron reactivity analysis and gel permeation chromatographic profiles using EM PGM-2000, a trade name for a polyethylene glycol dimethacrylate-type steric exclusion gel. A typical EM PGM-2000 gel permeation chromatogram of a 10% ACH solution aged for 5 weeks at room temperature is shown in Fig. 39. These workers reported that two chromatographically distinct $Al^{c'}$ complexes (denoted by an H and L doublet in Fig. 39) may be classified as slow-reacting Al^c complexes, i.e., as high (H) and low (L) molecular weight $Al^{c'}$ complexes. These complexes had GPC diffusion constants in the 0.20–0.65 region (more specifically, K_d = 0.30–0.55) normally observed for Al^b-type complexes. The preparation of various ACH' solutions which contained at least 70% $Al^{c'}$ complexes was found to be quite dependent on heating conditions as noted in Table 26. In the patent, these workers concluded that the production of $Al^{c'}$ complexes involved depolymerization of larger poly-

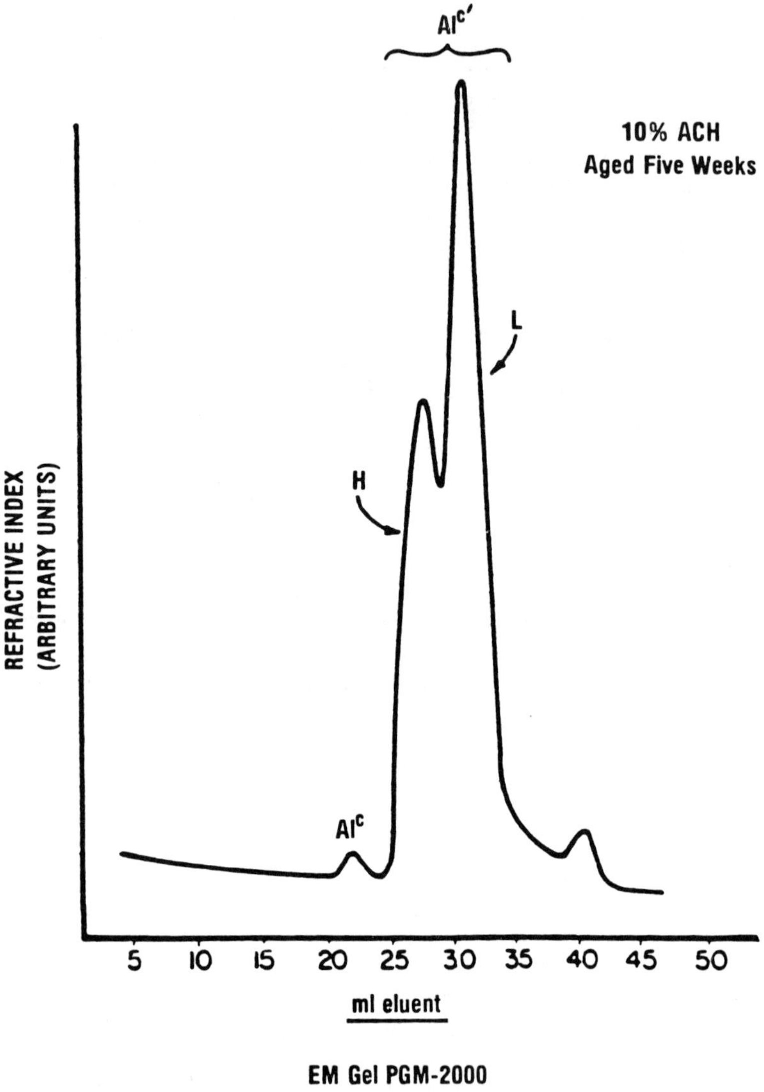

Figure 39 EM gel PGM-2000 GPC profile of activated ACH' prepared from 10% ACH solution by thermal aging for 5 weeks. (From Fitzgerald, 1984, Fitzgerald, Phipps, McClean, and Wu, 1980.)

Table 26 Conditions for Preparation of ACH' Solutions Containing at Least 70% $Al^{c'}$ Complexes

ACH concentrations (%)	Time of reaction at various temperatures		
	50°C	65°C	80°C
10	1 week	1 day	8 hours
15	—	4 days	16 hours
25	—	3 weeks	1 week

Source: Fitzgerald et al. (1980).

meric species (at K_d = 0.00 of the Al^c type) originally present in ACH solutions, similar to the depolymerization process described previously for aged ACH solutions. The rates of the depolymerization process described in the patent are consistent with the depolymerization rates observed from kinetic studies of these ACH processes by Sephadex G-25 GFC analysis. In contrast to the patents of Gosling et al. these workers indicated that activated ACH' solutions containing $Al^{c'}$ complexes had less than 2% polymers greater than 100 Å in molecular size.

The reported physical-chemical studies of activated ACH' systems by GFC (GPC), ferron, and other techniques as well as the patents reported in the 1978–1982 period have spurred the development of a variety of activated ACH' systems. Nelson (1985) described a series of activated ACH' systems, prepared by various proprietary processes, which may be distinguished by such techniques as GPC and ^{27}Al NMR. Figure 40 shows LiChrosorb RP-2 GPC profiles for normal ACH and a series of activated ACH' systems reported by Nelson. According to these GPC results, three major aluminum polymeric species, labeled A, B, and C, are present in varying amounts in ACH and ACH' systems. The highest-MW species, A, is found in highest abundance in normal ACH, Reach 401 and 501 contain lesser amounts, and Reach 101 and 201 contain primarily low-MW polymeric species B and C. Although it is inviting to correlate the intensity and relative elution position of these GPC peaks A, B, and C with the three peaks (the Al^c and two $Al^{c'}$, an H and an L component) reported by Fitzgerald and co-workers (1980, 1981), further research is necessary to more fully evaluate the relationships of these different chromatographic analyses. Clearly, increased analysis of ACH and activated ACH' by these techniques and others will provide a more complete understanding of their chemistry.

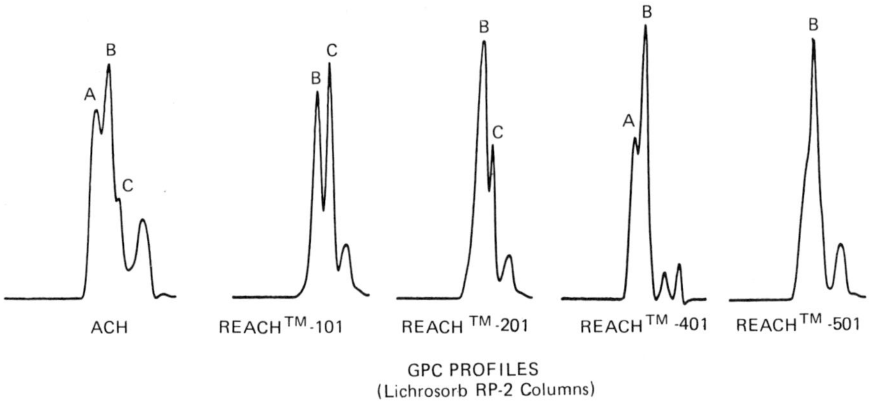

Figure 40 GPC profiles of various activated ACH' systems obtained on Lichrosorb RP-2 columns. (From Nelson, 1985.)

I. Redox Synthesis of ACH

The synthesis of concentrated aqueous aluminum chlorohydrate solutions has been described in the bulk of the patent literature using the following balanced oxidation-reduction reaction:

$$10Al(s) + 2AlCl_3(aq) + 30H_2O \rightarrow 6Al_2(OH)_5Cl(aq) + H_2(g) \quad (36)$$

to form a 5/6 basic aluminum chloride. This idealized reaction involves the dissolution of metallic aluminum in acidic aqueous aluminum chloride solution with the production of a viscous clear solution of the empirical formula $Al_2(OH)_5Cl$ and the evolution of hydrogen gas. The patent of Huehn and Haufe (1940) stresses that this method of synthesis of clear, highly concentrated (above 5 g Al_2O_3/liter) BAC solutions results in truly water-soluble colloidal solutions, whereas in preparing such solutions in dilute aluminum chloride, insoluble BAC salts are also formed. The original Huehn and Haufe patent describes a broader variation on the above reaction, including

1. The use of any aqueous soluble aluminum salts such as halides, nitrates, or even sulfates
2. The use of a variety of metallic aluminum forms including finely divided dust, pellets, shots, or ingots
3. The preparation of a soluble aluminum salt such as aqueous $AlCl_3$ by either dissolution of Al_2O_3 with HCl or dissolution

of aluminum metal with HCl prior to reaction with aluminum metal

The reaction stoichiometry may be varied from a 10/2 ratio of $Al/AlCl_3$ to even an 11/1 ratio using a 25% by weight aluminum chloride solution. The temperature may likewise by varied from 35 to 100°C, with a 70–100°C range preferred. Modification of this process using various catalysts such as mercury and thallium have been described by Jones.

The fundamental reaction for the synthesis of ACH given in Eq. 36 defines theoretical ratios including an $Al/AlCl_3$ ratio of 5/1, an initial Al/Cl solution ratio of 0.33, a molar $H_2O/Al^{3+}(aq)$ ratio of 15/1, and a molar H_2O/Al metal ratio of 3/1. Commercial syntheses normally use an excess Al metal/$Al^{3+}(aq)$ ratio on the order of 10–20/1.

Although numerous investigators have synthesized ACH solutions by this reaction to characterize their hydrolysis complexes relative to those found in commercially available ACH solutions, the only detailed examination of the redox synthesis reported to date has been the work of Fitzgerald and Brooks (1980). Using aluminum prills and aqueous aluminum chloride solutions, Fitzgerald and Brooks examined the ACH redox synthesis to more fully describe the polymer nucleation reactions involved. The experimental solution parameters examined included $[Al^{3+}]$, $[Cl^-]$, $\bar{n}$, pH, and solution density. A summary of the initial and final solution parameters up to 22.5 hours of the synthesis is given in Table 27, although the redox synthesis was carried out at 92–94°C for approximately 30.5 hours. Complete

Table 27 Redox Synthesis Solution Parameters

Solution parameter	Initial	Final
$[Al^{3+}]$, M	0.95	6.69
$[Cl^-]$, M	2.59	2.65
$[Al^{3+}]/[Cl^-]$	0.37	2.53
$\bar{n}$	0.00	2.60
pH	1.78	3.87
Density, g/ml	1.11	1.37

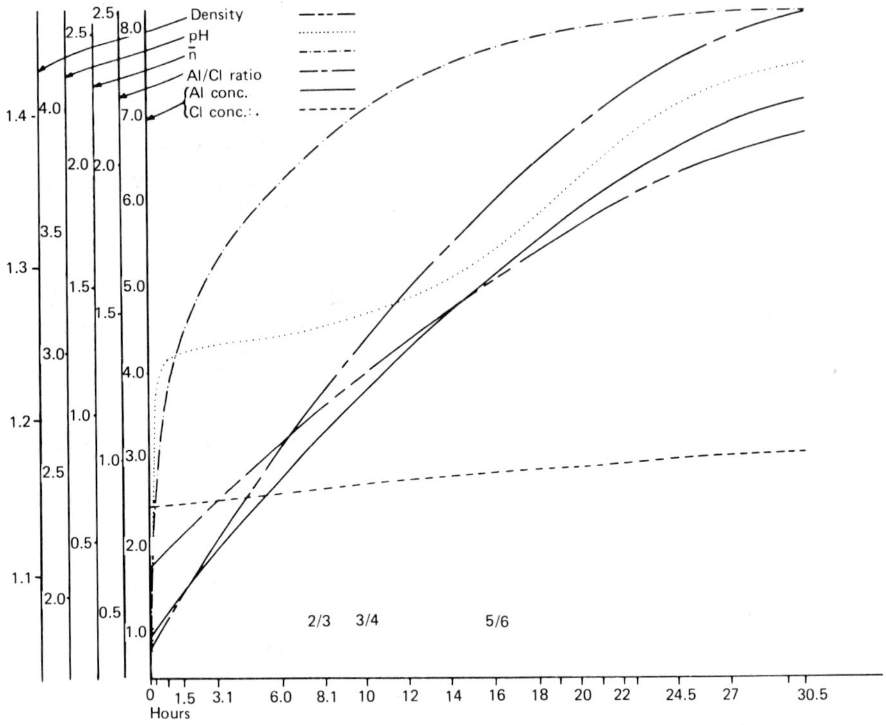

Figure 41 Various solution parameters obtained during the synthesis of basic aluminum chloride solutions from the Al metal-$AlCl_3$ solution reaction. (From Fitzgerald and Brooks, unpublished results.)

analytical results for these synthesis parameters are shown in Fig. 41 based on an initial reaction stoichiometry including an Al/Al^{3+} ratio of 17.5/1, a molar H_2O/Al^{3+} ratio of 52.6/1, and an H_2O/Al mole ratio of 3.0/1. Since aluminum metal was in excess, only 24.4% of the initial Al metal was consumed in the reaction process.

The overall analytical parameters show that the dissolution and oxidation of Al metal results in increases in $[Al^{3+}]$, Al/Cl ratios, pH, and $\bar{n}$ values, with only slight increases in solution $[Cl^-]$. After 22.5 hours, a clear viscous product ACH solution of [Al] = 6.69 M, $\bar{n}$ = 2.61, and pH = 3.87 is obtained. When the reaction is continued beyond 22.5 hours, an ACH solution of 7.15 M Al is obtained. Various points along the reaction time coordinate in Fig. 41 are noted which would correspond to the $\bar{n}$ value of Al/Cl ratio of a 2/3, 3/4, or 5/6 BAC solution, e.g., 16 hours for 5/6 BAC.

Examination of aliquots taken during the course of this reaction included Sephadex G-25 gel filtration chromatography and ferron assays. These measurements provided a means of interpreting changes in the aluminum species distributions occurring throughout the reaction. The Sephadex G-25 GFC results for this redox synthesis are summarized in Fig. 42 and Table 28 for some of the samples taken during the synthesis. The experimental GFC results show a very dramatic progression of species distribution changes throughout the reaction. The final reaction product solution (Sample 18)

Table 28 GFC Sephadex G-25 Results, Redox Synthesis

Sample No.	Time (hours)	GFC peak 1 %	GFC peak 1 K_d	GFC peak 2 %	GFC peak 2 K_d	GFC peak 3 %	GFC peak 3 K_d
1	0.00	—	—	—	—	100.0	0.90
2	0.25	—	—	—	—	100.0	0.94
3	0.75	—	—	5.8	0.42	94.2	0.93
4	1.50	—	—	10.6	0.46	89.4	0.94
5	3.10	—	—	16.4	0.40	83.6	0.95
6	6.03	—	—	47.3	0.53	52.7	0.96
7	8.06	—	—	56.7	0.45	43.3	0.96
8	10.00	—	—	61.7	0.43	38.3	0.93
9	12.00	—	—	62.1	0.39	37.9	0.94
10	14.00	0.5	0.00	67.2	0.39	32.3	0.96
11	16.00	0.6	0.00	76.7	0.41	22.7	0.94
12	18.00	3.3	0.00	74.7	0.38	22.0	0.97
13	19.00	12.1	0.00	70.9	0.35	17.0	0.94
14	20.00	30.1	0.00	55.4	0.42	14.5	1.00
15	21.00	51.2	0.00	35.8	0.44	13.0	1.00
16	22.00	64.3	0.00	26.4	0.44	9.3	1.00
17	22.50	70.4	0.00	21.1	0.44	8.5	0.96
18	23.00	75.8	0.00	14.5	0.34	9.7	0.90

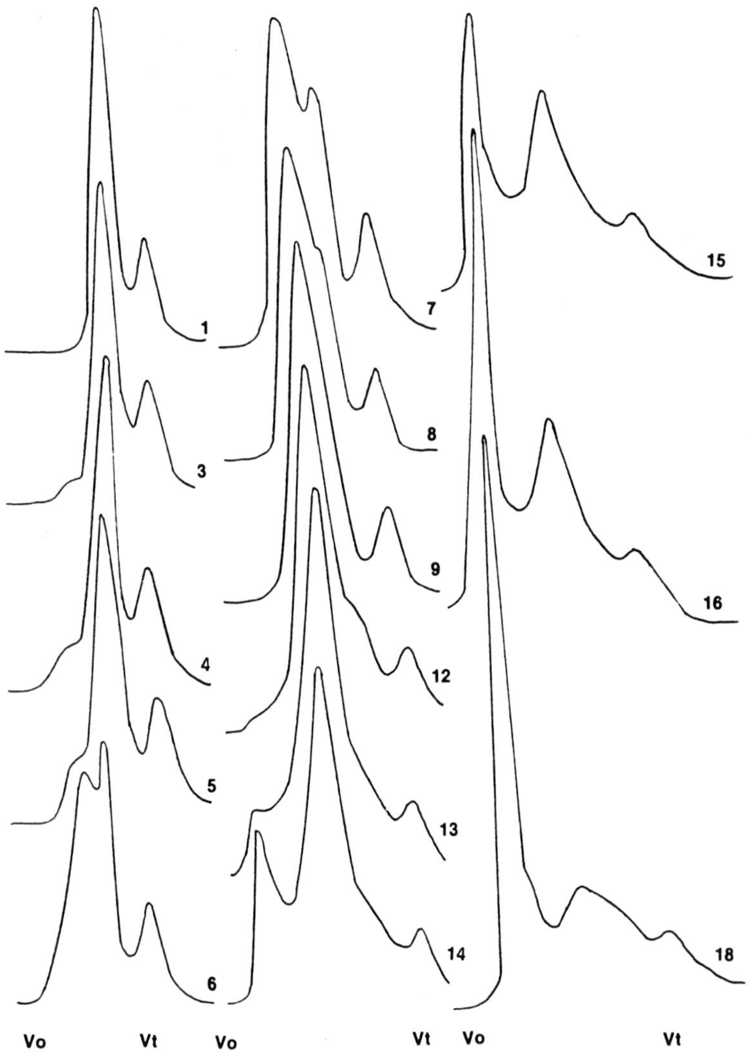

Figure 42 Sephadex G-25 GFC profiles for BAC solutions obtained during the reaction of Al metal with an aqueous aluminum chloride solution. (From Fitzgerald and Brooks, unpublished results.)

exhibits a GFC profile with a peak 1/2 distribution of 76%/15%, which is characteristic of commercial ACH solutions. As the reaction progresses, the solution chromatograms exhibit increases in the molecular weight distribution, including the appearance of peak 2 solutes at sample 3, the growth in peak 2 solutes throughout the middle samples 4–13, and finally the appearance of peak 1 solutes at samples 12 and 13. The % peak height values for the three major peaks are shown more clearly in Fig. 43. Also noted in Fig. 43 are the times along the reaction coordinate where 2/3, 3/4, and 5/6 basic solutions are obtained based on the solution Al/Cl analysis ratios. Interpretation of these GFC data suggests that the polymerization sequence

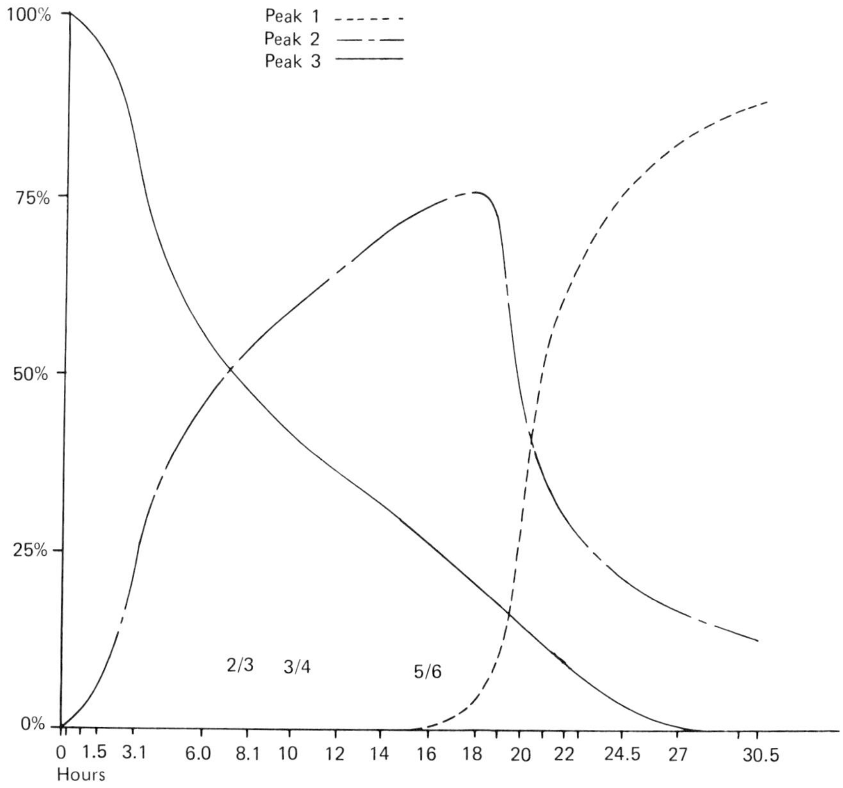

Figure 43 Profiles of the intensities of various GFC Sephadex G-25 peaks over the time course of the Al /$AlCl_3$ reaction. (From Fitzgerald and Brooks, unpublished results.)

Table 29 Ferron Results, Redox Synthesis[a]

Sample No.	Time (hours)	% Al of sample		
		Al^a	Al^b	Al^c
1	0.00	96.1	1.9	2.0
2	0.25	92.0	5.9	2.1
3	0.75	80.0	4.8	15.2
4	1.50	68.4	5.1	26.5
5	3.10	55.9	7.0	37.1
6	6.00	40.0	8.2	51.8
7	8.00	28.7	8.7	62.6
8	10.00	21.3	8.9	69.8
9	12.00	16.4	8.4	75.2
10	14.00	14.0	8.9	73.4
11	16.00	12.0	8.2	78.1
12	18.00	10.5	8.4	81.1
13	19.00	9.0	6.5	83.5
14	20.00	8.2	9.1	82.7
15	21.00	7.5	6.7	85.8
16	22.00	6.4	8.1	85.5
17	22.50	4.2	8.1	87.7
18	23.00	5.4	9.4	85.2

[a]Initial: 96% Al^a, 2% Al^b, 2% Al^c. Final: 4% Al^a, 8% Al^b, 88% Al^c.

peak 3 to peak 2 to peak 1 solutes occurs in a stepwise fashion. This is clearly eivdenced by the fact that the majority of peak 3 solutes are incorporated into medium-MW peak 2 solutes prior to aggregation of these solutes into the larger-MW peak 1 solutes in the range of Al_{80-160} units. The K_d value of peak 2 solutes varies from 0.35 to 0.53 over the reaction sequence. As expected, the content of peak 2a/2/2b solutes is not readily obtained because of lack of peak resolution in this region of the Sephadex G-25 chromatograms.

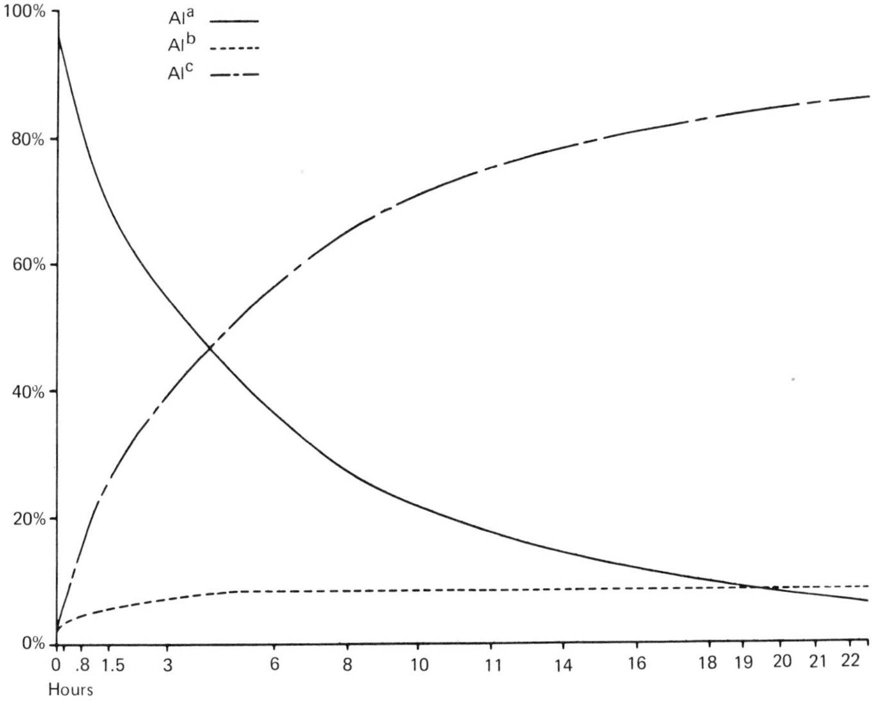

Figure 44 Percentages of aluminum species from ferron analysis of BAC solutions taken throughout the $Al/AlCl_3$ redox synthesis. (From Fitzgerald and Brooks, 1986.)

In an effort to clarify this question ferron assays were also conducted, and the results are summarized in Table 29. The tabulated ferron species contents are depicted graphically in Fig. 44. Ferron analysis of the starting aqueous solution indicated 96% Al^a content. The final reaction solution (sample 18) analyzed as 4% Al^a, 8% Al^b, and 88% Al^c, consistent with previously reported ferron assays of commercial ACH solutions. The ferron assays taken during the reaction show that the Al^b content of the synthesis solution rises rapidly to 8% in the initial 3 hours and then remains constant throughout the rest of the reaction. The Al^a contents decrease as the Al^c contents increase. This result suggests that the peak 2 solutes from GFC data are in fact $Al_{41\text{-mer}}$ complexes of the Al^c type rather than $Al_{13\text{-mer}}$ complexes of the Al^b type. Interpretation of the ferron and GFC data together suggests that one of the two polymerization sequences shown below occurs during the ACH redox synthesis:

Scheme 1

Al^{a} (peak 3) → Al^{b} (peak 2b) (37)

Al^{a} (peak 3) → $Al^{c'}$ (peak 2) (38)

$Al^{c'}$ (peak 2) → Al^{c} (peak 1) (39)

Scheme 2

Al^{a} (peak 3) → Al^{b} (peak 2b) (40)

Al^{b} (peak 2b) → $Al^{c'}$ (peak 2) (41)

$Al^{c'}$ (peak 2) → Al^{c} (peak 1) (42)

Scheme 1 involves simultaneous polymerization of $Al_{13\text{-mer}}$ and $Al_{41\text{-mer}}$ complexes, the reaction conditions favoring the latter peak 2 species. This is followed by aggregation of these complexes into higher-MW peak 1 complexes of Al_{80-160} units at the latter stages of the synthesis. Scheme 2 depicts stepwise polymerization from monomeric to $Al_{13\text{-mer}}$ to $Al_{41\text{-mer}}$ complexes, followed by aggregation to the higher-MW complexes. In this scheme, Al^{b} ($Al_{13\text{-mer}}$) complexes may be treated as an intermediate species which attains a steady-state concentration very early in the reaction process. The overall assembly of the building polymeric units such as the $Al_{13\text{-mer}}$ or the $Al_{41\text{-mer}}$ or both probably involves either monomeric hydroxy Al(III) complexes or dehydroxy bridged dimer complexes. More detailed study of this redox synthesis is necessary to clarify the structural organization occurring during these processes which lead to the production of the final solute species in commercial ACH solutions.

J. Molecular Model of Aluminum Chlorohydrate

This review of the chemistry of aluminum chlorohydrate solutions as derived from detailed physical-chemical investigations reported over the past decade has been most comprehensive. Aluminum chlorohydrate occupies a unique position among the numerous commercially important basic aluminum chloride systems since it is the parent solution from which the majority of the other available salts are produced. Aluminum chlorohydrate solutions are chemically unique in comparison with other aluminum hydrolysis systems in three fundamental ways. First, the solutions are polydisperse and contain larger and less reactive polymeric species than are observed in less dilute aluminum hydrolysis solutions. Second, the species in these solutions are metastable and are not in chemical equilibrium, but are subject to dramatic kinetic changes on dilution and aging. Third, the chemical species in these solutions are significantly influenced by Cl^- ion, which may act as a structural component in these systems, or at least is involved in outer-sphere complexation with the various aluminum hydrolysis complexes, thus influencing the aqueous chemistry and stability of ACH solutions.

The macromolecular characteristics of ACH solutions suggest that at least three polymeric species from 10,000 to 1000 daltons are contained in ACH. Freshly dissolved ACH solutions have been shown from Sephadex G-25 GFC results to contain three polymeric species with K_d values of 0.00, 0.25, and 0.40, in addition to a $K_d = 0.80$ peak. The relative intensity of the various polymeric species is given by the following contents: 78%, 20%, and 2% for peaks 1, peak 2a/2, and peak 3, respectively. The resolution of peaks 2a and 2 is best observed using 10% ACH solutions or less or following the aging of these dilute solutions. All three polymeric ACH species appear to be Al^c-type complexes, which are unreactive with the ferron reagent, with only 10% Al^a and Al^b solutes. In essence, freshly diluted ACH solutions represent a quite simple system containing two or three polymeric Al^c complexes at high concentration which are in a metastable state. The highest-MW species (identified as Al^c species) undergoes depolymerization on dilution or aging, producing a narrow molecular weight range of complexes that consists principally of two intermediate-sized polymeric complexes (denoted as $Al^{c'}$ species) in the MW 2500–3500 range. The fragmentation or dissociation process is accompanied by the production of Al^b oligomeric species of about MW 1000, the $Al(H_2O)_6^{3+}$ ion, and the release of Cl^- and H^+ ions. Experimental evidence for this species composition in ACH solutions has been obtained from GFC, UC, ferron kinetic assays, pCl and pH measurements, as well as ^{27}Al NMR spectroscopy. The NMR measurements have provided direct evidence for the presence of the monomer, the $Al_{13\text{-mer}}$ (an Al^b species), and an intermediate-sized polymeric species proposed to be an $Al_{20\text{-mer}}$ or $Al_{41\text{-mer}}$ cation. The latter complexes have molecular weights in the 2500–3500 range, consistent with the molecular size of peak 2a/2 solutes in ACH solutions. It has also been proposed that complexes such as these represent the principal building unit of polymeric species in ACH solutions and that the higher-molecular-weight peak 1 species are aggregates of these complexes formed at higher solution concentration.

Studies of the redox synthesis of ACH solutions support these conclusions, since the formation of intermediate-MW polymers is presumed to proceed from monomer and dimeric Al^a complexes to $Al_{13\text{-mer}}$-type Al^b species to $Al^{c'}$ complexes to Al^c complexes. This sequence of reaction steps leads to the production of principally high-MW aggregate complexes under the higher concentration conditions used in manufacturing ACH solutions.

Aging of dilute ACH solutions, at 10% or lower concentration, has been shown to involve depolymerization of higher-MW species to two intermediate-MW species, with accompanying formation of Al^b and Al^a complexes and release of H^+ and Cl^- ions. The depolymerization reaction is inhibited by high solution chloride concentrations. A kinetic scheme which accounts for the species compositions of fresh ACH solutions over a range of concentrations and for the observed aging processes accompanying dilution of ACH solutions is given below:

$$Al^c\ (Cl^-\text{-containing}) \rightleftarrows Al^c\ (\text{non-}Cl^-\text{-containing}) + Cl^- + H^+ \qquad (43)$$

peak 1 peak 1

$$Al^c\ (\text{non-}Cl^-\text{-containing}) \rightarrow Al^{c'}\ (\text{non-}Cl^-\text{-containing}) + Al^b + Al^a \qquad (44)$$

peak 1 peak 2 peak 2b peak 3

$$Al^{c'}\ (\text{non-}Cl^-\text{-containing}) + Cl^- \rightleftarrows Al^{c'}\ (Cl^-\text{-containing}) \qquad (45)$$

peak 2 peak 2a

In this kinetic model, ACH solutions are proposed to consist of higher-MW GFC peak 1 solutes which are both chloride-containing and non-chloride-containing, depending on solution concentration as governed by the dissociation of chloride ions from these cationic aluminum polymeric complexes. Following release of chloride ion, these larger polymeric species depolymerize to peak 2 solutes with release of Al^b fragments and the monomer. The presence of both peak 2/2a solutes in aged ACH solutions is accounted for by assuming a third reaction involving the association of free chloride ion with these intermediate-sized polymeric complexes. The above three-step reaction sequence thus accounts for much of our current chemical, macromolecular, and structural knowledge of the species contained in ACH solutions over a wide range of solution conditions.

VII. CHEMISTRY OF ALUMINUM SESQUICHLOROHYDRATE (ASCH) AND ALUMINUM DICHLOROHYDRATE (ADCH)

A. Chemistry of ASCH and ADCH—Macromolecular Nature

The macromolecular nature of less basic aluminum chlorides such as ASCH and ADCH has received only limited study by gel filtration chromatography, ultracentrifugation, and vapor phase osmometry

Table 30 GFC Parameters for 10% BAC Solutions, Sephadex G-25

BAC system (%)	[Al] (M) ($\bar{n}$)	Peak No.	K_d	% Normalized peak height
ACH (10%)	0.97 (2.49)	1	0.00	57.4
		2	0.40	32.2
		2b	0.69	10.4
		3	—	—
ASCH (10%)	0.99 (2.24)	1	0.00	10.4
		2	0.49	45.6
		2b, 3	0.74	44.0
ADCH (10%)	0.92 (2.05)	1	0.00	10.9
		2	0.52	36.6
		2b, 3	0.71	52.5

measurements. *Gel filtration chromatography* studies of these aqueous systems using Sephadex G-25 and G-50 gels have been reported by Fitzgerald and Johnson (1980) and Fitzgerald (1981). Sephadex G-25 chromatograms of 10% ACH, ASCH, and ADCH solutions are shown in Fig. 45. Experimental K_d values and percent normalized peak height data are summarized in Table 30. The GFC data for the three BAC solutions indicate the polydisperse nature of all three solutions as exemplified by the appearance of three resolved polymeric species distributions from K_d = 0.00 to 0.74 on Sephadex G-25 columns. No GFC evidence is observed for the presence of the monomeric $Al(H_2O)_6{}^{3+}$ ion at K_d = 0.81, although this chromatographic peak is probably hidden between peak 2b and the ghost peak (peak 4). All three commercial systems are extensively hydrolyzed, containing a variable composition of three polymeric species distributions of K_d = 0.00, 0.40–0.52, and 0.69–0.7 corresponding to peaks 1, 2, and 2b according to the assignments for ACH.

The major differences in the GFC profiles for 2/3, 3/4, and 5/6 basic solutions are the K_d variations of peak 2 and the larger

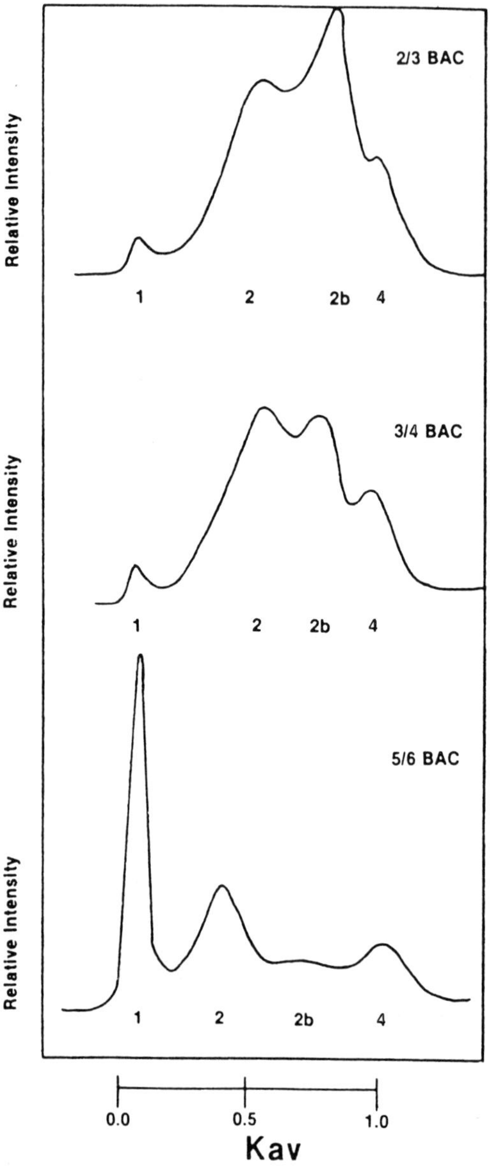

Figure 45 Sephadex G-25 GFC profiles of various 10% BAC solutions. (From Fitzgerald and Johnson, 1980.)

percent relative peak height differences between the three polymeric peak distributions. The K_d values of peak 2 shift from 0.40 to 0.49 to 0.52 for these species of the solute samples. The K_d shift may be interpreted as a reduction in molecular size of 300--800 daltons.

The most dramatic difference in the GFC profiles of the commercial BAC solutions is the percent normalized peak height values. ACH consists of 60% high-MW peak components and is more polydisperse than the two less basic systems. The GFC data for various freshly diluted ACH solutions show intensity variations of peak 1 from 60 to 90% of the aluminum contents. The less basic ASCH and ADCH systems, by contrast, contain about 10% peak 1, with the majority of the polymeric solutes in the intermediate to low molecular weight range. The major difference observed between these lower-MW systems is the differences in the percent contents of peak 2/2b (46/44% for ASCH and 37/53% for ADCH solutions). The degree of hydrolysis or the $\bar{n}$ value (2.24 versus 2.04) for these solutions is primarily responsible for the observed shift to a lower-MW species distribution in comparing the GFC profiles of ASCH to ADCH in solution. Clearly, these less basic systems contain quite different MW ranges of aluminum species than are observed for GFC profiles of ACH solutions.

It should also be noted that the K_d shift of peak 2b for ASCH (0.74) and ADCH (0.71) in comparison with ACH (0.69) is probably a consequence of incomplete resolution of peak 2b components and peak 3 monomeric peak solutes at K_d = 0.80. The K_d shifts of this GFC peak are quite large and originate from real MW differences caused by the presence of appreciable concentrations of monomeric and dimeric species in the ASCH and ADCH systems.

The polydisperse nature and variable species content of these less basic BAC systems are best observed from deconvolution Gaussian peak analysis of their chromatograms as reported by Fitzgerald and Johnson (1980). Figures 46 and 47 show a computer-generated deconvolution of the Sephadex G-25 chromatograms of 10% ASCH and ADCH solutions, respectively. In comparison with the similar deconvolution analysis of ACH shown in Fig. 18, these simulations show the presence of a large peak distribution in the monomer/dimer region of the Sephadex G-25 K_d region corresponding to peak 3. Using these simulated GFC chromatograms of ACH, ASCH, and ADCH, Fitzgerald and Johnson (1978) calculated the statistical molecular weight parameters summarized in Table 31, together with reference data obtained for the $Al_{13\text{-mer}}$ system of Aveston (1965). The results are consistent with the observed polydispersity of the BAC series from GFC data of Fig. 45. The weight-averaged molecular weights decrease with $\bar{n}$ within the series from 5950 to 1174. This MW parameter shows a significant contribution for ACH in particular due to high-molecular-weight species components. The number-averaged MW values decrease from 1122 to 885, showing the contributions

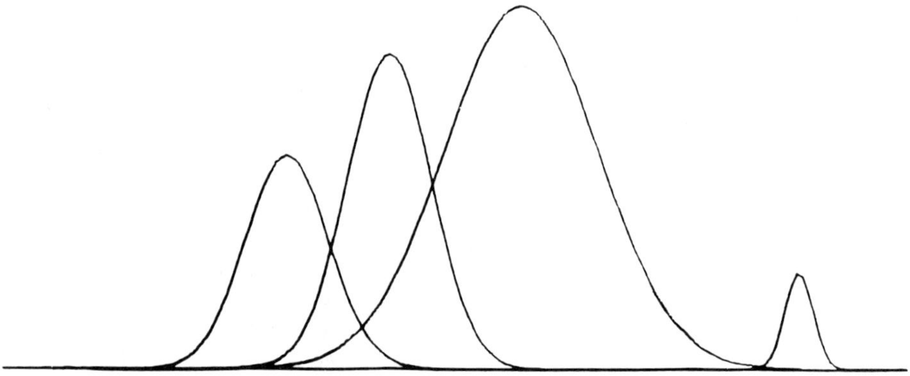

Figure 46 Deconvolution analysis of Sephadex G-25 chromatogram of 10% ADCH solution. (From Fitzgerald and Johnson, 1980.)

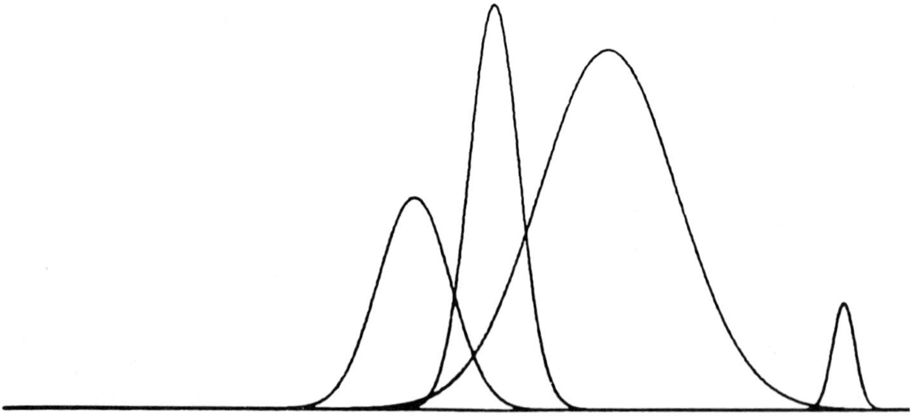

Figure 47 Deconvolution analysis of Sephadex G-25 chromatogram of 10% ADCH solution. (From Fitzgerald and Johnson, 1980.)

Table 31 Molecular Weight Data for BAC Solutions (Sephadex G-25)

BAC system	$\bar{n}$	MW_w	MW_n	I^a (polydispersity)
ACH	2.49	5950	1122	5.57
ASCH	2.24	1858	1069	1.74
ADCH	2.05	1174	885	1.33
$Al_{13\text{-mer}}$	2.43	950	940	1.02

[a] I equals polydispersity index MW_w/MW_n.

of the small molecular species to this statistical parameter. The low values for the MW parameters for both ASCH and ADCH reflect the contributions due to the presence of monomeric and dimeric species.

The determination of MW parameters for ASCH and ADCH has also been reported in the literature from *vapor phase osmometry* measurements by Bretschneider et al. (1975). An MW_n value of 975 was reported for ADCH solutions at infinite dilution, in good agreement with the MW_n value of 885 noted in Table 31 from gel filtration measurements and related calculations. Langerman (1986) also carried out *ultracentrifugation* studies of dilute ASCH and ADCH solutions at 0.1% wt/wt. Weight-averaged molecular weights of 2820, 630, and 500 were obtained for these dilute ACH, ASCH, and ADCH solutions. The much lower MW_w values obtained from ultracentrifugation than from GFC calculations for 10% ASCH and ADCH suggest that the contributions due to monomeric and oligomeric complexes are quite significant at 0.1% concentration. Additional polydispersity indices (MW_w/MW_n) of 1.07, 2.60, and 3.60 were obtained for these very dilute ACH, ASCH, and ADCH solutions respectively. The polydispersity values confirm that dilute ACH solutions probably depolymerize rapidly to an essentially monodisperse solution with an MW_w value of 2850. This is consistent with the depolymerization processes observed for dilute ACH solutions. A molecular weight range of 2850 is estimated for the major species in diluted, aged ACH solutions corresponding to the peak 2 components seen in Sephadex G-25 chromatograms. The rather high polydispersity indices for ASCH and ADCH solutions at this very low concentration support a variable composition of a range of low-molecular-weight species, while the low MW_w values of 630 and 500 indicate that only small concentrations of the higher-MW polymers are present in these dilute solutions.

Table 32 Ferron Analysis Results, BAC Solutions at Time Zero

BAC system	Concentration (%)	% aluminum of total		
		Al^a	Al^b	Al^c
ACH	25	2	3	95
	10	0	2	98
	5	8	5	87
	2	3	5	92
ASCH	25	31	15	54
	10	30	12	58
	5	30	14	56
	2	32	14	54
ADCH	25	41	13	46
	10	36	20	44
	5	49	14	37
	2	50	14	36

B. Ferron Studies of Fresh ASCH and ADCH Solutions

Ferron kinetic studies of solutions of freshly dissolved ACH, ASCH, and ADCH powders over a wide range of concentrations from 25 to 2% have been carried out by Fitzgerald and Johnson (1980) to assess the nature of the species contents of these less basic systems. The percent of aluminum species types for these three aluminum systems over the concentration ranges studied are shown in Table 32. A typical ferron reaction profile over 4 hours is also given for the three 10% BAC systems as shown in Fig. 48. As shown previously, ACH solutions contain 87–98% Al^c species over the 10-fold concentration range. By contrast, ASCH and ADCH solutions show large contents of Al^a solutes (primarily monomeric and dimeric species) from 30% Al^a for ASCH solutions to 40–50% Al^a for ADCH solutions. The ferron assay results over the concentration range studied do not show an appreciably dependence on dilution for ACH and ASCH solutions. The ADCH solutions, by contrast, show an increase in

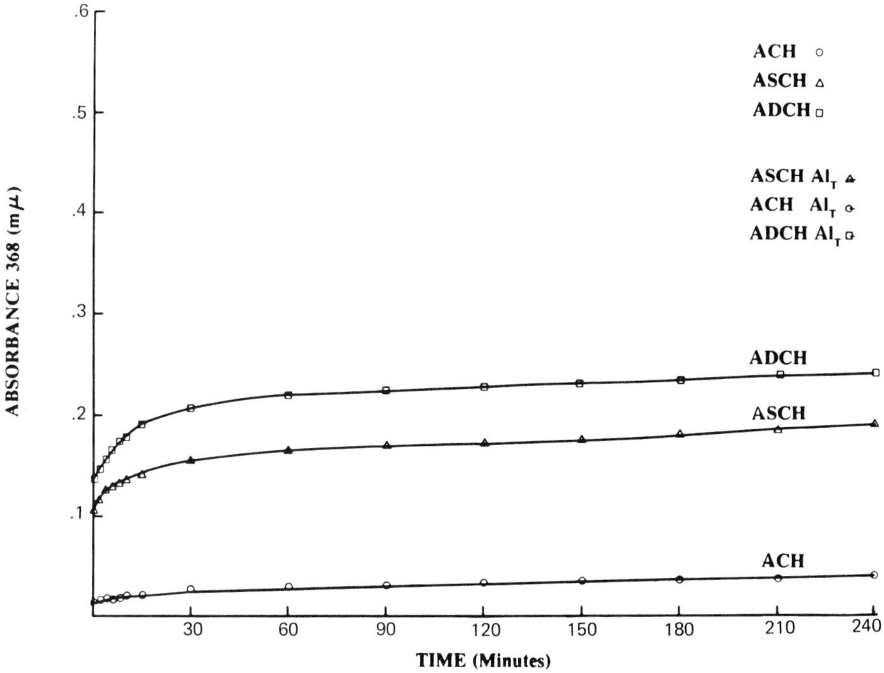

Figure 48 Ferron reactivity profile of 10% BAC solutions over a 4-hour reaction time period. (From Fitzgerald and Johnson, 1980.)

Al^a species content upon dilution from 41 to 50% with a decrease in the Al^c species contents from 46 to 36% over the 25–2% concentration range. The Al^c contents of the ASCH and ADCH solutions are 55 and 40%, respectively, indicating that the slower-reacting polymeric species account for about half of the aluminum in these systems in comparison with ACH solutions.

C. GFC and Ferron Relationship for ASCH and ADCH

A comparison of the GFC and ferron data for ASCH and ADCH solutions provides a means of interpreting the relationship between ferron reactivity and molecular size of the various species in these solutions. As noted in Table 19, Sephadex G-25 peak 1, 2a, and 2 solutes (K_d = 0.00–0.55) behave as Al^c species from studies using ACH solutions. The GFC data of Table 30 indicate that the combined % peak heights for peaks 1 and 2 for ASCH and ADCH solutions are 56% (10/46%) and 48% (11/37%), respectively. The ferron data for these solutions shows that 10% ASCH and ADCH solutions contain 55% and 36–46% Al^c species, respectively. The peak 1/2 solutes from the Sephadex G-25 chromatograms thus contain Al^c solutes, in agreement with the ferron-GFC correlation established from studies of ACH solutions. The remaining Al^a and Al^b species are found in the broad Sephadex G-25 peak at K_d = 0.71–0.74 for ASCH and ADCH solutions and consist of unresolved species corresponding to oligomers like the $Al_{13\text{-mer}}$, the dimer, and the monomeric complexes. The ferron data provide a means of quantifying these species, showing that ASCH solutions contain a 2/1 ratio of monomer and dimer versus $Al_{13\text{-mer}}$ (30% Al^a and 15% Al^b), whereas ADCH solutions contain a 3/1 monomer and dimer versus $Al_{13\text{-mer}}$ ratio (45% Al^a and 15% Al^b). This result is consistent with the GFC-ferron correlation for the lower-molecular-weight oligomeric species reported in Table 19 from experiments using ACH solutions.

Table 33 GFC Parameters for Aluminum Sesquichlorohydrate, Sephadex G-25 Column, 25% ASCH Solution

Time		Normalized height (%)		
Days	Hours	K_d = 0.0	K_d = 0.45 ± 0.03	K_d = 0.65 ± 0.01
0.08	2	25.6	36.2	38.2
1.08	26	16.5	47.2	36.3
1.17	28	16.3	46.9	36.8
3.88	93	10.2	47.8	42.0
3.92	94	12.4	48.1	39.5
7.08	170	9.2	53.9	36.9
21.21	509	3.7	62.3	34.0
21.25	510	3.7	63.0	33.3
35.00	840	0.9	70.0	29.1
35.04	841	0.9	72.6	26.5
78.13	1875	1.8	69.9	28.3
78.17	1876	0.3	70.3	29.4

D. GFC and Ferron Studies of Aged ASCH and ADCH Solutions

Various workers have suggested that less basic ASCH and ADCH solutions are subject to extensive changes in their hydrolysis chemistry upon aging dilute solutions. The large pH changes observed by Fitzgerald and Johnson (1980) support the idea that these systems undergo extensive species reorganization or hydrolysis reactions following dilution and aging. In an effort to characterize the details of these interspecies conversions, extensive GFC and ferron assays of aged ASCH and ADCH solutions have been carried out by Fitzgerald and Johnson (1980) over the concentration range 25 to 2%. Although these studies are too extensive to be reported here, representative GFC and ferron results are given to provide a basis for understanding the complexity of these systems upon dilution and aging.

Figures 49 and 50 show representative Sephadex G-25 gel filtration chromatograms of 25% ASCH and ADCH solutions aged for a 3-month period. The experimental GFC parameters for these two solutions are also summarized in Tables 33 and 34. The relative peak height data

Table 34 GFC Parameters for Aluminum Dichlorohydrate, Sephadex G-25 Column, 25% ADCH Solution

Time		Normalized height (%)		
Days	Hours	$K_d = 0.0$	$K_d = 0.45 \pm 0.03$	$K_d = 0.69 \pm 0.02$
0.08	2	31.5	—	68.5
1.08	26	23.2	—	76.8
1.17	28	25.8	—	74.2
4.13	99	14.5	33.7	51.8
4.17	100	13.7	35.3	51.0
7.29	175	10.6	37.3	52.1
7.37	177	6.7	42.4	50.9
21.17	508	3.6	46.5	49.9
21.21	509	1.6	48.7	49.7
35.00	840	0.4	49.6	50.0
35.04	841	0.7	50.9	48.4
78.00	1872	1.2	53.8	45.0
78.04	1873	0.3	55.7	44.0

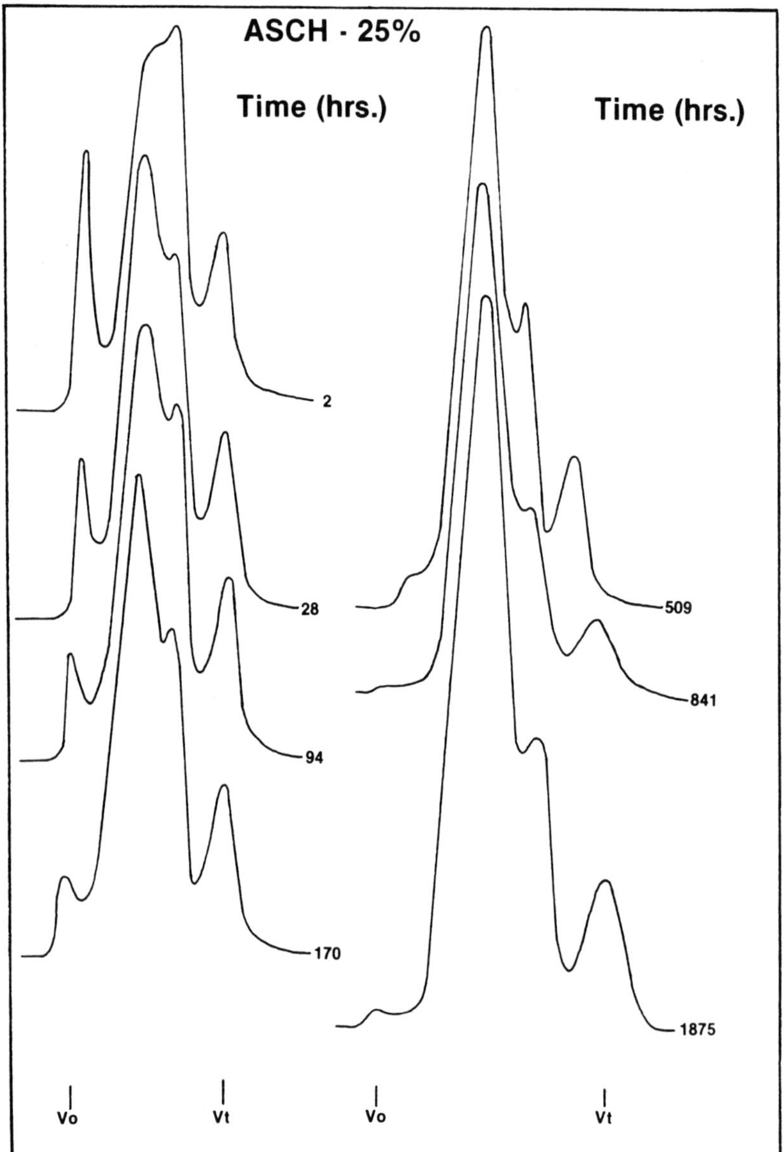

Figure 49 Sephadex G-25 GFC profiles of 25% ASCH solutions aged over 1875 hours. (From Fitzgerald and Johnson, 1980.)

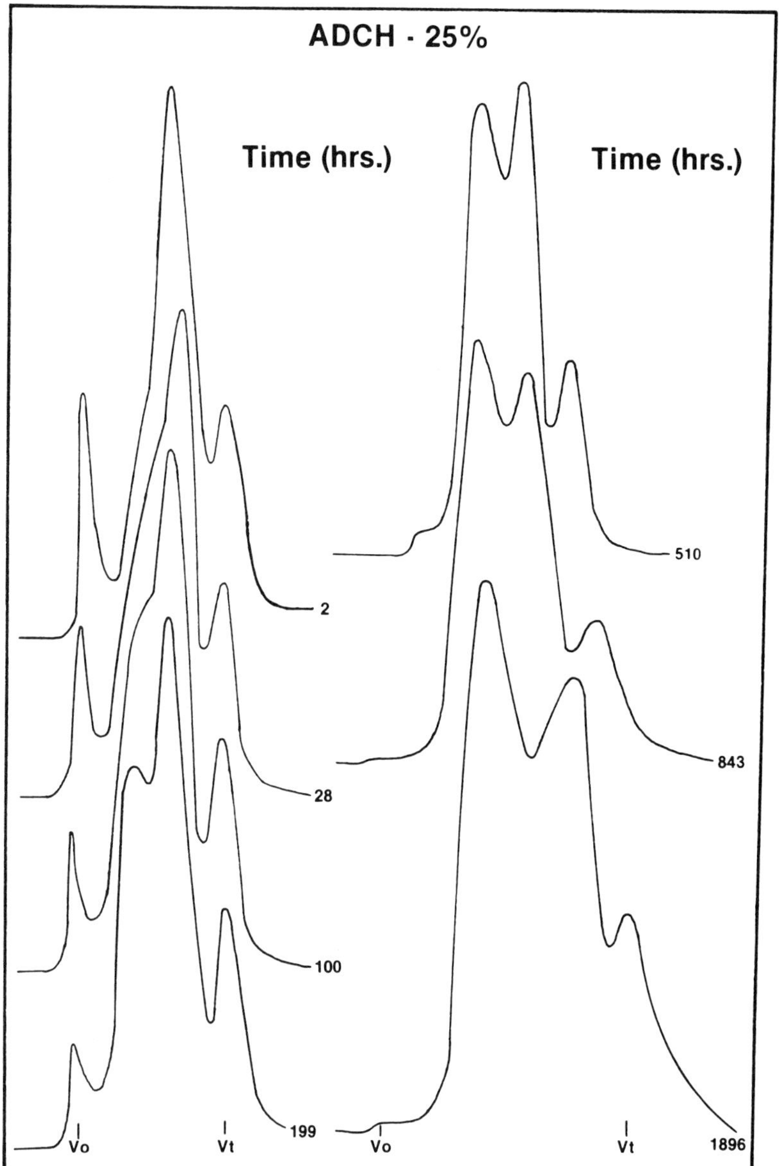

Figure 50 Sephadex G-25 GFC profiles of 25% ADCH solutions aged over 1896 hours. (From Fitzgerald and Johnson, 1980.)

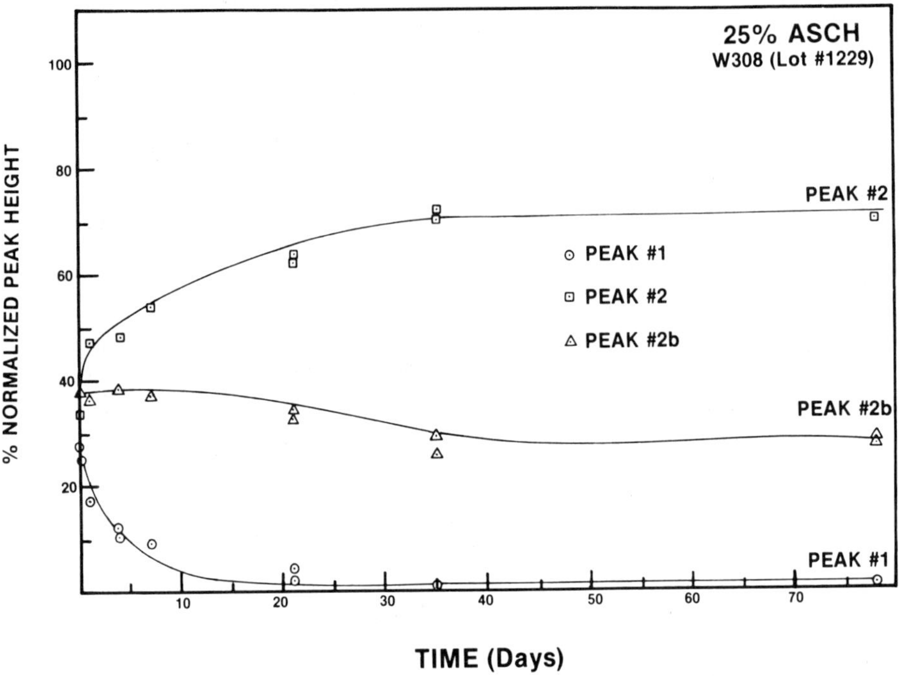

Figure 51 Aging profile for 25% ASCH solution depicting relative peak heights of Sephadex G-25 peaks 1, 2b, and 2 over an 80-day period. (From Fitzgerald and Johnson, 1980.)

from these tables have been used to construct plots of percent normalized peak height versus time for each peak observed in the 25% ASCH and ADCH solution chromatograms as shown in Figs. 51 and 52, respectively. Similar experiments and aging profiles have been reported for more dilute solutions over a similar 90-day period. The results of these aging studies reveal new information regarding these less basic BAC systems. First, the GFC data clearly demonstrate that these less basic solutions, like the aged ACH solutions discussed previously, are metastable systems containing polymeric, oligomeric, and monomeric species which undergo extensive species interconversions upon aging. Second, the aging phenomenon is concentration-dependent, showing an increase in the rates of the various kinetic processes as the sample concentrations decrease. Third, while the species interconversions for ACH solutions involve primarily a peak 1 to peak 2/2a depolymerization process, as described earlier,

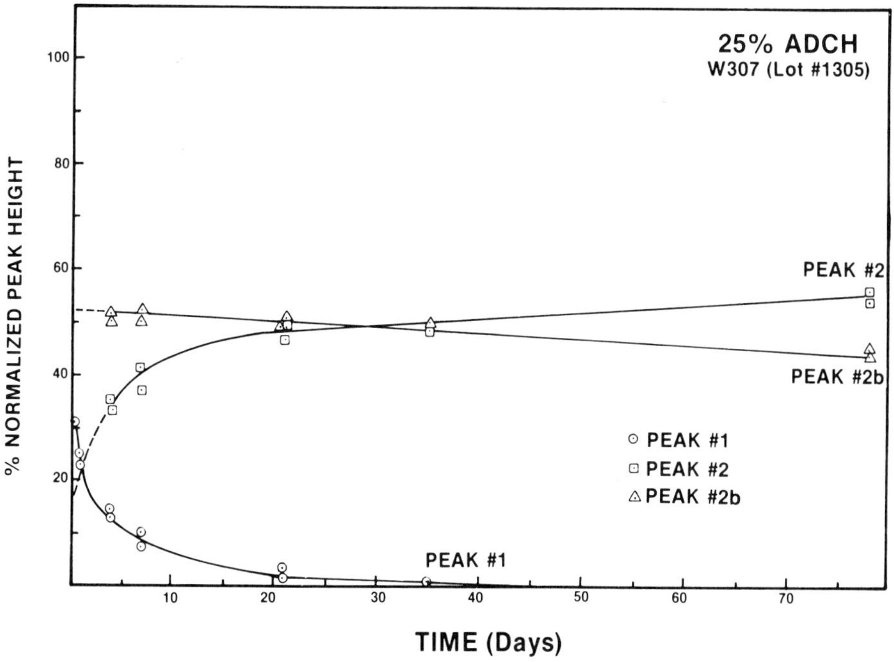

Figure 52 Aging profile for 25% ADCH solution depicting relative peak heights of Sephadex G-25 peaks 1, 2b, and 2 over an 80-day period. (From Fitzgerald and Johnson, 1980.)

both depolymerization and polymerization processes occur for ASCH and ADCH systems.

The more complex aging processes for 25% ASCH and ADCH solutions are most readily seen in Figs. 49 and 50, which show the presence of three peak distributions at K_d = 0.00, 0.45, and 0.65–0.69 assigned as peaks 1/2/2b, although the later peak contains unresolved oligomeric/dimeric/monomeric species. For 25% ASCH solutions (see Fig. 49 and Table 33), the peak 1/2/2b contents correspond to 28/34/38% at time zero. In 3 months, the composition changes to 0/70/30% (Table 33), indicating depolymerization of peak 1 to peak 2 and a further polymerization process involving some species to peak 2b or peak 2 solutes (Fig. 51). Approximately 8% of peak 2b and 28% of peak 1 solutes would account for the total increase in peak 2 from 34 to 70% (a net 36% change). This interpretation is consistent with monomeric/dimeric/$Al_{13\text{-mer}}$ species in peak 2b being converted to peak 2 species. These species interconversions are concentration-dependent, as observed for 5% ASCH solutions, where these processes are accelerated by nearly 100-fold in comparison with 25% ASCH solutions. In fact, a 2% ASCH solution was shown to reequilibrate upon dilution to a monodisperse solution containing nearly 100% GFC peak 2 solutes in a matter of hours.

The Sephadex G-25 data for ADCH systems show equally complex and similar species interconversions on dilution and aging. Chromatograms of a 25% ADCH solution exhibit extensive polydisperse GFC profiles with a peak 1/2/2b content of 31/0/69% at zero time, which changes to 14/35/51% relative peak intensities within 4 hours (see Table 34). As in the case of ASCH solutions, depolymerization and polymerization occur concurrently. More dilute ADCH solutions also show an acceleration in the rates of these processes; however, these systems do not reequilibrate to a monodisperse solution at 2% concentration. For example, the peak 1/2/2b contents of a 5% ADCH solution are 3/36/61% at time zero but change to 0/67/33% at 3 months. For a 5% ASCH solution, the system approaches a monodisperse peak 2 chromatogram in 3 months of aging. The 3-month aging profiles for 5 and 2% ADCH solutions do not reach the same equilibrium species distribution as ASCH solutions since the initial GFC species distribution of these ADCH solutions contains about 10–12% more peak 2b components (probably monomeric/dimeric species). It was noted from the aging studies that ADCH solutions have an increased rate of conversion of peak 1 to peak 2, consistent with a faster depolymerization process in comparison with ASCH solutions. In addition, the polymerization of peak 2b solutes to peak 2 components is much slower, probably because of the higher acidity of the ADCH solutions since they have substantially higher initial monomer/dimer contents than ASCH solutions.

In summary, the GFC results for dilute ASCH and ADCH solutions aged over a 3-month period reveal complex species interconversions which are influenced by concentration and $\bar{n}$ value of the original solution. Dilution and increases in $\bar{n}$ of the original metastable concentrated solutions accelerate the peak 1 to peak 2 interconversion. Polymerization processes for ASCH solutions are also observed due to the appreciable concentrations of low-MW oligomers and monomer/dimer species originally found in these systems. Polymerization of these species is observed from Sephadex G-25 chromatograms as a peak 2b to peak 2 interconversion, which is accelerated by dilution but inhibited by increases in the monomeric and dimeric species contents of the basic aluminum chloride system.

Ferron studies of ASCH and ADCH solutions have been carried out by Fitzgerald and Johnson (1978) for various aqueous solutions from 2 to 25% aged over a 2-month time period. The dependence of percent species composition of these less basic systems shows

Table 35 Ferron Analysis Results for Aluminum Sesquichlorohydrate, 2% ASCH at Various Times

Time		% of total aluminum		
Hours	Days	Al^a	Al^b	Al^c
0	0.00	32	14	54
2	0.08	17	29	54
24	0.17	12	34	55
48	1.00	15	41	44
120	2.00	15	41	44
192	8.00	—	—	—
360	15.00	8	42	50
720	30.00	19	38	43
1176	49.00	12	36	52

dramatic species interconversions. Ferron analysis aging results for 2% solutions of ASCH and ADCH solutions are shown in Tables 35 and 36. Aging results for ACH solutions at these concentrations are given in Table 33. As noted previously, aging of dilute ACH solutions results in small decreases in Al^c contents with concomitant increases in Al^b contents consistent with fragmentation of higher-MW species to lower-MW species. This species interconversion (Al^b to Al^c process) involves only about 10% of the aluminum solutes and accompanies the peak 1 to peak 2 depolymerization observed in the GFC aging studies (an Al^c to $Al^{c'}$ process).

The ferron aging results for dilute ASCH and ADCH solutions suggest a quite different sequence of species interconversions. Aging of dilute ASCH and ADCH solutions produces principally an Al^a to Al^b polymerization process, as predicted from the GFC data and shown in Tables 35 and 36. The Al^c contents of 2 and 5% ASCH solutions are invariant throughout the aging studies. This is expected since the GFC peak 1 to peak 2 depolymerization process involves only two different Al^c species. Over 2 months, 2% ASCH solutions show a decrease from 32 to 12% Al^a, simultaneous with an increase in Al^b solutes from 14 to 36%. Aging of 2 and 5% ADCH solutions involves more complex species interconversions. Initially an Al^a to Al^b polymerization occurs, followed by an increase in Al^c

Table 36 Ferron Analysis Results for Aluminum Dichlorohydrate, 2% ADCH at Various Times

Time		% of total aluminum		
Hours	Days	Al^a	Al^b	Al^c
0	0.00	50	14	36
2	0.08	39	28	33
4	0.17	33	31	36
24	1.00	34	31	35
96	4.00	31	35	34
192	8.00	30	32	38
336	14.00	33	27	40
768	32.00	29	27	44
1152	48.00	28	25	47

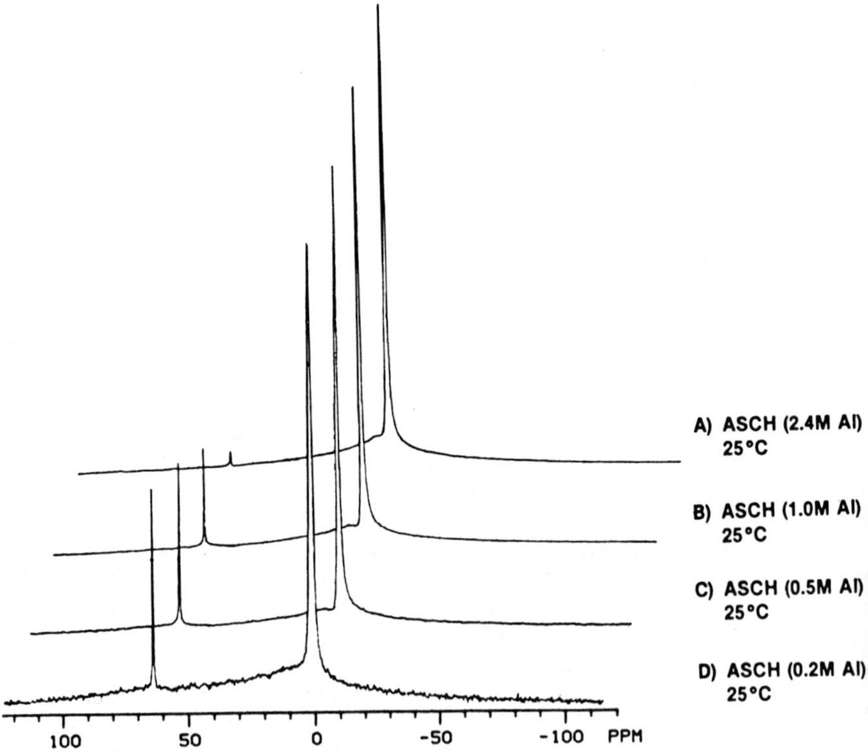

Figure 53 Aluminum-27 NMR spectra of ASCH solutions at various concentrations. (From Fitzgerald, 1983.)

species content probably originating from Al^b solutes as the Al^a concentrations reach an equilibrium with Al^b species. The ADCH data support an Al^a to Al^b to Al^c interconversion over the aging period studied.

E. ^{27}Al NMR Studies of ASCH and ADCH Solutions

The ^{27}Al NMR measurement reported by Fitzgerald (1980) at 39.10 MHz provides unique and direct evidence for the presence of the monomer and the bis-u-hydroxy dimer in these ASCH and ADCH solutions. The ^{27}Al NMR spectra of 2–25% ASCH and ADCH are shown in Figs. 53 and 54. Chemical shift data for the various NMR resonances observed in these solutions, ACH, and other reference

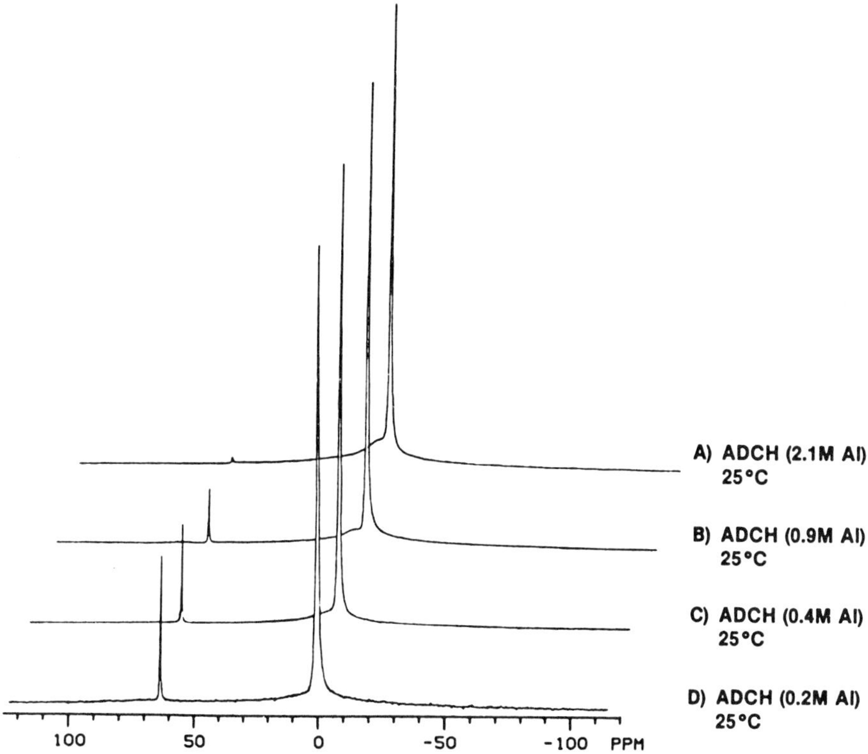

Figure 54 Aluminum-27 NMR spectra of ADCH solutions at various concentrations. (From Fitzgerald, 1983.)

Table 37 ^{27}Al NMR Data for Basic Aluminum Chlorides

Aluminum system	$\delta(^{27}Al)$, chemical shifts (ppm)[a]					
$Al(H_2O)_6^{3+}$ (0.2 M)	0.0	—	—	—	—	—
$Al_{13\text{-mer}}$ (0.2 M)	0.2	—	—	17.3[b]	62.8	—
ACH (1.0 M)	0.1	—	11.7	—	62.7	71 (i)[c]
ASCH (1.0 M)	0.2	4.6	12.0	—	62.9	71 (i)
ADCH (0.9 M)	0.1	4.5	12.5	—	62.8	71 (i)

[a]Relative to $Al(H_2O)_6^{3+}$ reference signal; 39.10 MHz.
[b]This $Al_{13\text{-mer}}$ signal is broadened at RT, but sharpens at elevated temperature
[c]Spectral features denoted by (i) indicate unresolved inflections where their positions are estimated.

compounds are given in Table 37. The dominant resonances observed in the NMR spectra of both ASCH and ADCH solutions are appearance of the monomer as an intense sharp resonance at 0.1–0.2 ppm and the broader dimer resonance at 4.50–4.65 ppm in higher concentration solutions (25–10%). Integration of these NMR resonances indicates that the monomer/dimer (Al^a solutes) represent up to 30% of the aluminum in ASCH and 40–45% of the solute species in freshly diluted ADCH solutions at higher concentrations. The presence of Al^b species such as the $Al_{13\text{-mer}}$ complex is confirmed on the basis of the sharp tetrahedral NMR resonance for the central aluminum in this complex observed at 62.8 ppm. An important feature of the concentration dependence of the RT NMR spectra of these less basic systems is the fact that the relative intensity of both the monomer and dimer resonances decreases concomitant with an increase in the resonance due to the $Al_{13\text{-mer}}$ complex at 62.8 ppm. This effect of dilution indicates that species interconversions involving polymerization of Al^a species to Al^b species, as suggested from GFC and ferron studies of these systems, are confirmed by this spectroscopic technique. Although the resonances due to polymeric Al^c species (peak 1/2 GFC components) are not readily seen in the spectra in Fig. 53 and 54, broad underlying resonances at 12.0–12.5 ppm and 71 ppm are observed in expanded intensity spectra of ASCH and ADCH solutions as summarized in Table 37. The ^{27}Al NMR resonance assignments for all of the signals

Table 38 ^{27}Al NMR Assignments for BAC Solutions

δ(^{27}Al), chemical shift (ppm)	Al site (geometry)	Sample type
0.00 (s)	Al-M (Oh)	$Al(H_2O)_6^{3+}$
0.20	Al-P_A (Oh)	ACH, ASCH, ADCH
4.5	Al-D (Oh)	Dimer, ASCH, ADCH
11.7	Al-P_B(Oh)	ACH, ASCH, ADCH
17.3	Al-P_C (Oh)	$Al_{13\text{-mer}}$, probaby BAC
62.8 (s)	Al-P_A (Td)	$Al_{13\text{-mer}}$, BAC
70.7	Al-P_B (Td)	ACH, ASCH, ADCH

observed in the various aluminum sites found in BAC solutions and reference aluminum hydrolysis species, including the monomer (M), dimer (D), and various octahedral and tetrahedral aluminum sites polymers (P), are given in Table 38. It is important to emphasize that the observation of the various species by ^{27}Al NMR spectroscopy represents the most direct means of identifying these complexes without diluting the solutions or modifying their chemistry.

F. Synthesis of ASCH and ADCH

The less basic aluminum chloride solutions are prepared on an industrial scale by neutralization of concentrated ACH solutions with acidic solutions of either aqueous $AlCl_3 \cdot 6H_2O$ or HCl. The definition of the basicity of less basic solutions produced by these reactions, such as ASCH and ADCH, is made purely on the basis of chemical analysis of the Al/Cl ratio of the resulting product solution. The major reaction variables reported to influence the final solute species composition following these syntheses include the choice of acid neutralization reagent (aluminum chloride or HCl), the concentration at which the neutralization is performed, and the temperature used to reequilibrate the resulting product solution. Examination of each of these synthesis variables is necessary to gain a chemical understanding of the species composition and macromolecular properties of the hydrolyzed aluminum species contained in acid-neutralized ACH solutions.

The two synthesis methods used to prepare ASCH and ADCH solutions are given in the following generalized reaction equations for a range of basic aluminum chloride solutions:

Route I: Neutralization with aqueous aluminum chloride

$$Al_2(OH)_5Cl + AlCl_3 \rightarrow Al_2(OH)_w(Cl)_p \tag{46}$$

where $2/p$ = Al/Cl ratio.

Route II: Neutralization with aqueous hydrochloric acid

$$Al_2(OH)_5Cl + HCl \rightarrow Al_2(OH)_w(Cl)_p \tag{47}$$

Control of ACH/$AlCl_3$ and ACH/HCl reaction stoichiometry is the basis for producing a final less basic system from ACH solutions as summarized in Table 39. For example, the synthesis of a BAC solution of $\bar{n}$ = 2.25 and Al/Cl ratio of 1.33 (ASCH or 3/4 basic) requires an ACH/$AlCl_3$ reaction stoichiometry of 9/2 moles or an ACH/HCl reaction stoichiometry of 2/1 ACH/HCl. By contrast, preparation of an ADCH system of $\bar{n}$ = 2.00 and Al/Cl ratio of 1.00 (2/3 basic) requires an ACH/$AlCl_3$ reaction stoichiometry of 4/2 moles or an ACH/HCl reaction stoichiometry of 1/1 ACH/HCl.

Detailed studies of the synthesis of ASCH and ADCH solutions prepared by the ACH/$AlCl_3$ and ACH/HCl reactions using Sephadex G-25 gel filtration chromatography have been carried out by Fitzgerald and Johnson (unpublished work, 1980). These two syntheses routes were examined starting with ACH solutions to obtain final ASCH and ADCH reaction solutions in the concentration ranges between 6.0 and 0.6 M Al. The ASCH and ADCH solutions were prepared by appropriate addition of either $AlCl_3$ or HCl in the required stoichiometry to obtain the required final solution Al/Cl ratios corresponding to the reaction equations given in Table 39 for these two systems. The initial reaction mixture produced following titrimetric neutralization of ACH with $AlCl_3$ or HCl was denoted as an "unheated" reaction product. The freshly reacted ASCH and ADCH solutions prepared in this manner were then heated at 50°C for 168 hours to obtain final equilibrated reaction products termed "heated" ASCH and ADCH solutions. Extensive GFC investigations of a range of "unheated" and "heated" ASCH and ADCH solutions produced at different reaction concentrations were carried out; however, this discussion will focus on GFC studies of ASCH and ADCH reaction products prepared to a final solution concentration of 1.2 M Al or approximately 10% wt/wt.

Table 39 Reaction Equations For Syntheses I and II

Reaction equation	Desired $\bar{n}$
Synthesis I ($ACH + AlCl_3$)	
$2AlCl_3 + 9Al_2(OH)_5Cl \rightarrow 10Al_2(OH)_{4.5}Cl_{1.5}$	2.25
$2AlCl_3 + 4Al_2(OH)_5Cl \rightarrow 5Al_2(OH)_4Cl_2$	2.00
$4AlCl_3 + 3Al_2(OH)_5Cl \rightarrow 5Al_2(OH)_3Cl_3$	1.50
$6AlCl_3 + 2Al_2(OH)_5Cl \rightarrow 5Al_2(OH)_2Cl_4$	1.00
Synthesis II ($ACH + HCl$)	
$HCl + 2Al_2(OH)_5Cl \rightarrow 2Al_2(OH)_{4.5}Cl_{1.5}$	2.25
$HCl + Al_2(OH)_5Cl \rightarrow Al_2(OH)_4Cl_2$	2.00
$2HCl + Al_2(OH)_5Cl \rightarrow Al_2(OH)_3Cl_3$	1.50
$3HCl + Al_2(OH)_5Cl \rightarrow Al_2(OH)_2Cl_4$	1.00

Sephadex G-25 chromatograms of the initial ACH solution (diluted to 1.2 M Al) used in the ACH/$AlCl_3$ reaction and the reaction product solutions of ASCH ($\bar{n}$ = 2.25) and ADCH ($\bar{n}$ = 2.00) produced by appropriate stoichiometric addition of aqueous $AlCl_3$ solutions are shown in Fig. 55. The product ASCH and ADCH reaction mixtures are denoted as unheated. Examination of the GFC profiles of these three solutions indicates that the species distributions of these samples are dependent on the final solution $\bar{n}$ value. The Sephadex G-25 chromatogram of the ACH solution shows two major species distributions at K_d = 0.00 (peak 1) and 0.40 (peak 2), typical of a 10% ACH solution. The GFC profiles for the reaction product ASCH and ADCH solutions show that the neutralization of the starting ACH solution results in a decrease in the intensity of the higher-MW peak 1, decreases in the intensity of peak 2 solutes at K_d = 0.44–0.48, and increases in the intensity of a broad distribution in the peak 2b/3 region (K_d = 0.69 for ASCH and 0.75 for ADCH) of the chromatograms. The addition of $AlCl_3$ to ACH solutions produces increasing depolymerization of peak 1 solutes with increases in the amount of $AlCl_3$ added. For example, the percent relative peak height of peak 1 changes from 80% for the ACH chromatogram

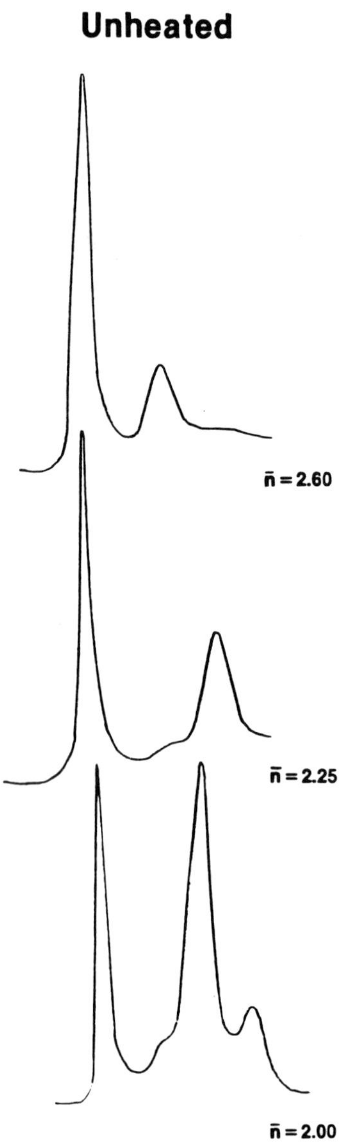

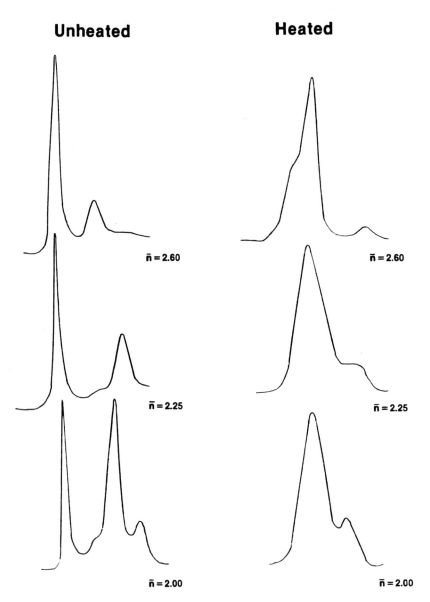

Figure 55 Sephadex G-25 GFC profiles of various BAC solutions prepared from ACH/$AlCl_3$ reaction (unheated and heated solutions). (From Fitzgerald and Johnson, 1980.)

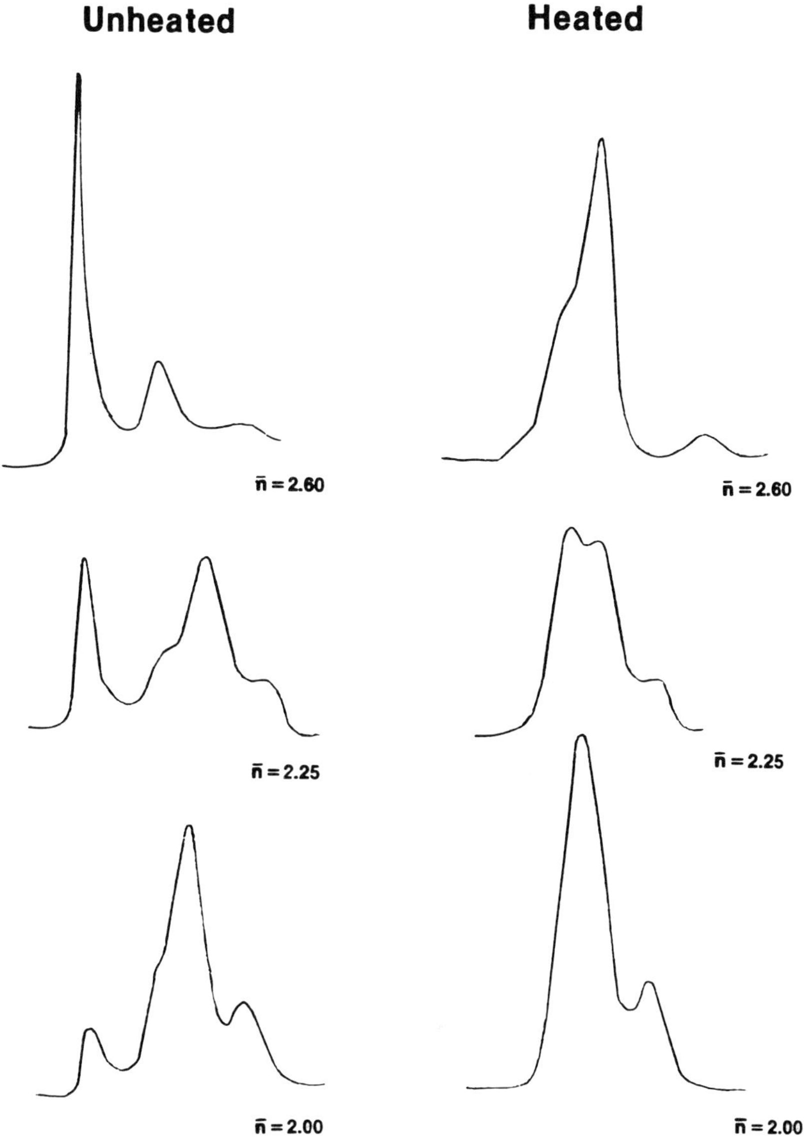

Figure 56 Sephadex G-25 GFC profiles of various BAC solutions prepared from ACH/HCl reaction (unheated and heated). (From Fitzgerald and Johnson, 1980.)

to 60 and 45% for the chromatograms of the ASCH and ADCH reaction products, respectively. The relative intensity of the peak 2 component of the chromatograms is about 20% for ACH and is reduced to about 8% for both the ASCH and ADCH reaction products. The major chromatographic change observed for ASCH and ADCH solutions is an increase in intensity in the peak 2b/3 region due to increases in the content of the $Al_{13\text{-mer}}$, dimer, and monomer complexes upon neutralization of the starting ACH solution. Addition of the monomer to ACH solutions produces only moderate depolymerization of the high- and intermediate-MW complexes, but also results in increased $Al_{13\text{-mer}}$ and dimeric contents of these less basic BAC solutions due to hydrolysis of appreciable amounts of the added aluminum hexaaquo complex. Up to 30% of the aluminum in the ASCH solution is a mixture of these species (as noted by the appearance of peak 2b/3 at K_d = 0.69), whereas the ADCH solution contains about 47% of the aluminum species in the form of these complexes in peak 2b/3 at K_d = 0.75.

Similar GFC studies of these three samples following heating at 50°C for 168 hours to obtain equilibrated heated ACH, ASCH, and ADCH solutions were also carried out, and the resulting Sephadex G-25 chromatograms are shown in Fig. 56. The GFC profiles of these samples are distinctly different from corresponding profiles of the unheated samples. As expected, the GFC profile of the heated ACH solution shows only a single two-component peak with a shoulder at K_d = 0.25 and an intense peak at K_d = 0.40. This chromatogram is characteristic of 10% ACH solutions following aging, as discussed previously. For the heated ASCH and ADCH solutions, it is apparent that heating the final reaction mixture produces extensive depolymerization of peak 1, with this species distribution being absent in both chromatograms of these less basic synthesized solutions. The chromatogram of the ASCH solution shows only a single very broad peak at K_d = 0.55 due to a mixture of both peak 2/2b components of intermediate molecular size. Thermal equilibration of the ADCH solution produces a final solution product with a GFC profile consisting of two peaks at K_d = 0.64 and K_d = 0.78, corresponding to peak 2/2b and peak 3 in a 65/35% relative intensity. The chromatograms of the heated ASCH and ADCH solutions prepared using the ACH/$AlCl_3$ reaction are very similar to those of commercial ASCH and ADCH solutions of comparable concentration, suggesting that the commercial process using this method must involve thermal reequilibration of the neutralization reaction mixtures. Of particular significance are the absence of any high-MW peak 1 polymeric species in these solutions, the substantial reduction in the peak 2 solute components in comparison with thermally equilibrated ACH solutions, and the appreciable content of the $Al_{13\text{-mer}}$ and dimeric and monomeric complexes in both synthesized heated samples.

Similar investigations of the ACH/HCl reaction were carried out to determine the species composition from the GFC profiles of the synthesized ASCH and ADCH solutions following neutralization (unheated) and following thermal treatment under similar conditions to the ACH/$AlCl_3$ synthesis. Sephadex G-25 chromatograms of six analogous samples from the ACH/HCl synthesis are shown in Fig. 53. The neutralized ASCH and ADCH solutions were prepared using the reaction stoichiometry shown in Table 39 for synthesis II. From inspection of the GFC profiles of the ASCH and ADCH solutions, it is apparent that their species compositions are markedly $\bar{n}$-dependent. Decreasing $\bar{n}$ causes a shift in the molecular weight distributions to smaller species, as observed in their corresponding chromatograms. Heating of the ASCH and ADCH solutions is also noted to shift the sample species compositions to a lower molecular weight distribution than observed for the fresh unheated solutions, as was observed for the synthesis I samples. Although similar trends are apparent in the GFC results for synthesis II samples and synthesis I samples, two significant differences are noted between the series. First, depolymerization of peak 1 polymeric solutes occurs to a greater degree following initial neutralization with HCl in comparison with $AlCl_3$ for both the unheated ASCH and ADCH reaction mixtures. The peak 1 relative intensities were 43% for ASCH and 16% for ADCH solutions following appropriate HCl neutralization to produce solutions of the correct Al/Cl or $\bar{n}$ ratios. Second, the molecular weight distributions from the GFC data for the unheated samples also show a decrease in the intensity of peak 2 components, suggesting that the addition of H^+ ion leads to depolymerization of these intermediate-MW Al^c-type polymeric complexes. This result is further supported by chromatograms of the heated ASCH and ADCH solutions, which show further shifts in the molecular weight distributions to smaller aluminum hydrolysis complexes. From the GFC data for the latter samples, it was observed that the addition of HCl leads to the formation of higher contents of the $Al_{13\text{-mer}}$, the dimer, and the monomer in the ASCH and ADCH solutions produced by this synthesis method than in those produced by the synthesis employing $AlCl_3$ neutralization. Acid-catalyzed depolymerization of both peak 1 and 2 polymeric complexes in the original ACH solution is accelerated in the HCl neutralization reaction in comparison with the $AlCl_3$ reaction. These investigations have thus provided a fundamental understanding of the origin of the hydrolysis species compositions of ASCH and ADCH solutions derived by these two synthetic patheways involving ACH as a starting material. They have also established that these two methods of synthesis produce different species compositions in the resulting BAC product solutions.

G. Molecular Model of ASCH and ADCH

The chemical nature and species composition of ASCH and ADCH solutions are best approached by considering two aspects: 1. the formation of these complex systems at high concentrations from ACH solutions, and 2. the description of the stability of the various species contained in ASCH and ADCH solutions following dilution and aging. ASCH and ADCH solutions are synthesized by reaction of either the monomeric $Al(H_2O)_6{}^{3+}$ ion or HCl with the various complexes contained in ACH solutions. The species compositions found in commercial ASCH and ADCH solutions are the result of reacting acidic reagents with the two major polymeric species in the molecular weight range 10,000 to 2500. The final species compositions of these synthesized solutions following reequilibration of the reaction mixtures by thermal treatment are dependent on the amount of added HCl or $AlCl_3$. Synthesis of ASCH, which uses a lower stoichiometric amount of these reactants, leads to a narrow range of polymeric complexes in concentrated ASCH solutions. Commercial ASCH solutions show the absence or a minimal content of the higher-MW GFC peak 1 polymers (up to 15%), since these complexes have undergone depolymerization due to the addition of appreciable amounts of acidic reactants. The final species distribution in these systems includes approximately 55% of two polymeric complexes between 2500 and 960 daltons corresponding to 45% $Al^{c'}$ (peak 2) solutes and 10% Al^{b} (peak 2b) solutes in addition to approximately 25–30% monomeric/dimeric Al^{a} species.

ADCH solutions, which are prepared from ACH solutions using at least twice the stoichiometric amounts of HCl or $AlCl_3$ reactants used in ASCH solutions, are distinctly different from ASCH solutions primarily in the fact that they contain appreciable monomer/dimer species after the neutralization reaction. ADCH solutions likewise contain limited amounts (10%) of higher-MW GFC peak 1 polymer due to acid-catalyzed depolymerization of these complexes. In addition, ADCH solutions contain less peak 2 intermediate polymers (about 30%) and higher levels of both Al^{b} (peak 2b species) and Al^{a} complexes (20 and 40%, respectively). Al^{b} complexes originate from two sources: fragmentation of $Al^{c}/Al^{c'}$ polymers and oligomerization of monomer/dimer complexes. The fragmentation of Al^{c} complexes, in particular, results in the release of H^{+} and Cl^{-} ions, while oligomerization of monomer/dimer complexes to $Al_{13\text{-mer}}$ complexes also releases hydrogen ion. In both ASCH and ADCH solutions, the presence of high levels of monomer/dimer complexes has been confirmed by direct observation using ^{27}Al NMR spectroscopy.

Dilution and aging of ASCH and ADCH produce species interconversions including further depolymerization of the remaining Al^{c}-type high-molecular-weight complexes. Furthermore, aging leads to hydrolysis of Al^a monomer/dimer complexes to Al^b species. For ASCH solutions aging increases the Al^b (peak 2) content to about 35% leaving about 10% Al^a solutes. ADCH solutions, by contrast, are more acidic, and less oligomerization of Al^a to Al^b species occurs on aging. The final solution composition is about 45% $Al^{c'}$ (peak 2), 25% Al^b (peak 2b), and 30% Al^a (peak 3) species.

A detailed chemical model of these very complex lower-MW systems is not presented, but the following series of reactions depicts some of the important interspecies reactions occurring in ASCH and ADCH solutions:

$$\underset{\text{peak 1/2}}{Al^{c}/Al^{c'}} + \underset{\text{peak 3}}{Al(H_2O)_6^{3+}} \rightarrow \underset{\text{peak 2a/2}}{Al^{c'}} + \underset{\text{peak 2b}}{Al^{b}} + \underset{\text{peak 3}}{Al^{a}} \text{ (unreacted)} + Cl^- + H^+ \qquad (48)$$

In this reaction the Al^b formed is a result of fragmentation of $Al^c/Al^{c'}$ complexes during the neutralization and of oligomerization of Al^a components, which occurs according to the reaction:

$$\underset{\text{peak 3}}{\text{monomer + dimer}} \rightarrow \underset{\text{peak 2b}}{Al_{13\text{-mer}}} \qquad (49)$$

Increases in $Al^{c'}$ content and in solution chloride ions are a result of the depolymerization of higher-MW Al^c complexes in the presence of hydrogen ion.

The above two reactions also account for the majority of the interspecies conversions occurring in diluted and aged ASCH and ADCH solutions. Further depolymerization of Al^c complexes occurs in addition to oligomerization of Al^a to Al^b complexes. The major difference between these two basic aluminum chloride systems is the increased amount of either $AlCl_3$ or HCl in the ADCH system, which accelerates both depolymerization/polymerization reactions in the synthesis reaction process and related reactions which occur on aging ADCH solutions.

VIII. SOLUTION CHEMISTRY OF BASIC ALUMINUM/ZIRCONIUM SALTS

A. Hydrolysis Chemistry of Zr(IV) Ion in Aqueous Solution

The hydrolysis chemistry of zirconium(IV) ion in aqueous solution has continued to be of interest because of the variety of commercially important zirconium compounds and the practical interest in understanding the separation behavior of hafnium(IV) and zirconium(IV) ions by ion exchange or solvent extraction techniques. The affinity of zirconium(IV) ion for oxygen dominates its aqueous chemistry, accounting for the formation of very stable oxide and hydroxide complexes, chelate complexes with ligands such as β-diketones, and other ligands containing carboxylate and hydroxy groups. In strongly acidic media, Zr^{4+} ion forms very stable anionic complexes such as the ZrF_6^{2-} ion, weaker cationic complexes with sulfate ion, and even weaker complexes with halides and nitrate ion.

Zirconium(IV)-oxygen compounds dominate the spectrum of zirconium compounds with industrial applications and include zirconium silicates and zirconium oxides, related to the most abundant zirconium minerals, baddeleyite ($ZrO_2 \cdot SiO_2$) and zircon (ZrO_2). Refractory zircon and numerous zirconium pigments are used in the ceramics industry. Aqueous zirconyl oxychloride and its derivatives are used as waterproofing agents for textiles and paper and as starting materials for antiperspirant complexes. The industrial applications of a wide variety of zirconium(IV)-oxygen compounds have been reviewed by Hoch (1974).

The aqueous chemistry of zirconium(IV), like that of hafnium(IV) ion, is dominated by increased coordination numbers in comparison with the exclusively 6-coordinate aluminum(III) ion. The large ionic radius (0.74 Å) facilitates bonding to seven or eight donor atoms in its coordination sphere, with square antiprism or dodecahedron geometries prevalent in 8-coordinate Zr-O compounds dissolved in or precipitated from aqueous solutions. The oxygen atoms are generally located at points of a square antiprism or at points of a triangular dodecahedron, sometimes being distorted.

B. pH Region of Zr(IV) Hydrolysis

Because of its high positive charge and relatively small ionic radius (due to the lanthanide contraction), Zr^{4+} ion is hydrolyzed strongly even in highly acidic aqueous solution. The aqueous hydrolysis chemistry of zirconium-(IV) ion is confined largely to the pH region 0–2 and has been reviewed by Baes and Mesmer (1976). The occurrence of soluble basic halides and sulfates such as $ZrOCl_2$ and $ZrOSO_4$ is the basis for the terminology "zirconyl" or "oxozirconium" ion used to denote the "ZrO" ion, although there is no evidence to show that this ion exists in either the solid state or solution. Rather, the mononuclear zirconyl ion may be thought of as a dehydrated form of the intermediate hydrolysis species, $Zr(OH)_2^{2+}$ ion, formed from the first two hydrolysis steps according to the reactions

$$Zr^{4+}(nH_2O) \rightleftarrows Zr(OH)[(n-1)H_2O)]^{3+} + H^+ \qquad (50)$$

$$Zr(OH)[(n-1)H_2O]^{3+} \rightleftarrows Zr(OH)_2[(n-2)H_2O]^{2+} + H^+$$

$$(ZrO^{2+}\cdot H_2O) \qquad (51)$$

Mononuclear hydrolysis equilibria have been examined using potentiometric methods, although compleximetric techniques empolying spectrophotometric analysis of the complexation of Zr^{4+} ion by thenoyltrifluoroacetone (TTA) in solution or extraction analysis of the resulting complexes have been most successful. Zielen and Connick (1956), using the TTA extraction technique, concluded that unhydrolyzed Zr^{4+} ion exists in acidic solutions below 10^{-4} M Zr. Noren (1973), using potentiometry, measured the first hydrolysis equilibrium and reported the constant $Q_{11} = 0.28$ ($\log K_{11} = 0.3$). The review of Baes and Mesmer (1976) indicates the following estimates for the equilibrium constants for the stepwise hydrolysis of Zr^{4+} ion in dilute solutions:

$$Zr^{4+} + H_2O \rightleftarrows Zr(OH)^{3+} + H^+ \qquad \log K_{11} = 0.3 \qquad (52)$$

$$Zr(OH)^{3+} + H_2O \rightleftarrows Zr(OH)_2^{2+} + H^+ \qquad \log K_{12}/K_{11} < -1.8 \qquad (53)$$

$$Zr(OH)_2^{2+} + H_2O \rightleftarrows Zr(OH)_3^{+} + H^+ \qquad \log K_{13}/K_{12} = -3.4 \qquad (54)$$

$$Zr(OH)_3^{+} + H_2O \rightleftarrows Zr(OH)_4 + H^+ \qquad \log K_{14}/K_{13} = -4.6 \qquad (55)$$

The formation of the anionic hydroxide form of Zr^{4+} ion in basic media (1–9 M NaOH) is given by the expression

$$Zr^{4+} + H_2O \rightleftarrows Zr(OH)_5^{-} + 5H^+ \qquad \text{where } K_{15} = -16.0 \qquad (56)$$

whereas the solubility of zircon, $ZrO_2(s)$, has been estimated to be $\log K_{s10} = -1.9$ or K_{14so} $(ZrO_2) = 10^{-12}$.

C. Macromolecular Nature of Zirconium in Dilute Solution

Polynuclear species have continually complicated measurements and interpretation of the equilibria involving mononuclear species in dilute solutions of 10^{-2} to 10^{-4} M Zr. However, in higher solution concentration, polynuclear complexes predominate in the pH range

0–3. The current understanding of the nature and composition of polynuclear complexes in zirconium(IV) hydrolysis chemistry has been obtained using ultracentrifugation, light scattering, small-angle x-ray scattering, TTA extraction studies of the TTA/Zr^{4+} complex, and potentiometric and coagulation methods. Zielen and Connick (1956) used the TTA extraction technique to study hydrolyzed Zr^{4+} ion solutions up to 0.02 M Zr in 1–2 M $HClO_4$ and first proposed the existence of the trimer $Zr_3(OH)_4{}^{8+}$ and the tetramer $Zr_4(OH)_8{}^{8+}$ ions. Such species are supported by the coagulation work of Matijevic et al. (1962) and the early pH measurements by Larsen and Gammill (1950). Simultaneous with the report of Zielen and Connick (1956), Johnson and Kraus (1956), using *ultracentrifugation* measurements of 0.05 M Zr^{4+} solutions in 0.5–2 M HCl and various Cl^- and $ClO_4{}^-$ supporting electrolytes, presented evidence for the presence of very stable small polymers (either the trimer or tetramer) with a charge of less than one per zirconium in these acidic solutions. Binding of Cl^- ion by these small polymers was also proposed. More acidic solutions were more polydisperse with an average degree of polymerization of less than three, suggesting the presence of unhydrolyzed or mononuclear hydrolysis species. *Light scattering* studies of $ZrOCl_2 \cdot 8H_2O$ dissolved in 0.01–0.1 M Zr in 2.8 M HCl by Angstadt and Tyree (1962) strongly supported the work of Zielen and Connick and Johnson and Kraus. These workers reported that a trimer species of charge +3 accounted for the principal species in these strongly acidic solutions and that the presence of species with an average degree of polymerization of six corresponding to a hexamer more readily fit the light scattering data under less acidic conditions of 0.75 M HCl. These results were consistent with the small-angle x-ray scattering studies of Muha and Vauhn (1960) on 0.5–2 M $ZrOCl_2$ solutions, which provided concrete evidence for the presence of a tetramer of the chemical formula $Zr_4(OH)_8(H_2O)_{16}x_z{}^{(8-z)+}$. The work of Muha and Vaughn demonstrated that chloride or bromide ion was involved in outer-sphere complexation with the hydrolyzed zirconium(IV) tetramer complex, based on the concentration dependence of the x-ray scattering of these solutions. Figure 57 shows the square planar arrangement of the four Zr^{4+} ions in the tetramer complex, in which each zirconium is coordinated to four bridging OH^- ions and four terminal water molecules in an eightfold Zr-O coordination of a slightly distorted square antiprism. The proposed structure of this solution tetrameric species was consistent with the x-ray diffraction studies reported for $ZrOCl_2 \cdot 8H_2O$ crystals by Clearfield and Vaughn (1956). This seemingly simple picture for the predominance of the tetramer species in acidic zirconium(IV) solutions is complicated by evidence for higher average degrees of aggregation of six reported by Angstadt and Tyree. Indeed, additional light

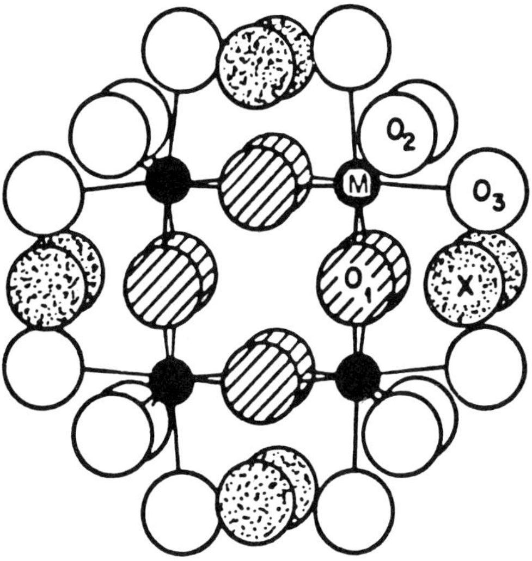

Figure 57 Square planar structure of zirconium(IV) ions in the tetrameric species, $Zr_4(OH)_8(H_2O)_{16}X_z^{(8-z)+}$, in solution, where X = Cl^-, O_1 = OH oxygen, O_2 and O_3 = H_2O oxygens. (Reproduced by permission from Muha and Vaughan, 1960.)

scattering measurements of aged Zr(IV) solutions (Tyree et al., 1968, 1969) and ultracentrifugation studies of lower-acidity solutions aged at an elevated temperature (Johnson and Kraus, 1956) revealed that less acidic zirconium(IV) ion solutions are more polydisperse and contain larger polymeric species. In solutions below 0.2 M HCl, zirconium(IV) ion solutions undergo further hydrolysis to form solutions containing 10 to 100 zirconium ions per aggregate. Johnson and Kraus, for example, reported that aged zirconium solutions of lower acidity do not reach equilibrium even in 10 days and that species of 10–40 zirconium atoms are formed. Copley and Tyree (1968) likewise showed the formation of zirconium polymers with 15–100 atoms per aggregate after 10 hours of aging at elevated temperature. Aging of lower-acidity or neutral $ZrOCl_2$-containing solutions thus showed further polymerization to higher-molecular-weight species of unknown size and structure. It is expected that such species involve structural changes including the formation of O^{2-} bridging from the OH^- bridging units during the polymerization process.

A summary of the major mononuclear and small oligomeric zirconium(IV) hydrolysis species proposed to exist in acidic solutions

Table 40 Summary of Zirconium(IV) Ion Hydrolysis Species, Mononuclear and Polynuclear Ions

Zirconium(IV) species	Species abbreviation	log K_{xy}
$Zr(OH)^{3+}$	1,1	-0.3
$Zr(OH)_2^{2+}$	1,2	-1.7
$Zr(OH)_3^{+}$	1,3	5.1
$Zr(OH)_4$ (aq)	1,4	-9.7
$Zr(OH)_5^{-}$	1,5	-16.0
$Zr_3(OH)_4^{8+}$	3,4	-0.6
$Zr_3(OH)_5^{7+}$	3,5	3.7
$Zr_4(OH)_8^{8+}$	4,8	6.0
ZrO_2 (s)	—	log Q_{s10} = -1.9

Source: Baes and Mesmer (1976).

based on a variety of physical-chemical evidence is given in Table 40, together with their corresponding species designation, (e.g., "3,4" referring to the trimer) and calculated equilibrium constants from Baes and Mesmer (1976). Appreciable evidence for the trimer and especially the tetramer is found in the solution literature. Baes and Mesmer also reported species distribution diagrams for these hydrolysis products of zirconium(IV) ion at 0.1 and 0.01 M Zr which support the predominance of the tetrameric complex in solution, as shown in Fig. 58. However, the experimental evidence needed to estimate the molecular size and structure of larger polymeric species in these solutions is rather limited even today, and plausible model species have not been proposed for these larger zirconium(IV) hydrolysis complexes in either low-acidity or neutral solutions.

D. Structure of Basic Zirconium(IV) Ion Species

The crystal structure of the zirconium(IV) tetrameric species obtained from concentrated acidic $ZrOCl_2$ solutions represents the most important starting point in considering the possible structural

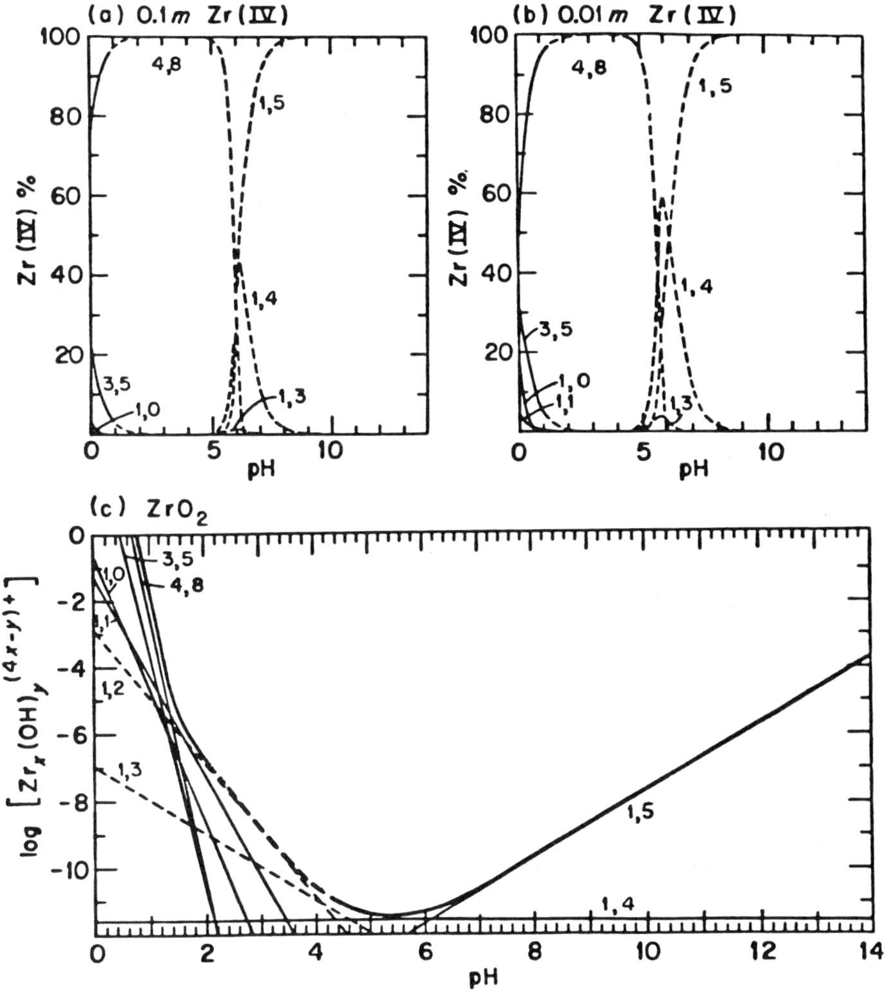

Figure 58 Distribution of zirconium(IV) hydrolysis species at (a) 0.1 M Zr(IV), (b) 0.01 M Zr(IV), and (c) solutions saturated with ZrO_2. (Reproduced by permission from Baes and Mesmer, 1976.)

means by which Zr(IV) ion polymerizes in aqueous solution. The formation of this complex involves sharing of two bridging hydroxide ions to facilitate oligomerization of Zr^{4+} ion in $ZrOCl_2 \cdot 8H_2O$ solutions. Continued hydrolysis of zirconyl solutions, which have an $\bar{n}$ value or $[OH^-]/[Zr^{4+}]$ ratio of 2.0, can be accomplished by aging these solutions as well as by neutralization with bases to form basic zirconium(IV) hydrolysis solutions such as ZrO(OH)Cl, which have an $[OH^-]/[Zr^{4+}]$ ratio of 3.0. Although the literature suggests that formation of higher-MW hydrolysis complexes occurs in the latter systems, research to clarify the possible modes of polymerization and the structural characteristics of such polymeric complexes is lacking. The work of McWhan and Lundgren (1966), however, does provide a basis for discussing some of the expected characteristics, both molecular shape and structure, of the potential species present in basic zirconium(IV) hydrolysis solutions.

McWhan and Lundgren reported the x-ray diffraction determination of the structure of a range of basic zirconium sulfates prepared by hydrothermal formation from the neutral salt $Zr(SO_4)_3(H_2O)_4$. The structures of three basic zirconium sulfates, $Zr_2(OH)_2(SO_4)_3(H_2O)_4$ (I), $Zr(OH)_2SO_4$ (II), and $Zr(OH)_2SO_4$ (III), obtained by thermal treatment of the neutral salt at 100, 200, and 300°C were reported, together with a discussion of the related structural characteristics of a number of other basic zirconium salts and the structure of the $ZrOCl_2 \cdot 8H_2O$ crystal. The neutral salt has a layered crystal structure with zirconium being 8-coordinate due to coordination to four sulfate oxygens and four water molecules in a square antiprism geometry. The structure of the basic zirconium salt (I) formed at 100°C is shown in Fig. 59. This zirconium complex still has a layered structure; however, two waters have been replaced by two bridging hydroxides to facilitate the bonding of each zirconium atom to four other zirconium units by the sulfate groups. The zirconium and hydroxide ions are in the form of a dimer of the formula $Zr_2(OH)_2{}^{6+}$, with the zirconium having a dodecahedral geometry. Formation of the 200°C basic sulfate (II), by contrast, results in the replacement of all coordinated waters with hydroxide ions to form an infinite zigzag chain with the composition of the doubly bridged unit $[Zr_2(OH)_2]_n{}^{2n+}$. The zirconium-oxygen coodination has a square antiprism geometry similar to that of the Zr^{4+} ion in the tetrameric species. The infinite chains are held together by sulfate groups. While these basic zirconium species are formed at elevated temperature and in the presence of sulfate ion, these structures suggest a number of common features to be expected for polymeric species present in basic zirconium chloride systems. First, oligomerization processes most likely require the formation of the doubly bridged hydroxide dimer or possibly the doubly bridged tetramer. Second, in the absence of sulfate ion, the formation of linear chains of dimeric species is likely from dimeric building units, whereas a more

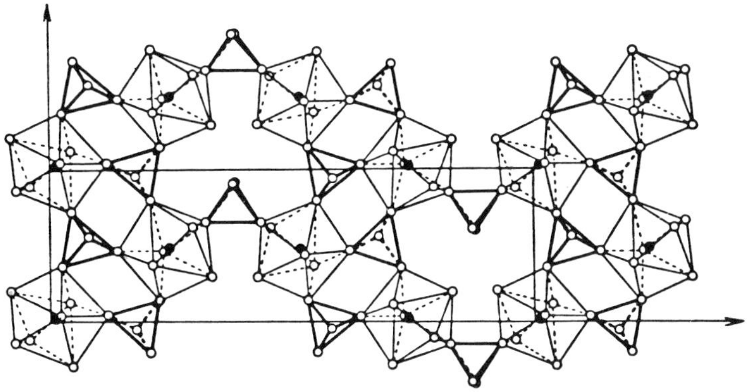

a

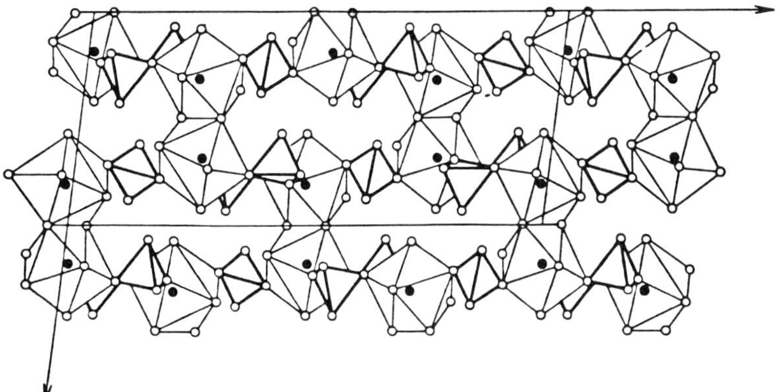

b

Figure 59 Structure of (a) $Zr_2(OH)_2(SO_4)_3(H_2O)_4$ and (b) $Zr_2(OH)_2(SO_4)_2(H_2O)_4$ from x-ray diffraction work of McWhan and Lundgren (1966). (Reproduced by permission.)

three-dimensional polymer is expected if this process involves a tetrameric building unit. Third, the zirconium ion is expected to be 8-coordinate, with a preference for the square antiprism geometry, although the dodecahedral geometry is also possible. Fourth, the degree of polymerization is expected to increase with the addition of hydroxide ion or following thermal aging of aqueous $ZrOCl_2$ solutions, resulting in the formation of oxo-bridges from hydroxide bridges. Clearly, the formation and isolation of polymeric zirconium

Table 41 Nomenclature and Analytical Specifications of Aluminum/Zirconium and ZAG Complexes

Name	Al/Cl ratio	Al/Zr ratio	Glycine
Aluminum zirconium trichlorohydrate (Al/Zr-3)	2.1—1.5	2—6	None
Aluminum zirconium tetrachlorohydrate (Al/Zr-4)	1.5—0.9	2—6	None
Aluminum zirconium pentachlorohydrate (Al/Zr-5)	2.1—1.5	6—10	None
Aluminum zirconium octachlorohydrate (Al/Zr-8)	1.5—0.9	6—10	None
Aluminum zirconium trichlorohydrex complex (ZAG-3)	2.1—1.5	2—6	Variable
Aluminum zirconium tetrachlorohydrex complex (ZAG-4)	1.5—0.9	2—6	Variable
Aluminum zirconium pentachlorohydrex complex (ZAG-5)	2.1—1.5	6—10	Variable
Aluminum zirconium octachlorohydrex complex (ZAG-8)	1.5—0.9	6—10	Variable

Source: FDA-OTC Monograph (1982).

species in basic zirconium(IV) ion hydrolysis systems in the presence of chloride ion will provide many interesting and exciting opportunities for obtaining new information regarding the hydrolysis polymerization processes of this 8-coordinate metal ion.

E. Basic Aluminum/Zirconium Complexes—Nomenclature

Commercial basic aluminum/zirconium solutions or solids used as antiperspirant materials constitute a large group of highly water-soluble systems prepared by the reaction of basic aluminum complexes with zirconium salts termed "Al/Zr complexes." If the preparation involves the use of the neutral amino acid glycine as a buffering agent, these systems are termed zirconium/aluminum/glycine or "ZAG complexes." The FDA-OTC tentative final monograph classification of antiperspirants embodies these two classes of compounds and include the following:

1. Al/Zr complexes such as aluminum zirconium trichlorohydrate, aluminum zirconium tetrachlorohydrate, aluminum zirconium pentachlorohydrate, and aluminum zirconium octachlorohydrate
2. ZAG complexes such as aluminum zirconium trichlorohydrex glycine complex, aluminum zirconium tetrachlorohydrex glycine complex, aluminum zirconium pentachlorohydrex glycine complex, and aluminum zirconium octachlorohydrex glycine complex.

The classification of a particular Al/Zr or ZAG complex on the basis of the FDA-OTC scheme is based on analysis of the aluminum, zirconium, and chloride contents of the solutions or spray-dried solids. A summary of the various Al/Zr and ZAG complexes is given in Table 41 based on the allowable Al/Cl and Al/Zr mole ratios of these systems.

The wide range of Al/Cl and Al/Zr ratios possible for each Al/Zr and ZAG complex makes a discussion of specific complexes and their chemistry exceedingly difficult. To simplify the nomenclature of these complexes in terms of specific Al/Zr or ZAG systems, specific synthesis approaches must be identified to give relevant examples of Al/Zr or ZAG complexes which fit these classifications. Commercial Al/Zr or ZAG complexes generally involve the reaction of either $ZrOCl_2$ or ZrO(OH)Cl solutions with a basic aluminum chloride system such as aluminum chlorohydrate according to the following two general expression:

$$ZrOCl_2(aq) + nAl_2(OH)_5Cl(aq) \rightarrow ZrOCl_2 \cdot nAl_2(OH)_5Cl(aq) \quad (57)$$

$$ZrO(OH)Cl(aq) + nAl_2(OH)_5Cl(aq) \rightarrow ZrO(OH)Cl \cdot nAl_2(OH)_5Cl(aq) \quad (58)$$

While it is apparent that less basic salts such as ASCH or ADCH or even $Al(OH)_3$ could be used as the source of aluminum, they are ignored here. The above two reactions of zirconyl chloride and zirconyl hydroxychloride with ACH may be carried out over a wide range of stoichiometries to produce various Al/Zr or ZAG complexes. Table 42A summarizes a series of reactions involving ACH/$ZrOCl_2$ stoichiometries of 1.0/1.0 to 5.0/1.0. Using these ratios, the elemental Al/Cl and Al/Zr mole ratios of the final solution products are summarized, and the FDA-OTC class in which the final product solution fits is identified on the basis of the elemental analysis specifications of these classes as summarized in Table 41. Two examples of Al/Zr systems produced by the reaction of zirconyl chloride with $Al(OH)_3$

Table 42 Reactions to Synthesize Various Al/Zr and ZAG Complexes

Reactant stoichiometry	Elemental ratios Al/Cl	Al/Zr	FDA-OTC Al/Zr or ZAG class
A. Using $ZrOCl_2$ and ACH or $Al(OH)_3$			
$ZrOCl_2 + 1.0Al_2(OH)_5Cl$	0.66	2	None
$ZrOCl_2 + 1.5Al_2(OH)_5Cl$	0.86	3	Al/Zr-4 or ZAG-4
$ZrOCl_2 + 2.0Al_2(OH)_5Cl$	1.00	4	Al/Zr-4 or ZAG-4
$ZrOCl_2 + 2.5Al_2(OH)_5Cl$	1.11	5	Al/Zr-4 or ZAG-4
$ZrOCl_2 + 3.0Al_2(OH)_5Cl$	1.20	6	Al/Zr-4 or ZAG-4
$ZrOCl_2 + 4.0Al_2(OH)_5Cl$	1.33	8	Al/Zr-8 or ZAG-8
$ZrOCl_2 + 5.0Al_2(OH)_5Cl$	1.43	10	Al/Zr-8 or ZAG-8
$ZrOCl_2 + 3.0Al(OH)_3$	1.50	3	Al/Zr-3 or ZAG-3
$ZrOCl_2 + 4.0Al(OH)_3$	2.00	4	Al/Zr-3 or ZAG-3
B. Using ZrO(OH)Cl and ACH or $Al(OH)_3$			
$ZrO(OH)Cl + 1.0Al_2(OH)_5Cl$	1.00	2	Al/Zr-4 or ZAG-4
$ZrO(OH)Cl + 1.5Al_2(OH)_5Cl$	1.20	3	Al/Zr-4 or ZAG-4
$ZrO(OH)Cl + 2.0Al_2(OH)_5Cl$	1.33	4	Al/Zr-4 or ZAG-4
$ZrO(OH)Cl + 2.5Al_2(OH)_5Cl$	1.44	5	Al/Zr-4 or ZAG-4
$ZrO(OH)Cl + 3.0Al_2(OH)_5Cl$	1.50	6	Al/Zr-4 or ZAG-4
$ZrO(OH)Cl + 4.0Al_2(OH)_5Cl$	1.60	8	Al/Zr-5 or ZAG-5
$ZrO(OH)Cl + 5.0Al_2(OH)_5Cl$	1.67	10	Al/Zr-5 or ZAG-5
$ZrO(OH)Cl + 1.0Al(OH)_3$	2.00	2	Al/Zr-3 or ZAG-3

are also given. Table 42B summarizes a similar set of reactions involving zirconyl hydroxychloride with ACH to give various product complexes with defined elemental Al/Cl and Al/Zr ratios, together with their corresponding designation in a specific FDA-OTC class. A single example of the reaction product complex formed from ZrO (OH)Cl and aluminum hydroxide is also given.

The summary shown in Table 42A illustrates a number of basic features of Al/Zr or ZAG complexes formed from the reaction of $ZrOCl_2$ with ACH (Eq. 57). First, a 1/1 $ZrOCl_2$/ACH reaction product solution does not fit under the FDA-OTC designations. Second, the preparation of an Al/Zr or ZAG complex by the reaction of a 1.5/1.0 through a 3.0/1.0 mole ratio of ACH/$ZrOCl_2$ produces a complex which is classified as a tetrachlorohydrate or tetrachlorohydrex system (Al/Zr-4 or ZAG-4 if the product also contains glycine). Reaction stoichiometries of 4.0/1.0 or 5.0/1.0 ACH/$ZrOCl_2$ lead to the formation of octachlorohydrate or octachlorohydrex complexes (Al/Zr-3 or ZAG-3). In essence, the reaction product stoichiometries limit the type of Al/Zr or ZAG complex which can be synthesized in the current FDA-OTC classes.

Analysis of the reactions of Table 42B for the preparation of Al/Zr and ZAG complexes from ACH and ZrO(OH)Cl leads to similar conclusions. Tetrachlorohydrate or tetrachlorohydrex complexes (Al/Zr-4 or ZAG-4) may be prepared by reactions where the ACH/ZrO(OH)Cl ratios vary from 1.0/1.0 to 3.0/1.0. At 4.0/1.0 or 5.0/1.0 ratios of ACH to ZrO(OH)Cl, the resulting products would be in the pentachlorohydrate or pentachlorohydrex class (Al/Zr-5 or ZAG-5). Similarly, a trichlorohydrate or trichlorohydrex complex can be prepared from ZrO(OH)Cl by reaction of 1 mole of $Al(OH)_3$ with 1 mole of ZrO(OH)Cl.

This analysis of the various synthetic options identifies the complications inherent in describing the nomenclature of commercial Al/Zr or ZAG complexes. Although an infinite number of systems appears possible based on the nomenclature and the reaction stoichiometries, the solution chemistry, economics of starting materials, and stability of the product solutions have limited the availability of Al/Zr complexes to a narrow range of systems. Future development of new Al/Zr or ZAG systems which are included within the current FDA-OTC classes must overcome some of these synthetic limitations to make these complex solutions or materials.

F. Basic Aluminum/Zirconium Complexes—Methods of Synthesis

Basic aluminum/zirconium or ZAG solutions currently available fall within a narrow range of the Al/Zr or ZAG complexes that are theoretically feasible based on the reactions summarized in Table 42A or 42B. The narrow range of available Al/Zr or ZAG complexes has been derived from extensive synthetic work, the results of which have been exclusively reported in the patent literature of commercial basic aluminum/zirconium chemistry. Except for a recent patent by Callaghan and Phipps (1985), the reported synthetic studies have not included the use of any physical-chemical methods of characterization of these systems except for elemental analysis of the aluminum, zirconium, and chloride contents. The patent literature pertaining to the synthesis of Al/Zr or ZAG complexes includes four key series of patents which describe specific advances to enable the synthesis of stable (nongelling) mixtures of aluminum/zirconium complexes.

Beekman (1959) is the first patent describing the preparation of stable aluminum/zirconium complexes and is one of a series of patents utilizing zirconyl chloride solutions prepared by dissolution of $ZrOCl_2 \cdot 8H_2O$ to obtain an Al/Zr system from reactions with various sources of aluminum. In this patent, stable nongelling Al/Zr solutions in the 2–8 Al/Zr mole ratio are prepared by heating the reaction mixture containing $ZrOCl_2$ and ACH, or $ZrOCl_2$ and $AlCl_3$(aq)/Al metal, or hydrous ZrO_2 and $AlCl_3$(aq)/Al metal. Heating of these reaction mixtures was required to produce stable nongelling product solutions with pH values above 3.0.

The patents of Daley (1957a) and Grad (1958) were the first to describe the synthesis of zirconium aluminum glycine or ZAG complexes. Although Beekman noted that a wide range of Al/Zr reaction stoichiometries could be used to prepare stable aqueous Al/Zr systems from $ZrOCl_2$, subsequent patent by Daley (1957a, 1957b) noted that the presence of glycine or urea was necessary to produce concentrated solutions from this reaction in the 15–35% wt/wt range which were stable beyond 2–3 weeks. The patent of Daley (1957a), which involves the use of glycine to stabilize Al/Zr solutions prepared by heating reaction mixtures of $ZrOCl_2$ and ACH with Al/Zr mole ratios of 1.5 to 3.5, is the first one involving the formation of ZAG complexes. Although a wide range of Zr/Al ratios is described, a 1.0/3.5 ratio with 1% glycine is preferred.

A subsequent patent by Grad (1958) was the first one to describe the synthesis of ZAG or zirconium aluminum urea complexes using a more basic zirconyl hydroxychloride solution of the form $ZrO(OH)Cl \cdot 3H_2O$ in place of $ZrOCl_2$. Because of the more basic characteristics of this starting zirconium reagent, the production of stable Al/Zr complexes in the presence of the buffering agent glycine required a reaction stoichiometry with a greater than 3/1 Al/Zr mole ratio to obtain stable nongelling solutions at high concentration. The preparation of ZAG complexes by heating mixtures of ACH and ZrO(OH)Cl in the presence of glycine (Grad, 1958) emphasizes the use of glycine/zirconium ratios near 1.0/1.0, although a range of glycine/zirconium ratios is reported. The patent of Daley (1957a) which describes the synthesis of ZAG complexes from $ZrOCl_2$ and ACH in the presence of glycine also includes a glycine/zirconium ratio near unity.

The patents of Grote (1961a, 1961b) encompass a range of alternative methods for synthesizing Al/Zr complexes in the absence of organic agents such as glycine or urea. The early patent of Grote (1959), following subsequent reactions as described by Beekman, involved the preparation of Al/Zr complexes which were unstable to gelling by the reaction of $ZrOCl_2$ with ACH. A later series of three patents by Grote (1961a, b) described the preparation of stable Al/Zr complexes from 2.0/1.0 to 12.0/1.0 Al/Zr ratios, and preferably in the 3.0/1.0 to 4.0/1.0 range, by heating a reaction solution containing $ZrOCl_2$ and $Al(OH)_3$ or various aluminum alkoxides of the form $Al(OR)_3$. Use of the aluminum alkoxides enables the in situ synthesis of aluminum hydroxide prior to its reaction with zirconyl chloride.

Since these early (1950–1960) patents on the synthesis of Al/Zr and ZAG complexes, numerous other patents involving the synthesis of aluminum and zirconium solution complexes have been reported. The Al/Zr and ZAG complexes currently available commercially, however, are probably prepared by the synthetic methods reported in the early 19602 or modifications of these approaches. Two important recent advances in the synthesis of Al/Zr or ZAG complexes must be noted. First, the preparation of zirconyl chloride, $ZrOCl_2$, or zirconyl hydroxychloride, ZrO(OH)Cl, has recently been carried out by the synthesis of their aqueous solutions from a basic zirconium carbonate past by reaction with HCl according to the following equations:

$$2ZRO(OH)(CO_3)_{0.5}(s) + 4HCl \rightarrow 2ZrOCl_2(aq) + H_2CO_3 + H_2O \quad (59)$$

$$2ZrO(OH)(CO_3)_{0.5}(s) + 2HCl \rightarrow 2ZrO(OH)Cl(aq) + H_2CO_3 \quad (60)$$

In these two reactions the synthesis of zirconyl chloride or zirconyl hydroxychloride solutions is controlled by the reaction mole ratio of zirconyl hydroxycarbonate to HCl.

The second major advance appears in the patent by Callaghan and Phipps (1985), which describes the synthesis of activated ZAG complexes, termed ZAG'. Callaghan and Phipps describe the thermal processing of aqueous ZAG solutions in the 2–20% concentration range prepared by heating ZAG mixtures containing ZrO(OH)Cl, ACH, and glycine in an Al/Zr/glycine ratio of 4/1/4 at 50°C to obtain zirconium aluminum glycine or ZAG complexes which are chemically distinct from those prepared by the method of Grad (1958a). This patent and the physical-chemical approaches used to characterize these complexes will be discussed separately.

In summary, studies of the synthesis of Al/Zr or ZAG complexes as described exclusively in the patent literature have provided a basis for discussing the synthetic origins of the principal Al/Zr and ZAG complexes commercially available today. Aluminum zirconium trichlorohydrates or aluminum zirconium trichlorohydrex complexes containing glycine may be synthesized by reactions of either $ZrOCl_2$ or ZrO(OH)Cl solutions with $Al(OH)_3$ in the absence or presence of glycine and have their origins in the early patents of Grote (1961a, 1961b). The Al/Zr tetrachlorohydrates may be synthesized using reactions of either $ZrOCl_2$ or ZrO(OH)Cl with ACH over a wide range of reaction stoichiometries; however, these systems have only limited stability toward gellation, as described in the patents of Beekman, Daley, and Grad. The zirconium aluminum tetrachlorohydrex glycine (ZAG-4) complexes represent the most commonly used commercial ZAG materials and have their origin in the Daley or Grad patents, depending on the choice of zirconium starting reagent, either $ZrOCl_2$ or ZrO(OH)Cl. Commercial ZAG-4 complexes are generally found to have a zirconium/glycine mole ratio near unity, but the Al/Zr ratio varies from 3.0/1.0 to 4.0/1.0. Although Al/Zr-5, Al/Zr-8, ZAG-5, and ZAG-8 systems are not widely used commercially, these complexes can readily be prepared from either ACH/$ZrOCl_2$ or ACH/ZrO(OH)Cl reaction mixtures in the presence or absence of glycine. These materials have their synthetic origins primarily in the patents of Beekman, Daley, and Grad. Finally, the recently reported preparation of activated ZAG systems, while involving a thermal processing step, is indirectly derived from the ZAG method of synthesis of Grad (1958) involving the preparation of zirconium aluminum glycine complexes from mixtures of glycine, ZrO(OH)Cl, and ACH.

G. Recent Investigations of Aluminum/Zirconium Systems

The wealth of published information on the chemistry of basic aluminum chloride solutions, particularly aluminum chlorohydrate, may be contrasted with the very limited published scientific investigations describing aqueous aluminum zirconium or zirconium aluminum glycine complexes. Two reports which can be cited include that by Oertel and Rush (1979) on the use of Raman and infrared spectroscopy to study a ZAG complex with an Al/Zr/glycine ratio of 3.3/1.0/1.3 and the report on the Sephadex G-50 gel filtration chromatographic behavior of aqueous ZAG and ZAG' complexes with an Al/Zr/glycine ration of 4.0/1.0/1.0 by Callaghan and Phipps (1985) in their patent on activated ZAG complexes.

The Raman and infrared spectroscopic studies of a ZAG complex of Al/Zr/glycine ratio 3.3/1.0/1.3 by Oertel and Rush were carried out in an effort to clarify how the glycine component in ZAG salts is incorporated in the structure of these complexes. These workers showed that the Raman and infrared bands for glycine were consistent with existence of glycine in the dipolar ion form, namely $^{+}NH_3CH_2COO^-$ (α-glycine), in both concentrated 50% aqueous ZAG solutions and the solid ZAG complex, as shown in the Raman and infrared spectra of glycine, the solid ZAG complex, and the 50% aqueous ZAG solutions in Fig. 60 and 61, respectively. Comparison

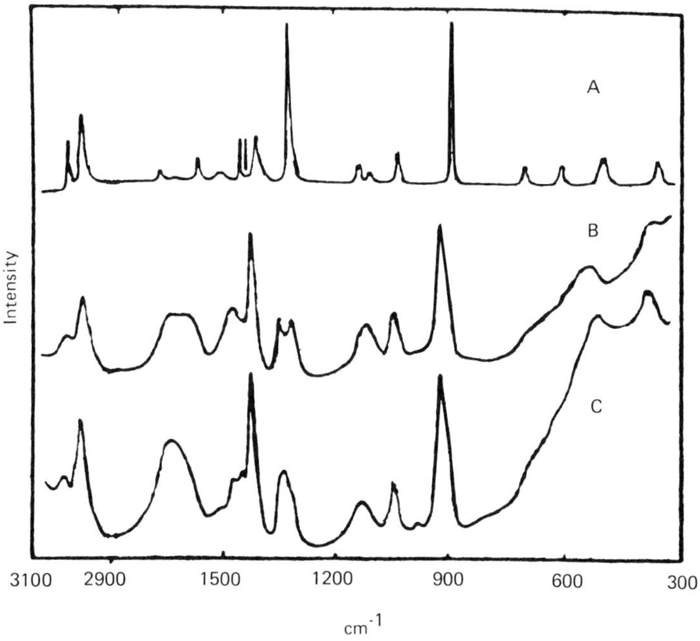

Figure 60 Raman spectra of (A) glycine, (B) solid ZAG, and (C) 50% ZAG solution from work of Oerter (1979). (Reproduced by permission.)

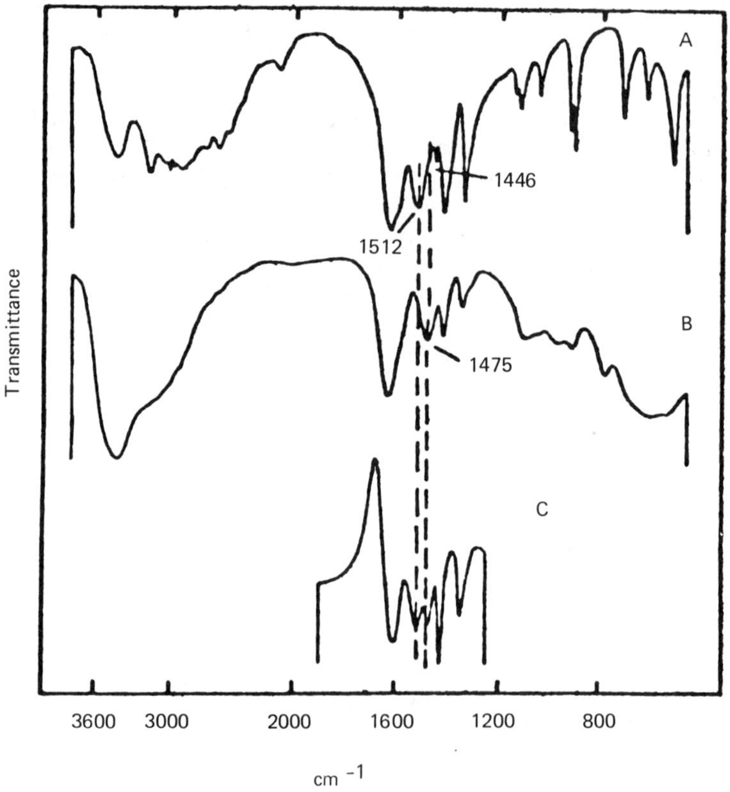

Figure 61 Infrared spectra of (A) glycine, (B) solid ZAG, and (C) 50% ZAG solution. (Reproduced by permission, Oertel and Rush, 1979.)

of the infrared and Raman spectra of these ZAG solutions and solids with the infrared spectra of other metal-glycine complexes led these workers to conclude that either the Zr(IV) ion or the Al(III) ion or both were coordinated to the carboxylate group of the dipolar glycine ion, and that the NH_3^+ end of the glycine was involved in hydrogen bonding in some fashion with molecular constituents in ZAG solids and solutions. In ZAG solids, almost all of the glycine was involved in complexation of this nature, whereas in a 50% ZAG solution from 50 to 75% of the glycine was bonded to various metal ion components of the aluminum and zirconium complexes. Based on the relative shifts of the infrared and Raman bands due to coordinated COO^- groups, it was proposed that stronger bonding to the zirconium(IV) ions was more likely and that this complexation involved bidentate coordination to the Zr(IV) ions in polymeric hydrolysis complexes or coordination as a bridging ligand to two metals ions. The complexation behavior exhibited by the dipolar glycine ion to various Zr(IV) and or Al(III) ions in the polynuclear hydrolysis complexes of these ZAG systems led Oertel and Rush to conclude that this binding of glycine was responsible for the gel-inhibiting properties of glycine in Al/Zr systems. Such complexation would interfere with the extended hydroxy and oxy bridging of various polymeric species in these systems, thus limiting further polymerization or aggregation of mixed Al/Zr complexes which would be unstable toward gelation. Throughout their discussion, the authors emphasized that ZAG systems most likely consisted of mixed Al-Zr polymeric complexes, although no evidence to support this assumption was given.

While the work of Oertel and Rush (1979) provides significant information about the role of glycine in ZAG systems, the work of Callaghan and Phipps (1985) has contributed enormously to the understanding of the macromolecular behavior of aqueous ZAG solutions in the 10% concentration range. In their patent describing the conditions used to prepare new activated ZAG salts such as a 4.0/1.0/1.0 Al/Zr/glycine complex, these workers reported Sephadex G-50 gel filtration chromatographic studies of normal and activated 10% ZAG solutions. Figure 62A shows a Sephadex G-50 chromatogram of an aqueous solution of a normal 4/1 mixture of ACH and ZrO(OH)Cl termed a normal ZAG solutions, and Fig. 62B shows a Sephadex G-50 chromatogram of the same solution heated for 2 hours at 100°C. The very complex chromatogram of normal ZAG solutions includes six peaks, with peak 1 containing zirconium polymeric species greater than 30,000 daltons and five chromatographic peaks attributed to polymeric, oligomeric, and monomeric aluminum hydrolysis complexes. These chromatograms provide significant information about the complex nature of normal Al/Zr and ZAG solutions. First, the authors indicate that peak 1 consists of polymeric zirconium(IV) hydrolysis complexes larger than MW 30,000. This result is consistent with the extensive hydrolysis anticipated to occur in basic zirconyl hydroxychloride solutions on the basis of previous light scattering and ultracentrigation work on aged $ZrOCl_2$ solutions. Second, the authors assigned chromatographic peak 2 to high-MW Al^c complexes found in normal ACH, and peak 3 to the $Al^{c'}$ complexes found in normal and aged or thermally treated ACH solutions. Peak 4 is assigned to a new species not observed in ACH solutions, while peaks 5 and 6 are assigned to Al^a species or low-MW oligomers. The Sephadex G-50 profile for the aluminum species components in Al/Zr solutions thus suggest that Al^c and $Al^{c'}$ complexes found in normal ACH solutions are depolymerized further by reaction with more acidic

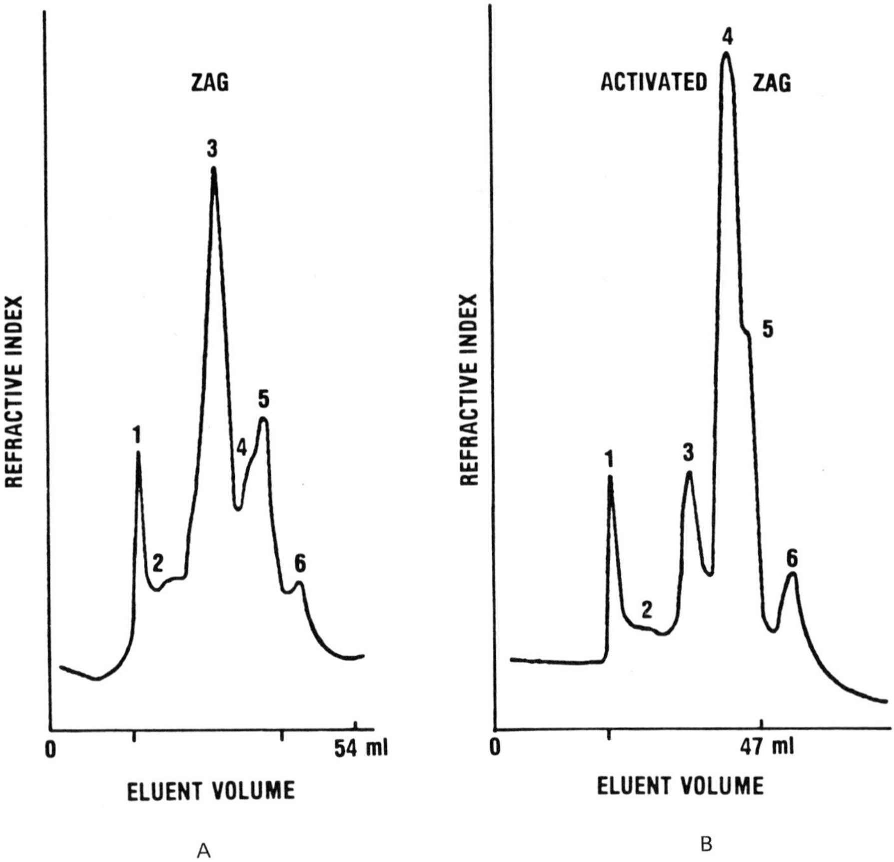

Figure 62 Sephadex G-50 gel filtration chromatogram of ZAG solution (A) before and (B) after thermal treatment. (Reproduced by permission from Callaghan and Phipps, 1985.)

ZrO(OH)Cl solutions to produce lower-MW species of unknown composition. The chromatogram of the thermally treated Al/Zr solution shown in Fig. 62B indicates that high-temperature treatment of Al/Zr solutions leads to further depolymerization of these aluminum complexes, as evidenced by the increase in intensity of peaks 4 and 5 at the expense of the peak 2 and 3 solute species. In fact, continued thermal treatment for up to 100 hours results in GFC profiles containing primarily peak 4 and 5 aluminum species. No experimental evidence is presented, however, to determine the nature of the zirconium hydrolysis species changes occurring throughout these thermal treatments. In addition, this work does not provide GFC results for specific glycine-containing ZAG solutions to establish the fate of these complex species distributions in the presence of glycine. The authors do note that the addition of glycine Al/Zr mixtures such as these does not alter the thermally induced depolymerization of the aluminum species from peak 2/3 species to peak 4/5 and 6 complexes. Despite the absence of GFC data for ZAG solutions and the limitations on the Al/Zr and ZAG systems studied in this patent, it is apparent that this work has already added greatly to the current understanding of the species distributions and complexities of the solution chemistry of basic aluminum zirconium complexes.

H. Future Approaches

The recent scientific reports of specific aspects of the chemistry of basic aluminum/zirconium and zirconium aluminum glycine complexes just described indicate that investigations of the aqueous chemical behavior of these complexes is in its infancy. The study of Al/Zr and ZAG complexes must of necessity involve the use of a number of physical-chemical approaches which will address some of the most important questions about the macromolecular properties, stability, and structure of these important but complex inorganic systems. Figure 63 is a generalized schematic illustrating the complexity of a typical ZAG salt produced by the thermal reaction of ACH with ZrO(OH)Cl in the presence of glycine. This schematic provides a perspective for considering some of the fundamental questions which still remain unanswered regarding the chemistry of this ZAG system as well as the chemistry of basic aluminum/zirconium complexes in general. Some of the most significant unanswered questions regarding ZAG and Al/Zr complexes are as follows:

1. What is the nature of the two reactant basic salts, particularly the polymeric nature and structure of Zr(IV) ion hydrolysis species in ZrO(OH)Cl solutions?
2. What is the nature of the interactions of glycine with Zr(IV) hydrolysis complexes prior to reaction with ACH solutions?
3. How have the species distributions and structure of aluminum hydrolysis complexes in ACH been altered following reaction under thermal conditions with a basic zirconium glycine solution?
4. What is the macromolecular and structural nature of the zirconium/glycine complexes as well as aluminum/zirconium/glycine complexes in aqueous ZAG solutions?
5. Are aluminum/zirconium/glycine complexes affected by solution aging, dilution, and various thermal treatments?

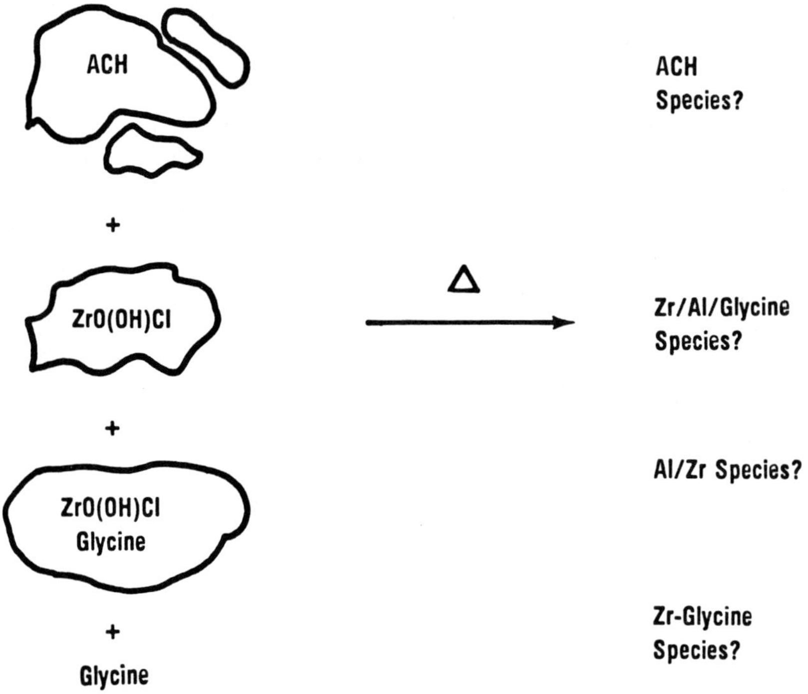

Figure 63 Depiction of the complexities of solution complexes in ZAG solutions prior to reaction and following thermal treatment. (From Fitzgerald, unpublished result.)

6. How are the polymeric nature and structure of the various reactants as well as product Al/Zr and ZAG complexes influenced by synthesis conditions, reactant stoichiometry, and various other alterations of synthesis conditions used in the production of these complexes?

Answers to these and other important questions regarding the aqueous chemistry of basic Al/Zr and ZAG complexes may be obtained through fundamental investigations of these systems using many of the approaches which have proved successful in the past 15 years in characterizing basic aluminum chloride systems. Applicable approaches to investigating these basic aluminum/zirconium and ZAG systems include:

1. Macromolecular investigations of these complexes using light scattering and ultracentrifugation as well as gel filtration or gel permeation chromatography with Sephadex gels and other applicable media.
2. Development and utilization of a zirconium(IV) ion spectrophotometric reactivity assay analogous to the ferron method for aluminum(III) ion hydrolysis solutions.
3. Detailed multinuclear NMR studies of these systems.
4. Studies of a broad range of Al/Zr and ZAG complexes at various concentrations following aging at room temperature and elevated temperatures.
5. Detailed characterization of the composition, stability, and structure of specific hydrolysis species in Al/Zr and ZAG systems synthesized by a range of applicable synthetic methods.

Future investigations with these approaches will undoubtedly lead to a more complete understanding of the stability, structure, and polymeric properties of basic aluminum/zirconium systems, thus providing direction for the future development of various salt systems with improved and consistent pharmacologic behavior.

ACKNOWLEDGMENTS

Over the past 13 years the author has interacted with numerous academic and industrial scientists and cosmetic chemists in industry who have continually raised important questions regarding the chemical nature of commercial basic aluminum hydrolysis systems. Many of these individuals not only have contributed to our current understanding of the chemistry of these systems through their queries but also have been active in much of the research discussed in this chapter.

The author acknowledges the stimulating research environment at the Gillette Company, where he initiated his research interest in basic aluminum complexes using gel filtration chromatography. Under the leadership of Dr. Bernard Siegel, who emphasized the paramount significance of understanding the solution chemistry of aluminum chlorohydrate, the author became involved in a multidisciplinary research program on commercial basic aluminum chloride systems. Various researchers including Dr. Alan Phipps, Ms. Diane Riesgraf, and Dr. Robert Farmer not only made significant contributions to the chemical literature of these systems but also left a legacy of new approaches applicable to studying these complex inorganic solutions. The author acknowledges the contributions of the late Dr. S. Y. Tyree of the College of William and Mary, and Dr. Neal Langerman, formerly at Tufts University and Utah State University, for their collaboration, which has brought about a better understanding of the macromolecular properties of BAC solutions through light scattering and ultracentrifugation studies.

Interactions and collaboration with industrial scientists such as Dr. Allan Rosenberg, Dr. Ronald Markarian, and Dr. Lenonard Weintroup of Bristol-Myers and Dr. Mortan Barr, presently at Mennen, over the past 7 years have been vital to the author in maintaining a consistent direction and perspective in research efforts on the basic aluminum hydrolysis systems discussed here. The financial support of research by the author specifically directed toward understanding basic aluminum chloride systems by the Bristol-Myers Products Division of Bristol-Myers Company is gratefully acknowledged.

Most recently, through a sabbatical leave funded by the South Dakota School of Mines and Technology and the support of Dr. G. E. Maciel, director of the Colorado State University Regional NMR Facility, advances in our understanding of the solution and solid-state ^{27}Al NMR of commerical BAC systems have been made possible. The author acknowledges the National Science Foundation for its support of this NMR facility through NSF Grant CHE 82-08821, and Dr. James Frye, Dr. Bruce Hawkins, and Mr. Mark Seger for their contributions to various technical problems involving solution and solid-state ^{27}Al NMR techniques.

Finally, the author would like to credit the contributions of the following graduate and undergraduate research students: Mr. Loren Johnson, Ms. Carolyn Schwarz, Dr. Theodore C. Brooks, and Ms. Jan A. Burns, for their fine experimental efforts during the past 7 years. An expression of additional gratitude is extended to Ms. J. A. Burns for her excellent contributions in the form of technical review and word processing efforts during the completion of this manuscript.

REFERENCES

Aluminum Hydrolysis Studies

Baes, D. F., Jr., and Mesmer, R. E. (1976). Boron, aluminum and scandium, *The Hydrolysis of Cations*, Wiley-Interscience, New York, pp. 112–123.

Frink, C. R., and Peech, M. (1963). Hydrolysis of the aluminum ion in dilute aqueous solutions, *Inorg. Chem.*, *2*(3): 473.

Frink, C. R., and Sawhney, B. L. (1967). Neutralization of dilute aqueous aluminum salt solutions, *Soil Sci.*, *103*: 144.

Grunwald, E., and Fong, D.-W. (1969). Acidity and association of aluminum ion in dilute aqueous acid, *J. Phys. Chem.*, *73*(3): 650.

Holmes, L. P., Cole, D. L., and Eyring, E. M. (1968). Kinetics of aluminum ion hydrolysis in dilute solutions. *J. Phys. Chem.*, *72(1)*: 301.

Morgan, J. J. (1967). Application and limitations of chemical thermodynamics in natural water systems, *Equilibrium Concepts in Natural Water Systems*, Adv. Chem. Ser. No. 67, American Chemical Society, Washington, D.C., pp. 1–29.

Raupach, M. (1963a). Solubility of simple aluminum compounds expected in soils: II. Hydrolysis and conductance of Al^{3+}, *Aust. J. Soil Res.*, *1*: 28.

Raupach, M. (1963b). Solubility of simple aluminum compounds expected in soils: IV. Reactions of aluminum hydroxide under acid conditions, *Aust. J. Soil Res.*, *1*: 55.

Schofield, R. K., and Taylor, A. W. (1954). The hydrolysis of aluminum salt solutions, *J. Chem. Soc. (London)*, 4445.

^{27}Al NMR Spectroscopy Studies

Akitt, J. W. Aluminum-27 NMR Spectroscopy, (1972). *Annu. Rep. NMR Spectrosc.*, *5A*: 465.

Akitt, J. W., and Farthing, A. (1978). New ^{27}Al NMR studies of the hydrolysis of the aluminum(III) cation, *J. Magn. Reson.*, *32*: 345.

Akitt, J. W., and Farthing, A. (1981a). Aluminum-27 nuclear magnetic resonance studies of the hydrolysis of aluminum(III). Part 2. Gel-permeation chromatography, *J. Chem. Soc. Dalton Trans.*, 1606.

Akitt, J. W., and Farthing, A. (1981b). Aluminum-27 nuclear magnetic resonance studies of the hydrolysis of aluminum(III). Part 3. Stopped-flow kinetic studies, *J. Chem. Soc. Dalton Trans.*, 1609.

Akitt, J. W., and Farthing, A. (1981c). Aluminum-27 nuclear magnetic resonance studies of the hydrolysis of aluminum(III). Part 4. Hydrolysis using sodium carbonate, *J. Chem. Soc. Dalton Trans.*, 1617.

Akitt, J. W., and Farthing, A. (1981d). Aluminum-27 nuclear magnetic resonance studies of the hydrolysis of aluminum(III). Part 5. Slow hydrolysis using aluminum metal, *J. Chem. Soc. Dalton Trans.*, 1624.

Akitt, J. W., and Mann, B. E. (1981e). ^{27}Al NMR spectroscopy at 104.2 MHz, *J. Magn. Reson.*, *44*: 584.

Akitt, J. W., and Milic, N. B. (1984). Aluminum-27 nuclear magnetic resonance studies of the hydrolysis of aluminum(III). Part 6. Hydrolysis with sodium acetate, *J. Chem. Soc. Dalton Trans.*, 981.

Akitt, J. W., Greenwood, N. N. and Lester, G. D. (1969a). Aluminium-27 nuclear magnetic resonance studies of acidic solutions of aluminum salts, *J. Chem. Soc.*(A), 803.

Akitt, J. W., Greenwood, N. N., and Lester, G. D. (1969b). Hydrolysis and dimerization of aqueous aluminum salt solutions, *J. Chem. Soc. Chem. Commun.*, 988.

Akitt, J. W., Greenwood, N. N., Khandelwal, B. L., and Lester, G. D. (1972). ^{27}Al nuclear magnetic resonance studies of the hydrolysis and polymerization of the hexa-aquo-aluminum(III) cation, *J. Chem. Soc. Dalton Trans.*, 604.

Bottero, J. Y., Cases, J. M. Flessinger, F., and Poirier, J. E. (1980). Studies of hydrolyzed aluminum chloride solutions. 1. Nature of aluminum species and composition of aqueous solutions, *J. Phys. Chem.*, *84*(*22*): 2933.

Delpuech, J. J. (1983). Aluminum-27, *NMR of Newly Accessible Nuclei,* Vol. 1. P. Laszlo, ed., Academic Press, New York, pp. 153–193.

Fedotov, M. A., Krivoruchko, O. P., and Buyanov, R. A. (1978). NMR study of the hydrolytic polycondensation of aluminum(III) aquo-ions as an intermediate stage of the formation of hydrated aluminum(III) gels, *Russ. J. Inorg. Chem.*, *23*(*9*): 1282.

Fitzgerald, J. J. (1981). "Temperature Effects on the ^{27}Al NMR Spectra of Polymeric Aluminum Hydrolysis Species," 23rd Rocky Mountain Conference, Denver, Colorado, August 2–6.

Fitzgerald, J. J. (1983). "^{27}Al NMR Studies of Polymeric Basic Aluminum Chloride Complexes," National ACS Meeting, Inorganic Division, Seattle, Washington, March.

Fitzgerald, J. J. (1986). Effects of temperature of the ^{27}Al NMR spectra of basic aluminum chloride complexes, *J. Magn. Reson.*, in press.

Fitzgerald, J. J., and Maciel, G. E. (1984a). "High-Resolution Solid-State ^{27}Al NMR Studies of Aluminum Oxides and Hydroxides," National ACS Meeting, Inorganic Division, St. Louis, Missouri, April.

Fitzgerald, J. J., and Maciel. G. E. (1984b). "Correlation of Solution and Solid-State ^{27}Al NMR Studies of Aluminum Hydrolysis Systems," 1984 International Chemical Congress of Pacific Basin Societies, Sym. Metal Cluster Chemistry, Honolulu, Hawaii, December.

Fitzgerald, J. J., and Maciel, G. E. (1988). High-resolution solid-state ^{27}Al NMR studies of aluminum oxides and hydroxides, *J. Magn. Reson.*, (in press).

Fitzgerald, J. J., Johnson, L. E. Schwarz, C., and Brooks, T. C. (1981). Characterization of solutions of $Al_{13}O_4(OH)_{24}(H_2O)_{12}{}^{7+}$ and $Al(H_2O)_6{}^{3+}$ ions, *Proc. S.D. Acad. Sci.*, *60*: 179.

Gessner, V. W., and Winzer, M. (1979). Contribution to an explanation of equilibrium relationships between differently condensed Al-oxo cations in dilute alkaline solutions, *Z. Anorg. Allg. Chem.*, *452*: 157.

Henty, M. S., and Prescott, A. (1978). Investigations of the reaction of aluminum chlorohydrate($[OH^-]/[Al^{3+}]$ = 2.5) with hydrochloric acid, *J. Chem. Res.* (M), *427*: 5222.
Hinton, J. F., and Briggs, R. W. (1978). Group III—Aluminum gallium, indium and thallium, *NMR and The Periodic Table* (R. K. Harris and B. E. Mann, Eds.), Academic Press, New York, pp. 279–285.
Schonherr, V. S., Gorz, H., Gessner, W., Winzer, M., and Muller, D. (1981a). Basic aluminum salts and their solutions. VII. Influence of preparation, concentration and aging on the constitution of solutions of basic aluminum salts, *Z. Anorg. Allg. Chem.*, *476*: 195.
Schonherr, V. S., Gorz, H., Muller, D., and Gessner, W. (1981b). Basic aluminum salts and their solutions. VI. Preparation and characterization of a water-soluble $Al_{13}O_{40}$ chloride, *Z. Anorg. Allg. Chem.*, *476*: 188.
Schonherr, V. S., Gorz, H., Bertram, R., Muller, D., and Gessner, W. (1983). Basic aluminum salts. XII. Comparative studies of basic aluminum chloride solutions prepared in different ways, *Z. Anorg. Allg. Chem.*, *502*: 113.

Antiperspirant Salts

Bretschneider, E. S., Rubino, A. M., and Jones, J. L. (1975). Physical properties of antiperspirants as related to their efficacy, *Cosmet. Perfum.*, *90*: 27.
Food and Drug Administration, Department of Health and Human Services. (1982). Antiperspirant drug products for over-the-counter human use, Tentative Final Monograph, 21 CFR Part 350, *Fed. Regis.*, *47(162)*: 36492.
Govett, T., and deNavarre, M. G. (1947). Aluminum chlorohydrate, new antiperspirant ingredient, *Am. Perfum.*, *49*: 365.
Lansdown, A. B. G. (1974). Aluminum compounds in the cosmetic industry, *Soap Perfum. Cosmet.*, *47*(5): 209.
Margres, J. J., Rubino, A. M., and Bretschneider, E. S. (1974). "Stability of Selected Antiperspirant Materials," Society of Cosmetic Chemistry National Meeting, New York.
Margres, J. J., Rubino, A. M., and Bretschneider, E. S. (1975). Stability of selected antiperspirant materials, *Cosmet. Perfum.*, *90*: 27.
Nelson, R. E. (1985). Enhanced efficacy actives for aerosol antiperspirant products, *Aerosol Age*, *30(10)*: 1.
Reheis Chemical Company. (1977). Report to Antiperspirant CTFA Chemistry Subcommittee, "On the Nature of Species Present in Aqueous Solutions of Aluminum Ion," Reheis Chemical Company, Berkeley Heights, New Jersey.

Reller, H. H., and Luedders, W. L. (1977). Pharmacologic and toxicologic effects of topically applied agents on the eccrine sweat glands, *Dermatology and Pharmacology*, Vol. 4, Advances in Modern Toxicology Series (Margulli and Maibach, eds.), Wiley, New York, pp 1–54.

Tanabe, H. (1962). Basic aluminum compounds, *Am. Perfum. Cosmet.*, 77: 25.

Basic Aluminum Chloride Patents

Fitzgerald, J. J., Phipps, A. M., McClean, J., and Wu, M. S. (1980). Aluminum chlorhydroxide and preparation thereof, British UK Patent Application 2,048,229A.

Fitzgerald, J. J., McClean, J. L., and Wu, M. S. (1984). Processes for making novel aluminum chlorhydroxide complexes, Canadian Patent 1,163,418.

Garizio, J. E., Rubino, A. M., and Almquist, M. (1959). Aluminum alcogels and process for making same, U.S. Patent 2,876,163.

Gosling, K., Jackson, N. L., and Leon, N. (1978a). Antiperspirant agent, Ger. Offen. 2,700,711.

Gosling, K., Jackson, N. L., and Leon, N. (1978b). Inhibition of perspiration, South African Patent 77/0033.

Gosling, K., Jackson, N. L., and Leon, N. (1980a). Inhibition of perspiration, British UK Patent 1,568,831.

Gosling, K., Mulley, V. J., and Baldock, M. J. (1980b). Antiperspirant composition, Ger. Offen. 2,929,048.

Huehn, W., and Haufe, W. (1940). Water-soluble basic aluminum compounds, U.S. Patent 2,186,016.

Jones, J. L., and Rubino, A. M. (1969). Methods of making alcohol soluble complexes of aluminum and preparations employing the complexes, U.S. Patent 3,420,032.

Jones, J. L., and Rubino, A. M. (1975). Alcohol-soluble basic aluminum chloride, U.S. Patent 3,904,741.

Kastning, E. (1946). Production of basic aluminum chloride, U.S. Patent 2,392,153.

Shih, I. K., and Blaser, E. (1972). Poly aluminum hydroxy complexes and process for their production, U.S. Patent 3,655,329.

Basic Aluminum Chloride Salts

Breuil, H. (1965). Basic aluminum chlorides and bromides, *Ann. Chim. (Paris)*, *10*: 467.

Klenert, G., and Denk, G. (1959). Formation and decomposition of basic aluminum chlorides, *Z. Anorg. Allg. Chem.*, *301*: 171.

Levitskii, E. A., and Maksimov, V. N. (1961). The composition of hydrolysis products of aluminum chloride solutions, *Dokl. Akad. Nauk SSSR*, *141*: 865; *Chem. Abstr.* *56*: 13749h.

Levitiskii, E.A., Maksimov, V. N., and Marchenko, I. Yu. (1961). The polymeric nature of the 5/6 basic aluminum chloride and the possibility of a higher basicity in aluminum oxychlorides, *Dokl. Akad. Nauk SSSR*, *139*: 884; *Chem. Abstr.* *56*: 5616.

Teagarden, D. L., Kozlowski, J. F., White, J. L., and Hem, S. L. (1981a). Aluminum chlorohydrate I: Structure studies, *J. Pharm. Sci.*, *70*(*7*): 758.

Teagarden, D. L., Radavich, J. F., White, J. L., and Hem, S. L. (1981b). Aluminum chlorohydrate II: Physiochemical properties, *J. Pharm. Sci.*, *70*(*7*): 762.

Teagarden, D. L., White, J. L., and Hem, S. L. (1981c). Aluminum chlorohydrate III: Conversion to aluminum hydroxide, *J. Pharm. Sci.*, *70*(*7*): 808.

Teagarden, D. L., Hem, S. L., and White, J. L. (1982). Conversion of aluminum chlorohydrate to aluminum hydroxide, *J. Soc. Cosmet. Chem.*, *33*: 281.

Walter-Levy, L., and Breuil, H. (1957). Basic aluminum chlorides, *Compt. Rend.*, *244*: 2510.

Walter-Levy, L., and Breuil, H. (1962). Formation of basic aluminum chlorides and bromides at 25°C, *Compt. Rend.*, *255*: 1744.

Walter-Levy, L. L., and Breuil, H. (1965). Hydrolysis of the basic chloride $AlCl_3 \cdot 4Al(OH)_3 \cdot 7.5H_2O$, *Compt. Rend.*, *260*(*2*): 568.

Basic Aluminum/Zirconium Patents

Beekman, S. M. (1959). Zirconyl and aluminum halohydroxy complexes, U.S. Patent 2,906,668.

Callaghan, D. T., and Phipps, A. M. (1985). Antiperspirants, British UK Patent Application GB2,144,992A.

Daley, E. W. (1957a). Buffered antiperspirant compositions, U.S. Patent 2,814,584.

Daley, E. W. (1957b). Buffered antiperspirant compositions, U.S. Patent 2,814,585.

Grad, M. (1958). Zirconyl hydroxy chloride antiperspirant compositions, U.S. Patent 2,854,382.

Grote, I. W. (1961a). Zirconyl trichloro aluminate and method of preparing the same, U.S. Patent 3,009,360.

Grote, I. W. (1961b). Method of preparing zirconium aluminum chloro complexes, U.S. Patent 3,009,769.

Rubino, A. M. (1977). Antiperspirant complexes formed with alkali and ammonium zirconyl carbonates, U.S. Patent 4,025,615.

Siegal, B., Gray, H. H., and Shin, C. T. (1968). Antiperspirant compositions and method for the application, U.S. Patent 3,407,254.

Coagulation Studies

Matijevic, E., Mathai, K. G., Ottewill, R. H., and Kerker, M. (1961). Detection of metal ion hydrolysis by coagulation: III. Aluminum, *J. Phys. Chem.*, *65*: 826.

Matijevic, E., Janauer, G. E., and Kerker, M. (1964). Reversal of charge of lyophobic colloids by hydrolyzed metal ions. I. Aluminum nitrate, *J. Colloid Sci.*, *19*: 333.

Packham, R. F. (1962) . The coagulation process. II. Effect of pH on the precipitation of aluminum hydroxide, *J. Appl. Chem.*, *12*: 564.

Stryker, L. J., and Matijevic, E. (1969). Counterion complexing and sol stability. II. Coaguation effects of aluminum sulfate in acidic solutions, *J. Phys. Chem.*, *73*(5): 1484.

Ferron and Other Spectrophotometric Studies

Bersillon, J. L., Brown, D. W., Fiessinger, F., and Hem, J. D. (1978). Studies of hydroxyaluminum complexes in aqueous solution, *U.S. Geol. Surv. J. Res.*, *6*(3): 325.

Gessner, V. W., and Winzer, M. (1979). On the behavior of aluminum salts with differently condensed Al-oxo cations in the reaction with ferron [8-hydroxy-7-iodoquinoline sulfonic acid-5], *Z. Anorg. Allg. Chem.*, *452*: 151.

Gildea, D. R., Phipps, A. M., Ferguson, J. H., and Kustin, K. (1977). Problems in the study of aluminum(III) complexation kinetics, *Inorg. Chem.*, *16*(5): 1257.

Hem, J. D., (1968a). Aluminum species in water, *Trace Inorganics in Water*, Adv. Chem. Ser. No. 73, American Chemical Society, Washington, D.C., pp. 98–120.

Hem, J. D., (1968b). Graphical methods for studies of aqueous aluminum hydroxide, fluoride, and sulfate complexes. *U.S. Geol. Surv. Water-Supply Pap. 1827-B*.

Hem, J. D., and Roberson, C. E. (1967). Form and stability of aluminum hydroxide complexes in dilute solution, *U.S. Geol. Surv. Water-Supply Pap. 1827-A*, Government Printing Office, Washington, D.C.

Hsu, P. H., (1964). Formation of x-ray amorphous and crystalline aluminum hydroxides, *Mineral. Mag.*, *33*: 173.

Hsu, P. H., (1968). Interaction between aluminum and phosphate in aqueous solutions, *Trace Inorganics in Water*, Adv. Chem. Ser. No. 73, American Chemical Society, Washington, D.C., pp. 115–127.

Hsu, P. H., (1977). Aluminum hydroxides and oxyhydroxides, *Minerals in Soil Environments* (J. B. Dixon and S. B. Weed, eds.), Soil Science Society of America, Madison, Wisconsin, pp 99–143.

Hsu, P. H., and Bates, T. F. (1964). Fixation of hydroxy-aluminum polymers by vermiculite, *Soil Sci. Soc. Am. Proc.*, *28*(*6*): 763.
MacDonald, D. D., Butler, P., and Owen, D. (1973). Hydrothermal hydrolysis of Al^{3+} and the precipitation of boehmite from aqueous solution, *J. Phys. Chem.*, *77*(*20*): 2474.
Roberson, C. E., and Hem. J. D. (1969). Solubility of aluminum in the presence of hydroxide, fluoride, and sulfate, *U.S. Geol. Surv. Water-Supply Pap. 1827-C.*
Ross, G. J., and Turner, R. C. (1971). Effect of different anions on the crystallization of aluminum hydroxide in partially neutralized aqueous aluminum salt systems, *Soil Sci. Soc. Am. Proc.*, *35*: 389.
Schoen, R., and Roberson, C. E. (1970). Structures of aluminum hydroxide and geochemical implications, *Am. Mineral.*, *55*: 43.
Schonherr, V. S., and Frey, H. P. (1979). Separation and characterization of condensed aluminum cations from basic aluminum chloride solutions, *Z. Anorg. Allg. Chem.*, *452*: 167.
Smith, R. W. (1969). The state of Al(III) in aqueous solutin and absorption of hydrolysis products on Al_2O_3, Thesis, University Microfilms.
Smith, R. W. (1970). Relations among equilibrium and nonequilibrium aqueous species of aluminum hydroxy complexes, *Nonequilibrium in Natural Water Chemistry*, Adv. Chem. Ser. No. 103, American Chemical Society, Washington, D.C., p. 250.
Smith, R. W., and Hem, J. D. (1972). Effect of aging on aluminum hydroxide complexes in dilute aqueous solutions, *U.S. Geol. Surv. Water-Supply Pap. 1827-D.*
Turner, R. C. (1965). Some properties of aluminum hydroxide precipitated in the presence of clays, *Can. J. Soil Sci.*, *45*: 331.
Turner, R. C. (1968a). Conditions in solution during the formation of gibbsite in dilute aluminum salt solutins: I. Theoretical treatment of the effect of the formation of mono- and polynuclear hydroxyaluminum ions, precipitations and crystallization on curves representing the titration of aluminum chloride with a base, *Soil Sci.*, *106*: 291.
Turner, R. C. (1968b). Conditions in solution during the formation of gibbsite in dilute aluminum salt solution: II. Effect of time of reaction on the formation of polynuclear hydroxyaluminum cations, the substitution of other anions for hydroxide in amorphous aluminum hydroxide and the crystallization of gibbsite, *Soil Sci.*, *106*: 338.
Turner, R. C. (1969). Three forms of aluminum in aqueous systems determined by 8-quinolinolate extraction methods, *Can. J. Chem.*, *47*: 2521.
Turner, R. C. (1971). Kinetics of reactions of 8-quinolinol and acetate with hydroxyaluminum species in aqueous solutions. 2.

Initial solid phases, *Can. J. Chem.*, *49*: 1688.

Turner, R. C. (1975). Equilibrium constant for the formation of bis(hydroxyaluminum)(4+) in aqueous solution, *Can. J. Chem.*, *53(19)*: 2811.

Turner, R. C. (1976a). Effect of aging on properties of polynuclear hydroxyaluminum cations, *Can. J. Chem.*, *54(10)*: 1528.

Turner, R. C. (1976b). A second species of polynuclear hydroxyaluminum cation and some of its properties, *Can. J. Chem.*, *54(12)*: 1910.

Turner, R. C., and Ross, G. J. (1969). Conditions in solution during the formation of gibbsite in dilute aluminum salt solutions. III. Hydroxyaluminum products of reactions during the neutralization of aluminum chloride solutions with sodium hydroxide, *Can. J. Soil Sci.*, *49*: 389.

Turner, R. C., and Ross, G. J. (1970). Conditions in solution during the formation of gibbsite in dilute Al salt solutions. 4. Effect of Cl concentration and temperature and a proposed mechanism for gibbsite formation, *Can. J. Chem.*, *48*: 723.

Turner, R. C., and Sulaiman, W. (1971). Kinetics of reactions of 8-quinolinol and acetate with hydroxyaluminum species in aqueous solutions. 1. Polynuclear hydroxyaluminum cations, *Can. J. Chem.*, *49*: 1683.

Gel Filtration Chromatography Studies

Akitt, J. W., and Farthing, A. (1981). Aluminum-27 nuclear magnetic resonance studies of the hydrolysis of aluminum(III). Part 2. Gel-permeation chromatography, *J. Chem. Soc. Dalton Trans.*, 1606.

Bombaugh, K. J. (1971). The practice of gel permeation chromatography, *Modern Practice of Liquid Chromatography* (J. J. Kirkland, ed.), Wiley-Interscience, New York, pp. 237–286.

Determann, H. (1969). *Gel Chromatography*, 2nd ed., Springer-Verlag, New York.

Fitzgerald, J. J. (1977). "Macromolecular Properties of Aluminum Hydroxychloride Polymers," National ACS Meeting, Inorganic Division, New Orleans, Louisiana, March 24.

Fitzgerald, J. J. (1981). "Basic Aluminum Chloride Complexes. I. Physical-Chemical Characterization of Their Aqueous Solutions," Society of Cosmetic Chemists Meeting, New York, December.

Fitzgerald, J. J. (1987). Application of gel filtration chromatography to the understanding of basic aluminum polymers, *J. Colloid Interface Sci.*, in press.

Fitzgerald, J. J., and Johnson, L. (1978) "Computer Data Analysis System for Aqueous Gel Filtration Chromatography of Metal

Hydrolysis Polymers," South Dakota Academy of Sciences Meeting, Madison, South Dakota.
Fitzgerald, J. J., and Johnson, L. (1980). "Physical-Chemical Investigations of Concentrated Basic Aluminum Chlorides," National ACS Meeting, Inorganic Division, Houston, Texas, March.
Fitzgerald, J. J., and Rand, J. R. (1975). "Application of Gel Filtration Chromatography to Metal Ion Hydrolysis Studies," First Chemical Congress of North American Continent, Mexico City, December 5.
Fitzgerald, J. J., and Rand, J. (1988). Solute-gel interactions in gel filtration studies of aqueous aluminum solutes, *J. Chromatogr.*, in press.
Fitzgerald, J. J., Johnson, L. Schwarz, C., and Brooks, T. C. (1981). "Characterization of Solutions of $Al_{13}O_4(OH)_{24}(H_2O)_{12}$ Ion," South Dakota Academy of Sciences Meeting, Vermillion, South Dakota, April.
Kirkland, J. J. (1971). *Modern Practice of Liquid Chromatography*, Wiley-Interscience, New York.
Pharmacia Fine Chemicals. (1968). "Sephadex-Gel Filtration Theory and Practice," Pharmacia Fine Chemicals, Uppsala, Sweden.
Poole, C. F., and Schuette, S. A. (1984). *Contemporary Practice of Chromatography*, Elsevier, Amsterdam, pp. 290–304.
Rollins, J. E., Bose, A., Caruthers, J. M., Tsao, G. T., and Okos, M. R. (1983). Aqueous size exclusion chromatography, *Polymer Characterization: Spectroscopic, Chromatographic, and Physical Instrumental Measurements* (C. D. Craver, ed.), Adv. Chem. Ser. 203, American Chemical Society, Washington, D.C., pp. 345–362.
Schonherr, V. S., and Frey, H. P. (1979). Separation and characterization of condensed aluminum cations from basic aluminum chloride solutions, *Z. Anorg. Allg. Chem.*, *452*: 167.

Infrared and Raman Spectroscopy Studies

Fripiat, J. J., Van Cauwelaert, F., and Bosmans, H. (1965). Structure of aluminum cations in aqueous solutions, *J. Phys. Chem.*, *69*: 2458.
Lippincott, E. R., Psellas, J. A., and Tobin, M. C. (1952). Raman spectra and structures of aluminate and zincate ions, *J. Phys. Chem.*, *20*: 536.
Moolenaar, R. J., Evans, J. C., and McKeever, L. D. (1970). The structure of the aluminate ion in solutions at high pH, *J. Phys. Chem.*, *74*: 3629.
Riesgraf, D. A., and May, M. L. (1978). Infrared spectra of aluminum hydroxide chlorides, *Appl. Spectrosc.*, *32*(4): 362.

Waters, D. N., and Henty, M. S. (1977). Raman spectra of aqueous solutions of hydrolyzed aluminum(III) salts, *J. Chem. Soc. Dalton Trans.*, 243.

Light Scattering Studies

Patterson, J. H., and Tyree, S. Y. Jr. (1972). Paper given at the 163rd National ACS Meeting, Boston, April.

Patterson, J. H., and Tyree, S. Y., Jr. (1973). A light scattering study of the hydrolytic polymerization of aluminum, *J. Colloid Interface Sci.*, *43*(*2*): 389.

Riesgraf, D. A., Farmer, R. F., Watson, B. S., and Yin, N. E. (1979). "Characterization of Aluminum Hydroxide Chloride Polymers," National ACS Meeting, Washington, D.C., August.

Ruff, J. K., and Tyree, S. Y., Jr. (1958). Light-scattering studies on aqueous aluminum nitrate solutions, *J. Am. Chem. Soc.*, *80*: 1523.

Potentiometry Studies

Brosset, C. (1952). On the reactions of the aluminum ion with water, *Acta Chem. Scand.*, *6*: 910.

Brosset, C., Biedermann, G., and Sillen, L. G. (1954). Studies on the hydrolysis of metal ions, *Acta Chem. Scand.*, *8*: 1917.

DeHek, H., Stol. R. J., and DeBruyn, P. L. (1978). Hydrolysis-precipitation studies of aluminum(III) solutions. 3. The role of sulfate ion, *J. Colloid Interface Sci.*, *64*(*1*): 72.

Dezelic, N., Bilinski, H., and Wolf, R. H. H. (1971). Precipitation and hydrolysis of metallic ions. IV. Studies of the solubility of aluminum hydroxide in aqueous solution, *J. Inorg. Nucl. Chem.*, *33*: 291.

Grunwald, E., and Fong. D. W. (1969). Acidity and association of aluminum ion in dilute aqueous acid, *J. Phys. Chem.*, *73*(*3*): 650.

Hayden, P. L., and Rubin, A. J. (1973). "Systematic Investigation of the Hydrolysis and Precipitation of Aluminum(III)," NTIS PB-241, 318.

Mesmer, R. E., and Baes, C. F., Jr. (1971). Acidity measurements at elevated temperatures. V. Aluminum ion hydrolysis, *Inorg. Chem.*, *10*: 2290.

Rubin, A. J., and Hayden, P. L. (1973). "Studies on the Hydrolysis and Precipitation of Aluminum," Project Report No. 364X, U.S. Department of Interior Water Resources Center, Ohio State University, p. 1.

Sillen, L. G. (1954). On equilibria in systems with polynuclear complex formation. I. Methods for deducing the composition of

the complexes from experimental data. "Core plus links" complexes, *Acta Chem. Scand.*, *8*: 229.
Sillen L. G. (1959). Quantitative studies of hydrolytic equilibriums, *Q. Rev. Chem. Soc. (London)*, *13*: 146.
Stol, R. J., Van Helden, A. K., and DeBruyn, P. L. (1976). Hydrolysis-precipitation studies of aluminum(III) solutions. 2. A kinetic study and model, *J. Colloid Interface Sci.*, *57(1)*: 115.
Vermeulen, A. C., Geus, J. W., Stol, R. J., and DeBruyn, P. L. (1975). Hydrolysis-precipitation of aluminum(III) solutions. 1. Titration of acidified nitrate solutions, *J. Colloid Interface Sci.*, *51(3)*: 449.

Small-Angle X-Ray Scattering Studies

Rausch, W. V., and Bale, H. D. (1964). Small-angle x-ray scattering from hydrolyzed aluminum nitrate solutions, *J. Chem. Phys.*, *40(11)*: 3391.

Ultracentrifugation Studies

Ansevin, A. T., Roark, D. E., and Yphantis, D. A. (1970). Improved ultracentrifuge cells of high-speed sedimentation equilibrium studies with interference optics, *Anal. Biochem.*, *34(1)*: 237.
Aveston, J. (1965). Hydrolysis of the aluminum ion: Ultracentrifugation and acidity measurements, *J. Chem. Soc. (London)*, *111*: 4438.
Flory, P. (1953). Determination of molecula weights, *Principles of Polymer Chemistry*, Cornell University Press, Ithaca, New York, pp. 266–316.
Johnson, J. S., Scatchard, G., and Kraus, K. A. (1959). Use of interference optics in equilibrium ultracentrifugations of charged systems, *J. Phys. Chem.*, *63*: 787.
Langerman, N. (1988). "The Hydrodynamic Behavior of Aluminum Chlorohydrate and Basic Aluminum Chlorides," unpublished results.
Yphantis, D. A. (1964). Equilibrium ultracentrifugation of dilute solutions, *Biochemistry*, *3(3)*: 297.

X-Ray Diffraction Studies

Johansson, G. (1962a). The crystal structure of $[Al_2(OH)_2(H_2O)_8](SO_4)_2 \cdot 2H_2O$ and $[Al_2(OH)_2(H_2O)_8](SeO_4)_2 \cdot 2H_2O$, *Acta Chem. Scand.*, *16*: 403.

Johansson, G. (1962b). The crystal structure of a basic aluminum selenate, *Ark. Kemi, 20*: 305.

Johansson, G. (1962c). On the crystal structure of the basic aluminum sulfate $13Al_2O_3 \cdot 6SO_3 \cdot 6SO_3 \cdot xH_2O$, *Ark. Kemi, 20*: 321.

Johansson, G. (1966). On the crystal structure of the potassium aluminate $K_2[Al_2O(OH)_6]$, *Acta Chem. Scand., 20(2)*: 505.

Johansson, G., Lundgren, G., Sillen, L. C., and Soderquist, R. (1960). On the crystal structure of a basic aluminum sulfate and the corresponding selenate, *Acta Chem. Scand., 14*: 769.

Zirconium Hydrolysis Chemistry

Angstadt, R. L., and Tyree, S. Y. Jr. (1962). The nature of zirconyl chloride in strong hydrochloric acid: Light scattering, *J. Inorg. Nucl. Chem., 24*: 913.

Baes, C. F., Jr., and Mesmer, R. E. (1976). Titanium, Zirconium, hafnium and thorium, *The Hydrolysis of Cations*, Wiley-Interscience, New York, pp. 152–158.

Bilinski, H. and Tyree, S. Y. Jr. (1969). "Studies of the Kinetics of Hydrolysis of Zirconium Ion in 1 M NaCl," American Chemical Society Meeting, New York.

Bilinski, H., Branica, M., and Sillen, L. G. (1966). Precipitation and hydrolysis of metallic ions. II. Solubility of zirconium hydroxide in dilute solutions and in M $NaClO_4$, *Acta Chem. Scand., 20(3)*: 853.

Blumenthal, W. B. (1973). Hydrochlorides of zirconium oxide, *J. Less-Common Met., 30*: 39.

Clearfield, A., and Vaughan, P. A. (1956). The crystal structure of zirconyl chloride octahydrate and zirconyl bromide octahydrate, *Acta Crystallogr., 9*: 555.

Copley, D. B., and Tyree, S. Y., Jr. (1968). Time and temperature variations in the hydrolytic behavior of hafnium(IV) in aqueous chloride media, *Inorg. Chem., 7(7)*: 1472.

Hardy, C. J., Field, B. O., and Scargill, D. (1966). Bonding of ligands in hydrated nitrates and chlorides of Zr(IV), *J. Inorg. Nucl. Chem., 28(10)*: 2408.

Hoch, A. L. (1974). Zirconium compounds: The industrial importance of their aqueous chemistry, *Chem. Ind. London, 2*: 864.

Johnson, J. S., and Kraus, K. A. (1953). Hydrolytic polymerization of zirconium(IV), *J. Am. Chem. Soc., 75*: 5769.

Johnson, J. S., and Kraus, K. A. (1956). Hydrolytic behavior of metal ions. VI. Ultracentrifugation of zirconium(IV) and hafnium(IV); Effect of acidity on the degree of polymerization, *J. Am. Chem. Soc., 78(16)*: 3937.

Johnson, J. S., Kraus, K. A., and Holmberg, R. W. (1956). Hydrolytic behavior of metal ions. V. Ultracentrifugation of hafnium(IV), *J. Am. Chem. Soc., 78(1)*: 26.

Kovalenko, P. N., and Bagdasarov, K. N. (1961). Determination of the solubility of zirconium hydroxide, *Russ. J. Inorg. Chem.*, *6(3)*: 272.
Kraus, K. A. (1949). In *The Transuranium Elements*, IV-14B (G. T. Seaborg, J. J. Katz, and W. M. Manning, eds.), McGraw-Hill, New York, p. 245.
Larsen, E. M., and Gammill, A. M. (1950). Electrometric titrations of zirconium and hafnium solutions, *J. Am. Chem. Soc.*, *72*: 3615.
Mak, T. C. W. (1968). Refinement in the crystal structure of zirconyl chloride octahydrate, *Can. J. Chem.*, *46*: 3491.
Matijevic, E., Mathai, K. G., and Kerker, M. (1962). Detection of metal ion hydrolysis by coagulation, V. Zirconium, *J. Phys. Chem.*, *66*: 1799.
McWhan, D. B., and Lundgren, G. (1966). The crystal structure of $Zr_2(OH)_2(SO_4)_3 \cdot 4H_2O$, *Inorg. Chem.*, *5*: 284.
Muha, G. M., and Vaughan, P. A. (1960). Structure of the complex ion in aqueous solutions of zirconyl and hafnyl oxyhalides, *J. Phys. Chem.*, *33*: 194.

REFERENCES ADDED IN PROOF

Aluminum Hydrolysis Studies

Brown, P.L., Sylva, R. N., Batley, C. E., and Ellis, J. (1985). The hydrolysis of metal ions. Part 8. Aluminum, *J. Chem. Soc. Dalton Trans.*, *9*: 1967.

^{27}Al NMR Spectroscopy Studies

Akitt, J. W., and Farnsworth, J. A. (1985a). Nuclear magnetic resonance and molar-volume studies of the complex formed between aluminum (III) and the sulfate anion, *J. Chem. Soc. Faraday Trans. 1*, *81(1)*: 193.
Akitt, J. W., and Elders, J. M. (1985b). Aluminum-27 nuclear magnetic resonance studies of the hydrolysis of aluminum (III). Part 7. Spectroscopic evidence for the cation $[Al(OH)]^{2+}$ from line broadening studies at high dilution, *J. Chem. Soc. Faraday Trans.*, *8 (8)*: 1923.
Akitt, J. W., and Elders, J. M. (1985c). The hexa-aquo aluminum cation as a reference in aluminum-27 NMR spectroscopy. Possible detection of second sphere effects, *J. Magn. Reson.*, *63*: 587.
Akitt, J. W., Elders, J. M., and Fontaine, X. L. R. (1986). Hydrogen-deuterium isotope effects in the proton NMR spectra of hydrated aluminum (III) species, *J. Chem. Soc. Chem. Comm.*, *13*: 1047.
Bertran, R., Gessner, W., and Muller, D. (1986). Formation of

aluminum oxide $Al_{13}O_{40}$ cation by dilution of highly concentrated basic aluminum chloride solutions, *Z. Chem.*, *26*(*9*): 340.
Denney, D. Z., and Hsu, P. H. (1986). ^{27}Al nuclear magnetic resonance and ferron kinetic studies of partially neutralized $AlCl_3$ solutions, *Clay and Clay Minerals*, *34*(*5*): 604.
Kunwar, A. C., Thompson, A. R., Gutowsky, H. S., and Oldfield, E. (1984). Solid-state aluminum-27 NMR studies of tridecameric Al-oxo-hydroxy clusters in basic aluminum selenate, sulfate, and the mineral zunyite, *J. Magn. Reson.*, *60*: 467.

Basic Aluminum Chloride Patents

Elliotts, Ltd. (1953). Aluminum chlorohydrate and method of preparation, Australia Patent 150, 410.
Gosling, K., and Hyde, M. R. (1981). Antiperspirants, British UK Patent 1,597,498.
Gosling, K., Mulley, V. J., and Baldock, M. J. (1982a). Antiperspirant composition and methods for its preparation and use, European Patent 0006739.
Gosling, K., Mulley, V. J., and Baldock, M. J. (1982b). Inhibition of perspiration, British UK Patent 2,027,419B.

Zirconium Hydrolysis Chemistry

Bertin, F., Bouix, J., Hannane, S., and Paris, J. (1987). Raman study of zirconium oxychloride hydrolysis in aqueous solutions, *C. R. Acad. Sci. Paris 2*, *304*(*9*): 405.
Noren, B. (1973). Hydrolysis of zirconium(4+) and hafnium(4+) ions, *Acta. Chem. Scand.*, *27*: 1369.
Oertel, R. P., and Rush, M. R. (1979). Vibrational spectroscopic investigation of glycine binding in a zirconium-aluminum-glycine antiperspirant material, *Applied Spectroscopy*, *33*(*2*): 114.
Zielen, A. J., and Connick, R. E. (1956). The hydrolytic polymerization of zirconium in perchloric acid solutions, *J. Am. Chem. Soc.*, *78*(*4*): 5785.

7

Clinical Evaluation of Antiperspirants

Carl B. Felger *Gillette Research Institute, Gaithersburg, Maryland*

Janice G. Rogers *Gillette Medical Evaluations Laboratories, Gaithersburg, Maryland*

I. INTRODUCTION

Antiperspirants are regulated as over-the-counter (OTC) topical drugs by the Food and Drug Administration (FDA). Because of this regulation, and because the definition of an antiperspirant is based on its human clinical efficacy, the FDA requires a specified level of antiperspirant efficacy in a recognized hot-room or ambient temperature test (1). Such test procedures have found extensive use and are generally accepted by both the government and the technical community. A more detailed discussion of axillary antiperspirant testing will be presented later in this chapter.

Sweat glands can be divided into two types, eccrine and apocrine. The apocrine glands have a limited distribution over the body and their sparse secretions are believed to be a major factor in the development of body odor, especially in the axilla (2). Eccrine glands are distributed over the entire body surface and produce a watery liquid that is responsible for the undesirable wetness of the axillary vault as well as other parts of the body (3, 4). The role of an antiperspirant product is to reduce axillary sweating, hence the clinical demonstration of efficacy is a critical portion of the development program.

The development of new antiperspirant products is a costly, time-consuming activity. The majority of people under conditions of moderate temperature and little emotional stress produce small amounts of eccrine sweat; therefore for the purpose of antiperspirant testing it is necessary to induce sweating. This can be done in several ways, such as emotional stimulation (5), heat stimulation (6), or a combination of the two (7). Depending on the test procedure chosen, these studies can require from 4 days to 4 weeks to evaluate one test product (1, 8–10). Because of this, a large number of alternative methods for evaluating (screening) antiperspirant efficacy have been devised that will allow the rapid testing of several formulations and rank order them for efficacy before selecting a few for axillary testing under either hot-room or ambient conditions (11). These screening tests are categorized and discussed separately. Although not all such tests have been included, it is hoped that this chapter will provide the chemist with an idea of the options that are available to assist in development activities.

For the purposes of this discussion these tests can be grouped into three major categories:

1. Visualization (colorimetric) techniques
2. Instrumental (sensor) methods
3. Gravimetric techniques

II. TESTING METHODOLOGY

A. Visualization Techniques

The visualization techniques represent the oldest type of routine testing to determine sweat output. The most useful methods in this category are those that visualize the reaction of a relatively large number of sweat glands to an antiperspirant. Many such tests have been proposed and used extensively (12–21). One of the first of these methods was developed by Minor (12) and subsequently used and modified extensively. Minor's technique involved a colorimetric method in which a mixture of iodine, castor oil, and alcohol was painted on the skin. When the solution was dry the skin was powdered with starch. The skin, which then appeared white, became the background for the sweat droplets, which appeared as dark spots. The sweat droplets increased in number and size as sweating progressed until the entire area became uniformly dark. This classic method was later refined by Wada (22) to allow visualization of sweat secretion from individual glands and also to detect minute quantities of secretion by preventing evaporation. This refinement consisted of painting the skin with an iodine-alcohol mixture, which was allowed to dry completely. Then the area was painted again with a mixture of starch powder and castor oil. When sweating occurred, spots or rings of black-stained starch appeared at the pores of the functioning sweat glands and could be viewed through the transparent starch-castor oil layer. This technique remains (some 35 years later) a useful tool when rapid qualitative estimations of sweat output are desired or in certain mechanistic studies concerned with the size and number of individual firing sweat glands.

A second, relatively simple but important modification of Wada's method was described by Papa (19) to visualize sweat on dark skin. To overcome the difficulty in visualizing the blue-black dots in the standard starch-iodine method, he applied a suspension of cornstarch and castor oil over the skin and initiated sweating. The milk-white sweat droplets thus formed were easily observed on dark-skinned individuals.

Early on, however, the need for quantification of sweat output arose, and the methods used initially were modifications of the original Minor technique. In 1946 Randall (23) published a procedure which eliminated the necessity of applying a coating of starch powder on the skin. Instead he used a dilute solution of iodine in alcohol, which was first painted over the area and allowed to dry. He then placed a blank piece of ordinary bond paper (although any starch-containing paper would be acceptable) over the area for 20 seconds. As the paper was held in place, the sweat secreted solubilized the starch-iodine, making a solution which produced definite blue-black spots on the paper. Although the prints produced were not permanent for long periods of time, they did allow a period of several weeks for analyses and could be kept longer in a dark place, and truly permanent records could be obtained by photographing the test papers. Quantification of sweat output could then be done by actually counting the sweat droplets.

A variation of this technique was later described by Papa and Kligman (18), who basically repeated Wada's method of painting the castor oil-starch mixture over the iodine-alcohol-painted area to enable direct visualization of the sweat pattern. In order to obtain a permanent record, they simply took ordinary paper towels and smoothed them over the site. The towels were then gently peeled from the skin. They recommended coating the surface of the paper with a clear plastic aerosol spray to preserve the record. One of the main advantages of this method over Randall's was that the paper toweling was more flexible than the bond paper and thus could be molded onto areas of irregular contour such as the axillae.

A more modern version of these imprint techniques was described by Harris et al. (24) in which the starch-iodine method was used to visualize and trap the sweat droplets, but then, instead of using paper, a silicone rubber monomer was mixed with a catalyst and applied as a thin film over the area prior to the initiation of sweating. The material polymerized within 1 minute and the silicone imprint was then gently peeled from the skin. The sweat droplets appeared as small holes in the resulting impression, and these could easily be counted as in the earlier imprint methods. In addition to being more sensitive, this silicone technique afforded a truly permanent record which required no special storage. It also offered an improvement over earlier plastic impressions, which tended to cause localized irritation. The only difficulty noted in using the silicone impression technique was mastering the art of applying a thin uniform film of the silicone-catalyst mixture.

Along with the above methods, procedures were developed for directly visualizing sweat droplets by use of dyes such as bromphenol blue. Some investigators (25) used dye-impregnated paper, which, when placed on the skin, created a print of "blue dots" that could be processed as in the starch-iodine procedure. Others (26) injected the dye at the test site and simply viewed the droplets as they formed with the aid of a hand lens. The test areas were covered with castor oil to minimize evaporation and spreading of the sweat droplets. No attempt was made in this procedure to produce a permanent record for quantification.

A unique study was conducted by Brun (27) in the late 1950s. He used a Ping-Pong ball coated with bromphenol blue, sodium carbonate, starch, and tragacanth. He wrapped the ball with a piece of gauze and placed it into the axilla for 10 minutes. On removal, the gauze was unfolded and the size of the blue stain produced by the sweat was evaluated. He also described a Prussian blue test in which a strip of scotch tape was sprinkled with a finely ground mixture of $Fe_2(SO_3)_4$ and $K_3Fe(CN)_6$. The test paper, which appeared light gray when prepared, became colored in contact with water (sweat). He felt that this offered an advantage over the other methods in being able to detect fine droplets of sweat without painting the skin with iodine.

Daley (28) described a method, similar to the dye methods, in which the test area was coated with a mixture of sodium carbonate (finely ground in a ball mill) and phenolphthalein dispersed in an anhydrous base. As droplets of sweat appear on the skin, the sodium carbonate was dissolved and the resulting alkaline solution was colored dark red by the phenolphthalein indicator. The method was strictly used to visualize sweat patterns, and Daley advocated its use on the back, where a small area could be treated with an antiperspirant product. In this way, because of the large surface area of the back, a visual comparison could easily be made between the treated area and the untreated surrounding sites.

These colorimetric (visualization) methods have had many applications over the years, but are perhaps most beneficial even today in the study of emotional sweat patterns on the palms and feet. The preferred method for studying this type of sweating remains the use of a dye- or indicator-impregnated paper which allows good visualization of large quantities of sweat on these areas. A good presentation of this application for measuring "mental" sweat was described by Ferreira and Winter (29).

For a comprehensive review of the early colorimetric methods the reader is referred to Kuno's 1956 article on human perspiration (24).

B. Sensor (Instrumental) Methods

Although instrumental methods of measuring sweat output have never achieved popular use in routine screening of antiperspirants, they have nonetheless been widely used in mechanistic studies of antiperspirant activity as well as in the basic study of sweat physiology.

One of the first reported methods describing an "apparatus" for measuring water loss (30) used an infrared gas analyzer as a detector to determine the continuous change in vapor pressure of a known stream of air passed over the test subject. Later methods used a variety of devices: electrolytic cells containing a water-sensitive mixture of methanol-acetone and oxalic acid (31), electrical resistance types of humidity-sensing elements (32–35), and tared collecting coils which could be reweighed at the completion of the run and used weight gain as an indication of sweat output (36). A more recent method, described in (37), again uses water-sensitive electrodes but uses phosphorus pentoxide cells to minimize the problem of temperature control experienced in some of the other techniques.

Thermography, a more recent instrumental technique, was described by Davis and Rees-Jones in 1978 (38). The basis for this technique lies in recording variations in skin cooling caused by evaporation. The effect of antiperspirant action is to lessen this evaporative cooling process, thus resulting in a higher skin temperature. These authors also presented a comparative review of hygrometric, gravimetric, and thermographic methods for measuring sweat output. Although these methods have been used in a limited way to study antiperspirant efficacy (39), they remain, in general, a research tool rather than standard antiperspirant efficacy test methods.

C. Gravimetric Techniques

The use of gravimetric techniques arose out of a need and desire for a method more quantitative than the standard visualization method yet simpler than the complicated (and expensive) instrumental methods. Gravimetry quickly became and remains today the real workhorse of the antiperspirant efficacy testing field.

Before discussing specific gravimetric methods, it is important that we discuss some of the many factors which influence gravimetric testing. Understanding and consideration of these factors are critical when designing a study protocol.

One important consideration in this, as in any type of clinical study, is the selection of subjects. Above and beyond the basic requirements for any clinical test, i.e., reliability, good health, and ability to understand and carry out instructions, there are unique needs in the selection criteria for antiperspirant testing. Carabello (4) lists some of these requirements, such as abstinence from antiperspirant treatment for 2 weeks and a qualifying perspiration rate with consistent contralateral perspiration. The abstinence requirement is to ensure that no residual antiperspirant activity remains at the start of the study. The use of a 2-week hiatus as prescribed, however, should be carefully monitored in light of new and improved products which may extend the duration of effect and thus necessitate a longer washout period.

Reller (41) investigated several variables in test methods which he showed could affect the test results. He began with the obvious factors such as temperature and humidity and included more subtle factors such as air movement and mental or emotional stimuli. For example, he noted that sweating which occurs in the axilla at 70°F is mental or emotional sweating. As the temperature increases the sweat rate also increases, until at about 80°F the first thermal response is noted. This factor, coupled with the large effect that emotional stimulation alone has on axillary sweating, could have a tremendous effect on the resulting sweat output. Therefore, it becomes even more important to control emotional stimuli if only thermal sweat is desired. This is often difficult to accomplish—for example, when using a new panelist unfamiliar with the test procedure and perhaps somewhat embarrassed, or when new and unknown personnel enter the testing area. However, it is certainly more likely that unwanted factors can be eliminated or at least minimized if one is aware of their impact.

Reller discusses some other influencing factors which are also presented in the review article by Uttley (3). However, Uttley goes on to describe how these factors can influence sweat output and consequently efficacy data. These factors include such variables as metabolic rate, equilibration time, position of the body, sleeping, sex differences, skin area differences, and the influence of environmental conditions. Consideration of these factors should be an integral part of a protocol designed to measure antiperspirant effectiveness.

One of the earliest publications on the gravimetric method was that of Fredell and Read in 1951 (8). Their first experiments consisted of collecting perspiration from the axillae simultaneously on filter paper pads backed with heavy waxed paper while the subjects were seated in a hot room at 100°F wearing raincoats (why raincoats is not explained, but it conjures up an interesting image). They noted that sweating did not begin as quickly in the axilla as it did on the forehead, arm, or leg. Further, they found that the 2 × 1 in. pads were too small. After this first attempt, and a short regression back to the old starch-iodine techniques, they tried various types of pads and harnesses to hold the pads in the axillae. They finally settled on the use of an absorbent paper packing material folded to form a pad of $3\frac{1}{4} \times 4\frac{1}{2}$ in. backed by a layer of rubber dental dam of the same size. The pad and rubber dam were then wrapped in a single layer of gauze. The coded pads prepared in this way were placed in glass jars and the tare weight obtained. Postsweating weight gains of the pads were considered a direct measurement of the amount of sweat output. They found that no harness arrangement was required to keep the pads in place even when the subjects were allowed to continue their normal activities during the collection period.

Although this early pad technique has since been refined, it did establish many of the criteria on which later techniques were based. Fredell and Read determined that a minimum of 100 mg of sweat per axilla was required for a meaningful ratio to be calculated. (The only exception was in the case where the inhibition on the treated axilla neared 100%.) They noted that subject output varied from the required minimum of 100 mg to 2000 mg of perspiration per axilla in a 30-minute wearing of the pads under ambient conditions. They also established that the subjects' sweating ratios were the same under hot-room or ambient conditions, with or without exercise, and that the ratios were independent of the rate of sweating. Their findings regarding sweat ratios helped to form the foundation for the statistical methods used later to handle antiperspirant test data.

In 1958 Fredell and Longfellow (9) published the techniques of various laboratories in which the ratio method was used to evaluate antiperspirancy. In general, they noted that while elevated temperatures (i.e., hot-room conditions) increased the output of sweat, they did not alter the axillary ratio. Thus, according to their observations, using conditions which stimulate sweating would be useful only to facilitate (speed up) the collection of data.

A major contribution in the field of quantitative methodology was made in 1964 by Wooding et al. (42). Although their specific method of sweat collection, using containers of dehydrated silica gel in the axillae, was not generally adopted as a standard technique, their statistical evaluation method was and became an integral part of subsequent protocols.

One such method, published in 1974 by Majors and Wild (6), has with only minor modifications become a routine procedure employing the statistical methods of Wooding et al. (10, 42). The Majors and Wild technique was as follows:

1. Panelists were required to abstain from use of all antiperspirant materials for at least 1 week prior to initiation of the study.
2. Sweat collections were carried out in controlled-temperature rooms (100°F and 35% relative humidity).
3. Sweat collections were made during two successive 20-minute periods using tared Webril pads.
4. Sweat collections were preceded by a 40-minute equilibration period during which panelists held unweighed Webril pads in their axillae.

Majors and Wild expressed the percent reductions in sweating which were obtained by determining the shift in sweating ratios. Control sweating ratios were determined during the period prior to treatment, and posttreatment sweating ratios were adjusted to compensate for the deviation of individual mean control ratios from 1.000. The basic formula for this calculation was

$$(1.000 - \text{posttreatment ratio/mean control ratio}) \times 100 = \%\ \text{sweat reduction}$$

They reported no correlation between sweating rate and product efficacy (reductions in sweating) and presented data which indicated that the responses of panelists to effective antiperspirants followed an essentially normal distribution. They also noted that the percent reduction was comparable under hot-room and ambient conditions. Finally, as reported earlier, they found that the axillary sweating ratios were much more uniform than the sweating rates.

In 1975 Majors and Carabello (43) expanded and refined the previous method in their presentation to the OTC Panel for Antiperspirants and called it the "Hill Top research method," a name that continues to be used today. Their refinements included extending the washout period from 1 to 2 weeks, during which time the panelists are issued a marketed deodorant to use. They also required that the sweat collection pads be placed in and removed from the axillae by technicians as the panelists filed by the technicians at 15-second intervals. During the hot-room collections, panelists were now required to sit quietly with both feet flat on the floor and their arms close against their sides and to avoid any topic which might be considered controversial. The latter restriction was felt necessary to avoid any undue movement of the arms which might result in dropped pads.

Panelists for Hill Top's antiperspirant studies were cleared by a staff physician initially and on an annual basis. In addition, they had to meet the standard sweating criteria, i.e., production of at least 100 mg of sweat from an untreated axilla in a 20-minute collection under hot-room conditions and control sweating ratios with less than 25% coefficient of variation.

The method of sample application for various product forms was also described:

> Aerosol formulations were applied using a 2-second spray (timed by means of an audible metronome).

Roll-ons and sticks were applied directly to the axilla using individual applicators.

For each of these product forms the amount of sample used for each application was determined by weighing the containers before and after use.

Liquids, creams, and lotions were applied by swabbing measured amounts (0.5 g or ml) over the axillary surfaces by means of cotton swabs which had been saturated with the test solution to ensure transfer of the entire amount.

At this time most of Hill Top's studies were carried out with a balanced crossover design of two to six legs. Each leg followed the same procedure, and treatment groups were such that equal numbers of panelists received treatment on right and left axillae and the same number of panelists were treated with each sample in each test week. The design was also balanced so that application of each sample followed application of each other sample in the same number of panelists. An example of typical assignments is shown below for A, B, C, and D:

Panelist group	Axillae treated	Test weeks 1	2	3	4
I	R	A	B	C	D
II	R	D	C	B	A
III	R	B	D	A	C
IV	R	C	A	D	B
V	L	A	B	C	D
VI	L	D	C	B	A
VII	L	B	D	A	C
VIII	L	C	A	D	B

Typically, this type of test design used 12 panelists. Hill Top recommended that in a test involving only 12 panelists a standard reference be included as one of the test samples (i.e., 10% aqueous solution of aluminum chlorohydrate).

Majors and Carabello also presented an important finding based on data their laboratory had collected regarding subjective perception of sweat reduction (antiperspirancy). From their data they showed graphically that the ED_{50}* of antiperspirants required for noticeable effect resulted from 20% reduction in sweating when determined under hot-room conditions.

Other investigators in the mid-1970s evaluated antiperspirant products using the "normal activity" method. Jungermann (44) compared antiperspirant results between hot-room and ambient conditions. He concluded that the normal activity method was a more useful tool and gave more reproducible results because it more closely approximates what the consumer experiences in actual use. In addition, better correlations were obtained when test results were compared with observations reported in large-scale, carefully designed consumer tests. Steed (45) introduced the use of silica gel moisture-absorbing tins strapped to the axilla to collect sweat. He, too, used the normal activity method but made no comparisons with other collecting devices or environmentally controlled methods.

Later methods published have primarily been modifications of the existing gravimetric method designed to expedite the evaluation process (46,47).

D. Miscellaneous Methods

Specialized methods for antiperspirant testing have evolved primarily to meet the demands of unique test materials or to satisfy specific marketing claim. The form is illustrated by Kilmer et al. (48) in a classic paper on use of the forearm to test and compare the relative antiperspirancy of several topically applied antichlolinergics. The latter is addressed in a paper by Quatrale et al. (5) describing the methods used to study emotional sweating. Three techniques are described which were used to elicit emotional sweat: mental arithmetic, electric shock, and the word associaiton list. Of these, word association was the most successful. Further, their paper compared different methods of actual measurement of emotional sweat output. They found that although the water-sensing instrument was sensitive and accurate for measuring minute volumes of sweat output, it was impractical for obtaining data from subjects simultaneously. The Webril pad technique used in thermal studies was the preferred collection method.

---

*Effective dose perceived by 50% of the panelists.

E. OTC Guidelines

In August 1982 the OTC review panel recommended "Guidelines for Effectiveness Testing of OTC Antiperspirant Drug Products." These guidelines were based on the Food and Drug Administration's proposed antiperspirant test methods, also published in August of that year (1), and set forth the criteria needed to qualify an antiperspirant drug product, in finished product form, as effective.

1. Test subjects
   a. Test subjects must be sufficiently representative in amount of sweat produced as evidenced by a difference between the highest and lowest rates of sweating among the test subjects in excess of 600 mg in 20 minutes/axilla.
   b. Test subjects are required to abstain from use of all antiperspirant and deodorant materials (except as furnished by the investigator) for at least 17 days prior to the start of the study.
2. Test conditions. Either hotroom or ambient conditions be used to obtain gravimetric measurements of axillary sweat rate.
   a. Hot-room conditions are specified to be 100°F, plus or minus 2°, and a relative humidity of 35 to 40%, taking care to ensure that factors known to influence axillary sweating (i.e., air movement, position of the trunk, emotional stimuli, etc.) are controlled.
   b. Ambient conditions are simply stated as allowing test subjects to go about their normal daily routine during the collection period.
3 Test procedures
   a. Hot-room procedure:
      Treatment is to consist of application of the test formulation to one axilla and the control formulation to the other axilla of each test subject.
      Subjects are to be randomly divided so that half of them receive the test formulation under the left axilla and the control formulation under the right axilla, with the remaining subjects assigned oppositely.
      The quantity of formulation applied must reflect the amount a person would normally apply.
      Treatment is to be made once daily.

Subjects are then placed in a controlled environment for a 40-minute warm-up period.
Preweighed absorbent pads are placed in both axillae at the end of the warm-up period.
Subjects remain in the controlled environment for 20 minutes and the axillary pads are collected and reweighed at the end of the collection period. Sweat is collected during one of two successive 20-minute periods for evaluation.
If a pretreatment evaluation is to be made to determine right:left ratio, the control formulation will be applied to both axillae of each subject.

b. Ambient procedure: The procedure is the same as described above for the controlled environment except that the subjects go about their normal routines with a collection period of 3 to 5 hours.

4. Data treatment. If no pretreatment observation is made the unadjusted right:left ratio is defined as

$$\text{Unadjusted right-to-left ratio} = R/L$$

where R is the raw milligram weight measure of moisture from the right axilla and L the raw milligram weight of moisture from the left axilla.
To adjust for 20% reduction due to treatment, the measure of moisture for the control axilla is multiplied by 0.80. The hypothesis that reduction in perspiration exceeds 20% is tested statistically by comparing X (the adjusted value for a subject treated on the right axilla) to Y (the adjusted value for a subject treated on the left axilla) using the Wilcoxon rank sum test (also called the Mann-Whitney test). The test is of the null hypothesis

$$H_0\text{: median } X \text{ is less than median } Y$$

versus the alternative hypothesis

$$H_A\text{: median } X \text{ is less than median } Y$$

This is a test of H_0 against a one-sided alternative, H_A, at a predetermined level of significance, usually 0.05. Rejection of the null hypothesis at significance

level 0.05 will justify the conclusion that at least 50% of the target population will obtain a sweat reduction of at least 20% with an error in no more than 5% of the cases in which sweat reduction does not exceed 20%.
When pretreatment observations are recorded, the ratio of treated to control axilla adjusted for the ratio of right:left sweating rate is defined as

$$Z = \frac{PC \times T}{PT \times C}$$

where Z is the adjusted ratio, PC the treated measure of moisture for the control axilla, PT the pretreatment measure for the treated axilla, T the treated measure for the test axilla, and C the measure for the control axilla.
The study results are analyzed by comparing the adjusted ratio to 0.80, which corresponds to a 20% reduction in moisture due to treatment. The hypothesis that reduction in perspiration exceeds 20% is tested by subtracting 0.80 from A for all subjects and testing the resulting number with the Wilcoxon signed rank test. All other parameters, i.e., significance level and projected efficacy rate, apply.

The tentative final antiperspirant monograph, as proposed by the FDA, reflects much of the current status of antiperspirant testing methodology. Although the regulations have not as yet become law, industry has adopted their procedure as the method of choice for establishing that products meet the minimum efficacy requirements.

REFERENCES

1. Department of Health and Human Services, Food and Drug Administration. "Antiperspirant Drug Products for Over-the-Counter Human Use; Tentative Final Monograph," 21 CFR Part 350 (August 20, 1982).
2. Hurley, H. J., and Shelly, W. B. *The Human Apocrine Glands in Health and Disease*, Charles C. Thomas, Springfield, Illinois (1960).

3. Uttley, M. Measurement and control of perspiration, *J. Soc. Cosmet. Chem.*, *23:* 23 (1972).
4. Kuno, Y. Methods for the measurement of perspiration, *Human Perspiration*, Charles C Thomas, Springfield, Illinois (1956), p. 359.
5. Quatrale, R. P., Stoner, K. L., and Felger, C. B. A method for the study of emotional sweating, *J. Soc. Cosmet. Chem.*, *28:* 91 (1977).
6. Majors, P. A., and Wild, J. E. The evaluation of antiperspirant efficacy: Influence of certain variables, *J. Soc. Cosmet. Chem.*, *25:* 139 (1974).
7. Johnson, C., and Shuster, S. The patency of sweat ducts in normal looking skin, *Br. J. Dermatol.*, *83:* 367 (1970).
8. Fredell, W. G., and Read, R. R. Antiperspirant—axillary method of determining effectiveness, *Proc. Sci. Sect. Toilet Goods Assoc.*, *15:* 23 (1951).
9. Fredell, G. W., and Longfellow, J. Report on evaluating antiperspirant and deodorant products, *J. Soc. Cosmet. Chem.*, *9:* 108 (1958).
10. Wooding, W. M., and Finkelstein, P. A critical comparison of two procedures for antiperspirant evaluation, *J. Soc. Cosmet. Chem.*, *26:* 255 (1975).
11. Bakiewicz, T. A. A critical evaluation of the methods available for the measurement of antiperspirancy, *J. Soc. Cosmet. Chem.*, *24:* 245 (1973).
12. Minor, V. Ein Neues Verfahren zu der Klinischen Untersuchung der Schweissabsonderung, *Dtsch. Z. Nervenheilkd.*, *101:* 302 (1927).
13. Roth, G. M. A clinical test for sweating, *Proc. Staff Meet. Mayo Clin.*, *10:* 383 (1935).
14. Guttmann, L. Topographic studies of disturbances of sweat secretion after complete lesions of peripheral nerves, *J. Neurol. Psychiatry*, *3:* 197 (1940).
15. Silverman, J. J., and Powell, V. E. Studies on palmar sweating, *Am. J. Med. Sci.*, *208:* 297 (1944).
16. Wada, M., and Takagaki, T. *Tohoku J. Exp. Med.*, *49:* 284 (1947).
17. Tashiro, G., Wada, M., and Sakurai, M. Bromophenol blue method for visualizing sweat at the openings of the sweat ducts, *J. Invest. Dermatol.*, *36:* 3 (1961).
18. Papa, C., and Kligman, A. Modification of the Wada method for visualizing and recording eccrine sweating, *J. Invest. Dermatol.*, *36:* 167 (1961).

19. Papa, C. M. A new technique to observe and record sweating, *Arch. Dermatol., 88:* 732 (1963).
20. Juhlin, L., and Shelley, W. B. A stain for sweat pores, *Nature (London), 213:* 408 (1967).
21. Sarkany, I., and Gaylarde, P. A., A method for demonstration of sweat gland activity, *Br. J. Dermatol., 80:* 601 (1968).
22. Wada, M., Sudorific action of adrenalin on the human sweat glands and determination of their excitability, *Science, 111:* 376 (1950).
23. Randall, C., Quantification and regional distribution of sweat glands in man, *J. Clin. Invest., 25:* 761 (1946).
24. Harris, R., Polk, B., and Willis, I., Evaluating sweat gland activity with imprint techniques, *J. Invest. Dermatol., 58:* 78 (1972).
25. Herrman, F., Prose, P. H., and Sulzberger, M. B., Studies on sweating: V. Studies of quantity and distribution of thermogenic sweat delivery to the skin, *J. Invest. Dermatol., 18:* 71 (1952).
26. Hurley, H. J., and Witkowski, J., Dye clearance and eccrine sweat secretion in human skin, *J. Invest. Dermatol., 36:* 259 (1961).
27. Brun, R., Studies on perspiration, *J. Soc. Cosmet. Chem., 10:* 70 (1959).
28. Daley, E. W., Antiperspirant testing: A comparison of two methods, *Proc. Sci. Sect. Toilet Goods Assoc., 30:* 1 (1958).
29. Ferreira, A. J., and Winter, W. D., The palmar sweat print: A methodological study, *Psychosom. Med., 25:* 377 (1963).
30. Palmes, E. D., An apparatus and method for the continuous measurement of evaporative water loss from human subject, *Rev. Sci. Instrum., 19:* 711 (1948).
31. O'Malley, J., and Christian, E., The design of a continuous recording in vivo method of measuring sensible perspiration over a limited area, *J. Am. Pharm. Assoc., 49:* 398 (1960).
32. Custance, A. C., Cycling of sweat gland activity recorded by a new technique, *J. Appl. Physiol., 17:* 741 (1962).
33. Bullard, R. W., Continuous recording of sweating rate by resistance hygrometry, *J. Appl. Physiol., 17:* 735 (1962).
34. Gordon, B. I., and Maibach, H. I., Effect of systemically administered epinephrine on palmar sweating, *Arch. Dermatol., 92:* 192 (1965).

35. Adams, T., and Vaughan, J. A., Human eccrine sweat gland activity and palmar electrical skin resistance, *J. Appl. Physiol., 20:* 980 (1965).
36. Jenkins, J. W., Quellette, P. A., Healy, D. J., and Lana, C. D., A technique for perspiration measurement, *Proc. Sci. Sect. Toilet Goods Assoc., 42:* 12 (1964).
37. James, R. J., A new and realistic electronic approach to the evaluation of antiperspirant activity, *J. Soc. Cosmet. Chem., 17:* 749 (1966).
38. Davis, W. B., and Rees-Jones, A. M., Evaluating the performance of antiperspirants, *J. Soc. Cosmet. Chem., 29:* 413 (1978).
39. Zahejsky, J., and Rovensky, J., A comparison of the effectiveness of several external antiperspirants, *J. Soc. Cosmet. Chem., 23:* 775 (1972).
40. Carabello, F. B., Guidelines for the clinical study of antiperspirant and deodorant efficacy, *Cosmet. Toiletries, 95:* 33 (1980).
41. Reller, H. H., Factors affecting axillary sweating, *J. Soc. Cosmet. Chem., 15:* 99 (1964).
42. Wooding, W. M., Jass, H. E., and Ugelow, I., Statistical evaluation of quantitative antiperspirant data (I), *J. Soc. Cosmet. Chem., 15:* 579 (1964).
43. Majors, P. A., and Carabello, F. B., "Presentation to the OTC Panel for Antiperspirants of the Hill Top Research Method of Antiperspirant Evaluations and General Discussion of Results Obtained." Presented at an open meeting (August 14, 1975).
44. Jungermann, E., Antiperspirants: New trends in formulation and testing technology, *J. Soc. Cosmet. Chem., 25:* 621 (1974).
45. Steed, M. W., Evaluation of antiperspirant preparations under normal conditions of use, *J. Soc. Cosmet. Chem., 26:* 17 (1975).
46. Cullum, D. C., A rapid hotroom procedure for testing the performance of antiperspirants, *J. Soc. Cosmet. Chem., 29:* 399 (1978).
47. Hozle, E., and Kligman, A. M., Simplified procedure for evaluating antiperspirants: A method for rapid screening with subsequent assessment of axillary antiperspirant activity, *J. Soc. Cosmet Chem., 34:* 255 (1983).
48. Kilmer, F. S., MacMillan, Reller, H. H., and Synder, F. H., The antiperspirant action of topically applied anticholinergics, *J. Invest. Dermatol., 43:* 363 (1964).

8

Bacteriology of the Human Axilla: Relationship to Axillary Odor

James J. Leyden *University of Pennsylvania, Philadelphia, Pennsylvania*

I. INTRODUCTION

The bacterial flora of human skin is a relatively simple one, consisting of aerobic cocci of the Micrococcaceae family, aerobic diphtheroids, primarily *Corynebacterium* species, anaerobic diphtheroids of the genus *Propionibacterium*, yeast from the genus *Pityrosporum*, and occasional Gram-negative species. Significant variations in both the total number of bacteria and the composition of the bacterial flora exist for different body regions. These variations reflect differences in the amount of water and nutrients available to support bacterial growth (1). The human axilla provides ideal ecological factors for bacterial growth. The rich supply of eccrine sweat glands, which provide water, electrolytes, and minerals, apocrine sweat glands, which secrete a substance rich in protein and lipid, and sebaceous glands, which produce a mixture of lipids, contributes to the ecology of the axilla. The outermost region of the skin, the stratum corneum, is a compartment which is constantly being shed and provides a rich source of amino acids necessary for bacterial growth. Sweat glands deliver water to the surface and provide the critical moisture required for bacterial proliferation. In addition, sweat contains amino acids and minerals such as copper, iron, magnesium, zinc, and calcium which are important for bacterial growth and metabolism as well as toxin production by pathogenic bacteria. The presence of large amounts of water and the rich supply of proteins, lipids, and minerals coupled with the anatomy of the axilla, which produces a semioccluded environment that minimizes evaporation of water, results in an ideal ecosystem for bacterial growth.

Axillary bacteria have been shown to be responsible for generating undesirable odors as a result of their interaction with odorless apocrine sweat (2–4). In more recent studies, quantitative and qualitative changes have been correlated with differences in the quality of axillary odor.

II. AXILLARY MICROFLORA

The microbiology of the axilla has long interested skin microbiologists. However, systematic quantitative studies are of recent origin. The axilla supports one of the highest densities of bacteria on human skin and, because of the semioccluded anatomy, is relatively less prone to environmental contamination. Organisms recovered from the axilla are therefore representative of the indigenous flora of skin.

The majority of published studies to date have employed a nonquantitative methodology in which a sterile cotton swab moistened with buffer has been used to sample the axilla. More recent studies have employed Triton X-100 or other detergents to facilitate removal of organisms from the skin surface and to disperse macrocolonies and permit quantification.

In a multifacted classical study in which cultures were made from a small number of subjects ($n = 20$) and apocrine sweat was collected and inoculated with the various organisms isolated, Shelly et al. (2) regularly isolated *Staphylococcus aureus*, *S. epidermidis*, coryneforms, *M. luteus*, and Gram-negative bacteria. Strauss and Kligman (3) found coagulase-negative cocci and coryneforms to be the dominant organisms recovered in another small group of 29 subjects. In a study designed to determine the effects of topical antibacterial agents on the flora of the axilla and the production of odor, Shehadeh and Kligman (4) also found aerobic coagulase-negative cocci and coryneform bacteria to be the most abundant organisms. They reported that the number of bacteria recovered from the axilla was of the order of 10^6 organisms/cm^2.

Marples and Williamson (5) found that coagulase-negative cocci and diphtheroids, which they classified as lipophilic, made up 85% of the flora. Their purpose was to study the interaction of bacteria during the ecological pressure of systemic antibiotics rather than to survey the flora of the axilla and its relationship to odor. However, they did describe two distinct types of axillary flora, namely a flora dominated by lipophilic diphtheroids and a flora dominated by coagulase-negative cocci (5). In experimental manipulation of the cutaneous flora, excessive hydration favors proliferation of coryneform bacteria, and for this reason we have speculated that a major factor determining the composition of the axillary flora may be a difference in eccrine sweating (6).

More recent surveys have supported these earlier findings. Kloos and Musselwhite (7) found 70% of the axillary flora to be coryneforms and concluded that these organisms were predominantly *Corynebacterium* species. Aly and Maibach (8) found the axillary flora to be dominated by coryneforms but concluded that the majority (78%) were nonlipophilic in nature.

In a comprehensive study involving 163 males and 122 females, Jackman (9) also found that the axillary flora was dominated by aerobic cocci or by coryneforms. In his series, 64% of the males had an axillary flora dominated by coryneforms and 27% had a cocal flora. This ratio was reversed in the females. The predominant genus found in the axillary coryneforms was *Corynebacterium* (83%), with 5% *Brevibacterium* and 12% other coryneforms (9). In those with a coryneform-dominated flora a mean of 10^6coryneforms/cm^3 was recovered, compared to a fivefold lower density in those with a coccal-dominated flora.

We studied 205 subjects, employing quantitative techniques, and found the axillary flora to consist of a stable population of aerobic and anaerobic organisms with the total number ranging from 500,000 to 1,000,000 organisms/cm^2 (10). The day-to-day variation was minimal, with a coefficient of variation of 13%. Axillary hairs were found to have extremely low numbers of bacteria and were not viewed as a significant aspect of axillary bacteria. Comparison of the axillary flora in terms of right- and left-handness did not reveal any significant differences, nor did a comparison of the right versus the left axillary flora within subjects. Males and females had similar numbers of bacteria; significant compositional differences (described below) were found. Our finding that the axillary flora was a stable mixture of organisms is in agreement with previous work (11).

Table 1 Cell Wall Characteristics of Axillary Corynebacteria

Meso-diamino-pimelic acid	Major fatty acids
Arabinose, mannose, galactose	C16:1, 3%
Mycolic acids, 90% C_{30} to C_{36}	C16, 17%
	C18:1, 61%

All subjects had Micrococcaceae recovered from the axillae. *Staphylococcus epidermidis* was the most frequently recovered species (51%), *Staphylococcus saprophyticus* was recovered from 29%, while *S. aureus* was found in only 10% (10). In contrast to these findings, Prince and Rodgers (11) found a 38% prevalence of *S. aureus* in the winter months and 73% in the summer, and Shelly et al. (2) frequently recovered *S. aureus*. On the other hand, Aly and Maibach (8) found a 12% prevalence of *S. aureus*. These differences may be explained by population and/or climate differences.

Aerobic diphtheroids or coryneforms were frequently recovered. Until recently, these organisms have been poorly classified. Recent work in analyses of cell wall constituents has helped to strengthen their classification (12–14). It is now clear that there are aerobic diphtheroids belonging to two genera—*Corynebacterium* and *Brevibacterium*. The cell wall constituents of *Corynebacterium* are shown in Table 1. The presence of meso-diamino-pimelic acid, arabinose, mannose, galactose, and long-chain mycolic acids; a dominance of C18:1 cell wall fatty acids; and a strict nutritional requirement for lipid (lipophilia) characterize members of the *Corynebacterium* genus. The aerobic diphtheroids that do not require lipid (large colony diphtheroids) have not been as well characterized, but those recovered from the axilla also appear to belong to *Corynebacterium*. Significant differences in the carriage rate of lipophilic diphtheroids (*Corynebacterium lipophilicus* is the proposed name for this species) were found in males and females (Table 2). Males had a carriage rate of 85%, compared to 66% for females (10). However, the mean number of these organisms, when present, was the same for males and females. The non-lipid-requiring diphtheroids (large colony diphtheroids) were found in 26 and 25% of males and females and the mean number of these bacteria when present

Table 2 Prevalence and Density of Axillary Bacteria

	Males (N = 128)		Females (N = 77)	
	Prevalence (%)	Density[a]	Prevalence (%)	Density[a]
Total aerobic count	100	5.84 ± 0.06	100	5.95 ± 0.09
Micrococcaceae	100	5.51 ± 0.07	100	5.56 ± 0.11
Lipophilic diphtheroids	85	5.40 ± 0.09	66	5.36 ± 0.14
Large colony diphtheroids	26	4.43 ± 0.21	25	4.15 ± 0.24
Gram-negative bacteria	20	3.36 ± 0.25	19	3.32 ± 0.31
Propionibacteria				
P. acnes	47	3.86 ± 0.29	30	4.26 ± 0.38
P. avidum	34	3.62 ± 0.22	21	4.18 ± 0.43
P. granulosum	8	3.61 ± 0.32	5	3.65 ± 0.28

[a]Logarithm means per square centimeter.

was the same for both sexes. Prince and Rodgers (11) found diphtheroids (not classified further) in all subjects. Aly and Maibach (8) found higher numbers of diphtheroids, i.e., 10^6 to $10^7/cm^2$.

Gram-negative bacteria were found in approximately 20% of subjects with *Escherichia coli*, the most frequently isolated gram-negative organism; *Klebsiella*, *Enterobacter*, and *Proteus* species were also found.

Propionibacterium were found in 70% of males and 47% of females, with *P. acnes* and *P. avidum* the most frequently isolated species. The sex difference in carriage of *Propionibacterium* is most likely related to the correlation of *Propionibacterium* levels with sebum production, which is greater in males (15).

III. BACTERIAL FLORA OF AXILLA AND AXILLARY ODOR

Axillary odor is a mixture of many "notes," with the dominating notes identified as isovaleric acid and 5-androst-16-en-3-one and 5-androst-16-en-3-ol. The alcohol has a musky odor that is not unpleasant, while the ketone confers an extremely disagreeable odor that is often described as stale urine. Isovaleric acid has a "sweat" odor.

In studies comparing the axillary flora of those with a pungent disagreeable axillary odor to those with a "sweaty" odor, significant quantitative and qualitative differences were found. Those with a pungent unpleasant odor had a significantly higher number of bacteria, $10^6/cm^2$ compared to $10^5/cm^2$(10). More striking, however, were the marked differences in the composition of the flora. Every subject with a pungent axillary odor had lipophilic diphtheroids present in high numbers. The geometric mean count was 810,000 compared to a 55% prevalence with a mean of 53,000 in those with a sweaty odor. These results suggested that lipophilic diphtheroids (*C. lipophilicus*) were responsible for generating the offensive Δ^{16}-androgen steroid notes (10). In further studies, sterile odorless apocrine sweat was placed on the forearms of volunteers and subsequently incubated with aerobic cocci, coryneform bacteria, and Gram-negative organisms. Typical pungent axillary odors were produced only when lipophilic and large colony Corynebacteria were incubated with apocrine sweat (10). A sweaty, isovaleric acid type of odor was produced by micrococci and gram-negative bacteria.

Table 3 Apocrine and Axillary Lipid Profiles

	Glandular secretion		
	Apocrine	Sebaceous	Skin surface
Cholesterol	76.2%	3.4%	8.9%
Cholesterol esters	0.9%	—	8.8%
Wax esters	3.6%	21.8%	21.2%
Squalene	0.2%	19.0%	13.4%
Glycerides and fatty acids	19.2%	55.9%	47.4%
Total lipids	20 μg/μl	—	60 $\mu g/\mu m^2$
Protein	90 μg/μl	—	

IV. BIOCHEMICAL ASPECTS OF AXILLARY ODOR

Investigations utilizing radioimmunoassay techniques have demonstrated differences in the concentration of androstenone in male and female subjects (16). These results correlate well with differences in odor and the bacteriological differences described above. Apocrine secretion is odorless but does contain significant amounts of cholesterol and two sulfated steroids, i.e., androsterone sulfate and dehydroepiandrosterone sulfate (17) (Table 3). The cholesterol, other steroids, and proteinaceous substances present in apocrine secretion provide an ecosystem for bacterial growth and odor development. Other substances in the axilla originating from sebaceous and eccrine glands may also contribute to the total odor profile of the axilla. Sebum intermingles with apocrine secretion in the infundibulum of the follicle and contains approximately 10% squalene, a material that is used as a fixative in fragrances and may be important in prolonging axillary odor.

Investigation of the role of bacteria in generating axillary odor was poineered by Shelly et al. (2). Their studies demonstrated that apocrine sweat was sterile and odorless and that odor was produced by axillary microorganisms acting in apocrine sweat. Sterile eccrine sweat produced no odor when incubated with bacteria. Numerous studies now demonstrate that a decrease in axillary odor correlates with a decrease in axillary coryneforms (18–20).

In vitro incubation of sterile apocrine sweat and micrococci produces a sweaty odor, which has been identified with isovaleric acid by gas chromatography-mass spectroscopy (21). Incubation of apocrine sweat with coryneforms produces isovaleric acid and androstenone. The latter, however, while recognized by the human nose, appears to be at levels below the limit of detection of gas chromatography-mass spectroscopy.

The importance of bacteria in generating axillary odor has been further demonstrated in several studies in which topical antibodies and antimicrobial agents clearly reduce axillary odor. Several studies suggest that the use of these agents results predominantly in a reduction in coryneform bacteria (2,4,18,19). These findings correlate well with the conclusion that coryneform bacteria play the dominant role in generating axillary odors.

In conclusion, the presence of androgen steroids in apocrine sweat, the production of androgen steroids by the action of axillary lipophilic coryneforms of *Corynebacterium* species, and a strong positive correlation between axillary odor and the presence of an axillary flora dominated by coryneform bacteria have been firmly established. The biochemical pathways by which coryneform bacteria produce these malodorous materials have yet to be deciphered.

REFERENCES

1. Leyden, J. J., McGinley, K. J., and Webster, G. F., Cutaneous microflora *Biochemistry and Physiology of the Skin* (Goldsmith, ed.), Oxford University Press, New York (1983).
2. Shelly, W. B., Hurly, H. J., and Nicholas, A. C. Axillary Odor: Experimental study of the role of bacteria, apocrine, sweat and deodorants, *Arch. Dermatol. Suppl.*, *68:* 430 (1953).
3. Strauss, J. S., and Kligman, A. M. The bacteria responsible for axillary odor, *J. Invest. Dermatol.*, *27:* 67 (1956).

4. Shehadeh, N. H., and Kligman, A. M. The effect of topical antibacterial agents on the bacterial flora of the axilla, *J. Invest. Dermatol., 40:* 61 (1963).
5. Marples, R. R., and Williamson, P. Effects of demthylchlorotetracycline on human cutaneous microflora, *Appl. Microbiol., 18:* 228 (1969).
6. Kligman, A. M., Leyden, J. J., and McGinley, K. J. Bacteriology, *J. Invest. Dermatol., 67:* 160 (1976).
7. Kloos, W. E., and Musselwhite, M. S. Distribution and persistence of *Staphylococcus* and *Micrococcus* species and other aerobic bacteria in human skin, *Appl. Microbiol., 30:* 301 (1975).
8. Aly, R., and Maibach, H. Aerobic microbial flora of intertriginous skin, *Appl. Environ. Microbiol., 33:* 97 (1977).
9. Jackman, P. J. H. Taxonomy of aerobic axillary coryneforms based on electrophoretic protein patterns, Ph.D. Thesis, University of London (1981).
10. Leyden, J. J. The microbiology of the human axilla and its relationship to axillary odor, *J. Invest. Dermatol., 77:* 413 (1981).
11. Prince, H., and Rogers, J. A. Studies on the aerobic axillary microflora employing a standardized swabbing technique, *Cosmet. Perf.*, (October 1974).
12. Pitcher, D. G. Rapid identification of cell wall components as a guide to the classification of aerobic coryneform bacteria from human skin, *J. Med. Microbiol., 10:* 439 (1977).
13. Sharpe, M. E., Law, B. A., Phillips, B. A., and Pitcher, D. G. Methanethiol production by coryneform bacterial strains from dairy and human skin sources and *Brevibacterium* linens, *J. Gen. Microbiol, 55:* 433 (1977).
14. McGinley, K. J., Norstrom, K. M., Webster, G. F. and Leyden, J. J. Analysis of cellular components, biochemical reactions and habitat of human cutaneous lipophilic diphtheroids, *J. Infet. Dis.*, in press.
15. McGinley, K. J., Webster, G. F., and Leyden, J. J. Regional variations in the density of cutaneous propionibacteria. Correlation of *Propionibacterium acnes* population with sebaceous secretions, *J. Clin. Microbiol., 12(5):* 672 (1980).
16. Bird, S., and Gordon, D. B. The validation and use of a radioimmunoassay for 5-androst-16-en-3-one in human axillary collections, *J. Steroid Biochem., 14:* 312 (1981).

17. LaBows, J. N., Preti, G., Leyden, J. J., and Kligman, A. M. Steroid analysis of human apocrine secretion, *Steroids, 3:* 249 (1979).
18. Bibel, D. J. Ecological effects of a deodorant and a plain soap upon human skin bacteria, *J. Hyg. Comb., 79:* 1 (1977).
19. Evans, N. M. The classification of aerobic diphtheroids from human skin, *Br. J. Dermatol., 80:* 81 (1968).
20. McBride, M. E., Duncan, W. C., and Knox, J. M. The environment of the microbial ecology of human skin, *Appl. Environ. Microbiol., 33:* 603 (1977).
21. LaBows, J. N. Human odors, *Perf. Flavorist, 4:* 12 (1979).

9

Odor Detection, Generation, and Etiology in the Axilla

John N. Labows, Jr. *Colgate-Palmolive Company, Piscataway, New Jersey*

*An ugly rumour harms Your reputation. Underneath your arms They say you keep a fierce goat which alarms All comers—and no wonder, for the least Beauty would never bed with that rank beast. So either kill the pest that makes the stink Or else stop wondering why the women shrink.**

I. INTRODUCTION

Axillary odor has typically been characterized by subjective olfactory descriptions. One of the earliest literature references is a poem of Catallus dated 50 BC which describes the odor as goatlike and correctly points to the involvement of bacterial pests (1). Additional descriptions include the odors of ammoniated valerian, chlorinated urine, lamb, coumarin, burnt coffee beans, and heliotrope (2). These descriptions suggest a multifaceted nature of the odor, which can be related to the "primary odors" of sweaty, urinous, and musky (3). Other authors have commented on the sexual effects of axillary odors and why the axilla may be the best source of such attractive chemicals (4). This chapter will detail the research efforts which have

---

*From *The Poems of Catullus*, translated by James Michie. Copyright© 1969 by James Michie. Reprinted by permission of Random House, Inc., and Grafton Books, A Division of the Collins Publishing Group.

led to the identification of the chemicals responsible for axillary odor and to suggestions for their metabolic origins. It will also examine their reported psychological and physiological effects.

II. COMPOSITION OF APOCRINE SECRETION

In the axilla, secretions from all three skin glands, the apocrine, sebaceous, and eccrine glands, are necessary for odor production. The eccrine gland opens directly onto the skin surface, while the secretions from both the sebaceous and apocrine glands reach the surface through the hair follicle. The eccrine glands provide moisture for bacterial growth, and lipids, particularly long-chain fatty acids from the sebaceous glands, promote the growth of the coryneform bacteria (5). However, the key contributor to odor development is the substrate provided by the apocrine gland.

To understand the metabolic development of axillary odor it is necessary to describe the chemical nature of the substrate provided by the apocrine gland. Apocrine secretion can be obtained by stimulation of the gland with epinephrine (6,7). A milky liquid ($\sim 1 \mu l$) is obtained at the follicular opening on the skin surface. Thin-layer chromatographic (TLC) and gas chromatographic-mass spectrometric (GC-MS) analyses showed the presence of lipids, cholesterol, and C_{19} androgen steroids.* The lipid profile resembles the sebaceous lipids and, in fact, the major lipids come from sebum contamination because of the collection methods. However, the amount of cholesterol observed is clearly more than expected from sebaceous or epidermal lipid and is a feature of the apocrine secretion (7). One of the axillary bacteria, *Staphylococcus epidermidis*, is capable of efficiently hydrolyzing triglycerides at the pH of the axilla (6 to 8.5) and of esterifying cholesterol with the resultant fatty acids to form cholesterol esters (8). Thin-layer chromatographic studies of axillary surface lipids show a large increase in cholesterol and cholesterol esters over material obtained from other sebaceous-rich sites or from sebum itself (Table 1) (9).

*The C_{19} androgens are steroids having 19 carbons and possessing androgenic activity, e.g., DHA and androsterone. Androst-16-enes are C_{19} steroids with a double bond at carbon 16 (see Fig. 1).

Table 1 Comparison of Lipid and Cholesterol Profiles from Apocrine Secretion, Axillary Sweat, and Sebum

	Percent composition			
	Glandular secretion		Extract skin surface	
	Apocrine[a]	Sebaceous[b]	Axillae	Facial
Cholesterol	76.2%	3.4%	8.9%	1.5%
Cholesterol esters	0.9%[c]	21.8%	8.8%	3.0%
Wax esters	3.6%[c]		21.2%	26.0%
Squalene	0.2%[c]	19.0%	13.4%	12.0%
Glycerides and fatty acids	19.2%[c]	55.9%	47.4%	57.5%
Total lipid	20 μg/μl	—	60 μg/cm	100 μg/cm

[a]Stimulated and collected at the skin surface.
[b]Collected by microdissection of gland (8).
[c]Probably of sebaceous origin.
Source: From Ref. 9.

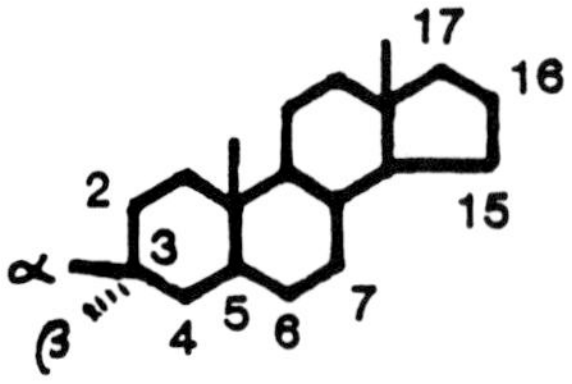

Figure 1 Carbon designations of androstane nucleus.

Histochemical studies also indicate cholesterol and cholesterol esters in the secretory cells of the apocrine gland (10).

Protein analysis showed approximately 10% protein in the secretion. Enzyme activity of apocrine secretion itself has not been studied; however, β-glucuronidase, 3-hydroxysteroid dehydrogenase, 5α-reductase, and esterase activities as well as lysozyme have been reported in excised apocrine glands (11,12).

The presence of steroids in the secretion is particularly intriguing. Previously, C_{19} androgens had been found by TLC analyses of extracts of axillary sweat (Table 2) (13). Ingestion of [^{14}C]progesterone or [^{3}H]pregnenolone led to the preferential secretion of radioactive steroids in the axillary regions, compared with general body sweat, suggesting that the apocrine gland was the source of these steroids (14). When apocrine secretion is injected directly into a gas chromatograph, androsterone, dehydroepiandrosterone (DHA), and the thermal breakdown products of their conjugates are observed (7). The pattern is identical to that found on injection of the corresponding sulfate derivatives. Enzymatic (for DHA) or acid hydrolysis of apocrine secretion followed by GC-MS confirms the presence of these two androgen sulfates. Since DHA is not an efficient precursor of the odorous androst-16-enes described below, one might expect to find other steroid conjugates in the secretion (15,16) (Figure 1).

III. METABOLISM BY AXILLARY BACTERIA

The original work with apocrine secretion showed that axillary bacteria produced unique odors when incubated with the secretion (6,17). A more recent extension of that work confirmed that specific axillary bacteria, lipophilic diphtheroids and micrococci, gave unique odors when incubated with this secretion

Table 2 Steroids Identified in Axilla

Steroids	Analytic method	Reference
Apocrine secretion:		
Cholesterol[a]	GC-MS	7
Androsterone (sulfate)[a]		
Dehydroepiandrosterone (sulfate)[a]		
Axillary sweat		
Androst-4-ene-3,17-dione	TLC, RT[b]	13
DHA-(sulfate)		13,70
Pregn-5-ene-3β-ol-20-one		14
Androst-16-en-3α-ol	GC-MS	b,29
Androst-16-en-3-one	RIA, GC-MS	c,15,35,36
Androsta-5,16-dien-3α-ol	GC-MS	c
Androsta-4,16-dien-3-one	GC-MS	c

[a]Also found in axillary sweat by TLC and/or GC-MS.
[b]RT, Radioactive tracer studies; other abbreviations as in text.
[c]Reported here.

either in solution or applied to the forearm (9). For example, the micrococcus *S. epidermidis*, when incubated on apocrine secretion, produces a sweaty odor, while the axillary diphtheroids give an odor judged as pungent/urinous. Thus, the role of bacteria and apocrine secretion can be demonstrated in experiments in vitro confirming the belief in and commercial exploitation of the bacterial involvement in odor production.

The distribution of the normal bacterial flora is detailed in Chapter 8 and consists of species of lipophilic and large colony diphtheroids and micrococci (18). Recent studies have characterized the axillary diphtheroids as lipid-dependent and differing in cell wall composition from the typed corynebacteria (5,19). Incubation of either bacteria isolated from the axilla with apocrine secretion gave isovaleric acid as the major compound identified by headspace analysis (9). The odor of the micrococcal cultures corresponded to this sweaty odor. The diphtheroid cultures had a more pungent odor, resembling that of the androst-16-enes, though these have not yet been identified in these cultures. However, the concentration level may be such that it is detectable olfactorily, having a low odor threshold, and not instrumentally (20). The relative amounts of these two bacterial genera differ among individuals, particularly by sex, with the diphtheroids being more prevalent in males (compare the quantitation of androstenone levels below). The in vitro studies are in accord with the observed in vivo odor (sweaty versus pungent), which can be correlated with the types of bacteria present in the axilla (18,19).

Since bacteria are involved in the production of axillary odor, it is important to know whether the axillary bacteria can transform nonodorous steroid substrates into the odorous androstenes. In studies at the Monell Chemical Senses Center, axillary isolates of the lipophilic diphtheroids were incubated in trypticase soy broth with the appropriate steroids and the culture extracts analyzed by GC-MS (21). The reactions observed were specific in terms of the geometry of the 3 position of the steroid nucleus. These included the hydrolysis and oxidation of 3β-sulfates but not 3α-sulfates and the oxidation of 3α-hydroxyls but not 3β-hydroxyls. Thus the sulfates of pregnenolone, DHA, and androstadienol, all with β-geometry, are converted to their corresponding 3-keto derivatives. Under the same conditons androstenol, but not its sulfate, is oxidized to androstenone.

The metabolism of pregnenolone, testosterone, and 5α-dihydrotestosterone (DHT) by axillary coryneform bacteria has also been examined (22). The first two steroids led to the formation of DHT, while testosterone also gave 17β-hydroxy-5α-androst-3-one. Evidence is suggestive but not conclusive that DHT gave androstenol in these 3-week cultures. Further experiments with isolated corynebacteria and with mixed axillary bacteria confirmed the metabolites of testosterone reported earlier but could not detect any metabolites with C-16,17 unsaturation (23,24). The ratio of metabolites changes according to the particular mixture of bacteria used in the cultures. These studies have been reviewed (24). The above series of experiments suggests that an androst-16-ene derivative may be secreted in the apocrine secretion and converted to the observed odorous steroids. Experiments with ovarian or testicular tissue indicated the presence of a dehydratase enzyme system which converts testosterone into androstadienone and androstenols (<1% yield) (25).

Measurement of androstenone levels in the axilla was used to directly demonstrate the effects of an antibacterial on axillary odor. Povidone-iodine was applied to the axilla that had previously been shown to have the higher level of the steroid by radioimmunoassay (RIA). After 3 days of treatment the level of androstenone decreased below that of the untreated axilla, confirming a bacterial pathway for formation of the androst-16-enes (26).

IV. ANALYSIS OF AXILLARY ODOR

Until recently, little work had been reported on the chemical analysis of axillary odor. Though bacteria and apocrine secretion were clearly involved, no attempts were made to characterize the metabolites from this interaction either in vivo or in vitro. Substances which interfered with bacterial growth, such as antibacterials and antiperspirants, or which involved odor masking, such as deodorants, were deemed sufficient to control odor. State-of-the-art analytical methods have made possible the detection and identification of trace levels (picograms to nanograms) of chemical compounds. Of particular importance has been the ability of GC-MS to analyze complex mixtures from biological sources (27,28). In 1974 the presence of androstenol in axillary sweat was observed in extracts from the combined axillary pads worn by 12 individuals for a period of 7 days (29). With improvements in column technology (bonded fused silica capillary columns) and mass spectral sensitivity, the same compound can be detected on an axillary pad worn overnight.

Early studies on human odor attempted to sweep the entire body or the axilla with an inert gas and collect the volatiles on a polymeric adsorbent (27,30,31). Olfactory evaluation of the eluting gas chromatographic peaks showed some peaks which were malodorous; however, the compounds were not identified. Subsequent GC-MS studies for direct analysis of "odors" in the axilla, either by drawing air over the axilla or by sweeping an axillary pad and collecting the volatiles on polymer traps, led to the identification of compounds from exogenous sources such as alkylbenzenes, limonene, and isopropyl esters and no "natural odors" (32). In retrospect, these approaches are unsuitable for the analysis of compounds of low volatility such as the short-chain aliphatic acids and androst-16-enes.

The androst-16-enes are mammalian pheromones and are directly involved in the mating behavior of the pig (16). Androstenone is also present in boar tissue and is responsible for the boar taint odor. Restrictions on the allowed levels of this compound in pork have been responsible for the development of sensitive methods for detecting the androst-16-enes, including RIA and isotope dilution GC-MS (33,34). Radioimmunoassay was subsequently used to quantitate the amount of androstenone in extracts of axillary pads. The level in males was 3 to 310 ng and in females was 3.5 to 11 ng (35). A second study reported production levels of 14 ng per armpit per hour (36). The RIA approach is very sensitive but requires a separation step to remove the lipids and a TLC procedure to separate the steroids since the antibody cross-reacts with other androst-16-enes, particularly androstadienone.

Gas chromatography-mass spectrometry has the sensitivity and specifity necessary for the separation and analysis of axillary pads for all the androst-16-enes. In 1972 the presence of androstenone in one subject was reported using GC-MS (15). In our studies, pads are extracted with methanol/acetone and, following concentration, the extract is redissolved in toluene and base-extracted. This step serves to remove the fatty acids and the steroid sulfates, which when injected decompose to androstene-17-ones, which elute in the same area of the chromatogram as the androst-16-enes. Figure 2 shows the computer-reconstructed ion chromatogram of a pad extract as well as the mass chromatogram of the ions corresponding to the molecular ion (intact ionized parent compound) of some of the steroids of interest. The mass chromatogram is a display of the intensity of a specific mass value versus time in scan numbers (1 scan = 1 second) and makes the mass spectrometer a specific and highly sensitive detector in this case for the androst-16-enes. The chromatogram was searched for these ions at the known retention times determined from standards. Androstane is used as a standard for quantitation and comparison of retention times. All four steroids, androstenone (molecular ion = 272), androstenol (274), androstadienol (272), and androstadienone (270), can be detected on pads worn overnight by a single individual using this approach. The presence of each steroid can be confirmed by examining its mass spectrum and comparing it with a standard spectrum from the computerized library compilation of the National Bureau of Standards or from an authentic sample. Under the GC conditions used (25 m × 0.25 mm CP Sil8 or DB-5 bonded fused silica capillary column), androstenol and androstadienol could not be separated by more than one scan (1 second) but are separated from β-androstenol) six scans). All four steroids were detected in only 2 of 10 subjects, and androstenol

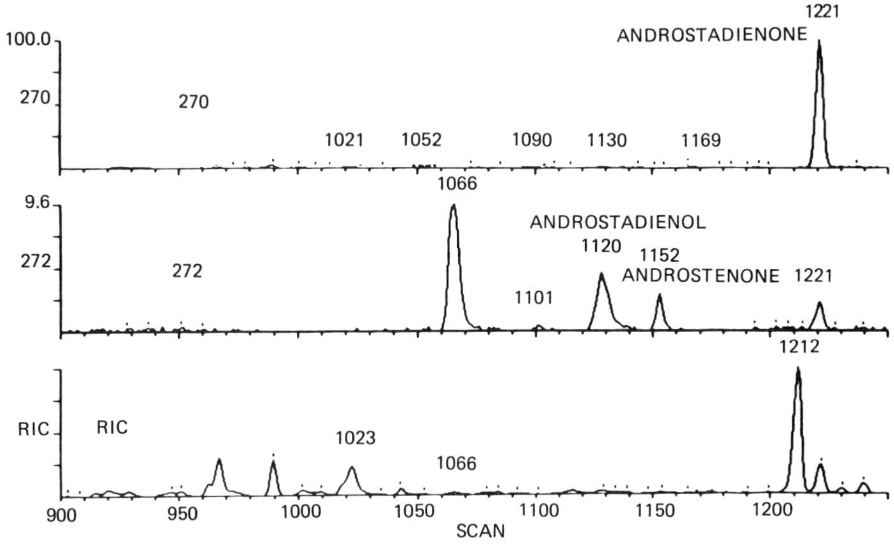

Figure 2 Reconstructed mass chromatogram of an axillary pad extract with monitoring for androst-16-enes.

was predominant (8 of 10 subjects). Levels of these steroids ranged from 10 to 150 ng. In all samples, DHA and androsterone were found free and as the sulfates (from base extract) and testosterone was found in one subject. In addition, studies at Monell have shown a significant change in levels of androstenol across the menstrual cycle (37).

A similar GC-MS approach was developed to separate and identify all the metabolites of pregnenolone in porcine testicular incubation extracts, including derivatives of the androst-16-enes (38). Androstadienol was separated from androstenol but not from β-androstenol when the steroids were run as their silyl derivatives.

An alternative method for searching the chromatograms is to use a reverse library search algorithm. Figure 3 shows a search of every scan in the chromatogram for the presence of androstadienone using the Incos software of the Finnegan GC-MS data system. A peak is shown where two successive scans contain the major ions found in the library spectrum. This corresponds to the correct retention time of this steroid and selects it from a complex background of peaks, as shown in the lower trace of

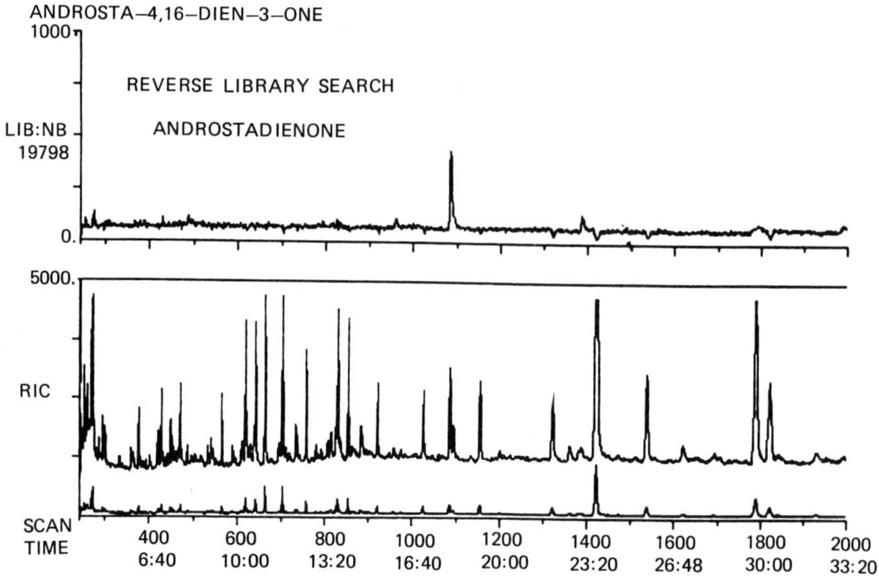

Figure 3 Reverse library search of reconstructed ion chromatogram for androsta-4,16-dien-3-one.

Fig. 3. Gas chromatographic analysis alone is insufficient to detect and quantitate the steroids at the levels present against this background.

V. ORIGIN OF AXILLARY ODOR—SWEATY

As indicated above, androst-16-enes clearly contribute to axillary odor. However, these odors have also been characterized as sweaty or hircine (goatlike). This implies the presence of either aliphatic acids such as isovaleric acid (the sweaty primary odor) or branched medium-chain acids. Isovaleric acid, as indicated above, can be produced by the axillary bacteria on apocrine secretion; however, the exact precursor is unknown. Characteristic of the hircine odor is 4-ethyloctanoic acid, which is responsible for the "soo-odor" of mutton and is the main odor constituent of the sebaceous gland secretion of mature male goats during the breeding season. Mature does in estrus respond specifically to the odor of 4-ethyloctanoic acid (39). This compound shows the lowest threshold value of the aliphatic acids at 1.8 parts per billion (ppb), compared to 1 part per million (ppm) for isovaleric acid, and elicits a mixed olfactory response from men while being disagreeable to women (40). Interestingly, there was 16% anosmia for this compound, which differed from the specific anosmia to isovaleric acid (4%). Since 4-ethyloctanoic acid is detected on goat hairs, it could originate from the triglyceride fraction of sebum after hydrolysis at the skin surface. A similar possibility for both isovaleric and 4-ethyloctanoic acids exists in human sebaceous secretions, since unspecified shorter-chain acids are reported in trace levels as part of the skin lipids (41). These acids have not been specifically reported in axillary sweat, although there are reports of volatile acids in body and axillary sweat (37,42,43).

One deodorant approach utilizes triethyl citrate, an ester of citric acid. A suggested mechanism for activity involves its hydrolysis (bacterial lipase/esterase) to citric acid, which would lower the axillary pH (44). Such an acidic pH would reduce the lipolysis of triglycerides for *S. epidermidis* and might also affect the steroid reactions described below.

VI. ORIGIN OF AXILLARY ODOR—PUNGENT

The metabolic origin of the androst-16-enes has been well studied using enzyme preparations from the boar testes (also adrenals or sow ovaries) (15,16,45,64). Pregnenolone is the prime substrate and is converted to androstadienol. This compound is sequentually modified to androstadienone, androstenone, and androstenol (Fig. 4). The stereochemistry of the reduction to androstenol (α or β) is controlled by the source of the enzyme system and the cofactor NADH or NADPH (16). In the sow nasal tissue, for example, the presence of NADPH results in formation of higher levels of the α-alcohol. This biochemical step is postulated as a mechanism for removing the more odorous androstenone from the olfactory receptor site and thus interrupting the olfactory signal.

The metabolic capabilities of axillary diphtheroids described earlier indicate their potential ability to metabolize pregnenolone to androst-16-enes. They also suggest that a derivative of androstadienol may be a primary substrate. This is supported by a report that a β-glucuronidase inhibitor, glucarolactone,

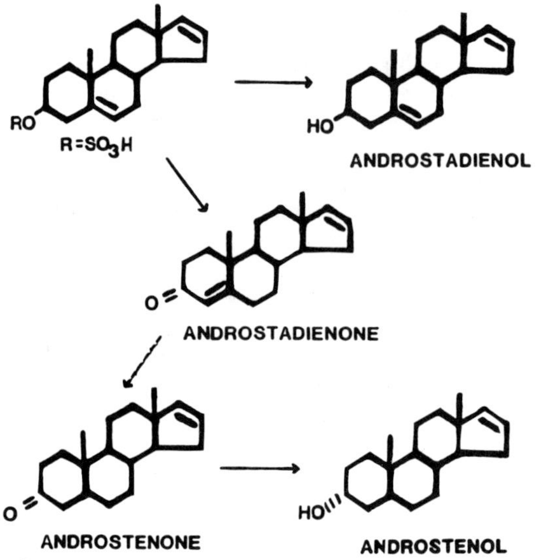

Figure 4 Structures and metabolism sequence of androst-16-enes.

is effective against axillary odor (42). In the boar, the conversion of pregnenolone to androstadienol takes place primarily in the testes. The androst-16-enes and their conjugates are in the blood and are further metabolized at the androgen target organ, i.e., the submaxillary gland (16,28). The human apocrine gland is also an androgen-dependent organ, and these androst-16-enes are present in the blood and saliva and excreted in the urine (Table 3) (28,47,48). It has not been established whether a sequence of events involving transport and modification of these compounds occurs in humans as in the boar.

VII. ODOR PERCEPTION/COMMUNICATION

Isovaleric acid, adrostenone, and androstenol represent the three "primary odors" of sweaty, urinous, and musky, respectively (3). These classifications are based on the concept of specific anosmias, i.e., the condition in which a person of otherwise normal olfactory acuity cannot perceive a particular compound at a concentration such that its odor is obvious to most other people. The fact that these compounds represent

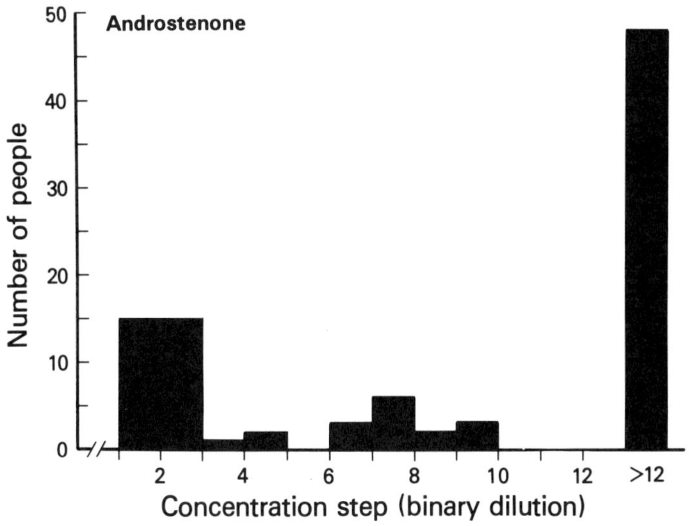

Figure 5 Distribution of detection thresholds for androst-16-en-3-one. (From Ref. 49.)

axillary odor suggests that individuals, i.e., panel members, may have different perceptions of the odor they are being asked to evaluate and should be screened for these anosmias. The olfactory properties of androstenone are particularly striking because it has an odor threshold of 0.18 ppb and a trimodal distribution in the normal population (Fig. 5) (20,49). Individuals who are very sensitive to androstenone find it to have a very unpleasant urinous, perspiration odor, while the middle group of low-sensitivity individuals describe it as having a perfume, sandalwood odor. A third group does not respond to androstenone even at full strength and represents approximately 46–50% of the adult population. In addition, the specific anosmia to androstenone has a genetic component. In a study of twins, all identical twins but only half of the nonidentical twins were alike in their sensitivity to androstenone (50). Interestingly, continued exposure to androstenone results in a change in sensitivity; increasing thresholds were also observed with age for males but not for females (51). In the same individual axillary levels of androstenone are not related to the anosmia for this compound (52).

Addition of androstenone to a perfume, i.e., oeillet, negatively affected the perception of the perfume only for individuals who where not anosmic to androstenone (53). Androstenol is noteworthy because it represents a natural musk odor arising from the body's scent gland. It gained notoriety by being added to a perfume (Andron) but no pheromonal effects have been demonstrated in humans for this or any other musk compounds.

Structure-odor correlations have been made for sterochemically related androst-16-enes (54). Alterations in the geometry—that is, α versus β substitution in the alcohols or changes in the ring fusion from trans to cis in rings A/B or C/D—reduce or in some cases eliminate the odor properties. Ohloff et al. (47) refer to this as an "enantioselective olfactory response." Interestinly, the 5β-androstenol and 5β-androstenone are both odorless to most humans but yet are very active in stimulating the sexual response in the sow.

As interesting as the unique odor characteristics of these steroids is the fact that they are mating pheromones for another mammalian species. They are secreted in the saliva of the boar and induce lordosis in the sow (16). Table 3 is a comparison of the occurrence of and response to these androstenes in pigs and humans. The olfactory acuity in the sow changes with the estrous cycle since the response to these compounds is only at the time of ovulation. In women a change in sensitivity to musk odors, which would include androstenol, and other odors has been reported to occur across the menstrual cycle (55). However, at least one study reports no changes in sensitivity (56).

In mammals it is well documented that sexual and social information is communicated by the transfer of chemicals. Both volatile and nonvolatile chemicals from saliva, skin gland secretions, and urine serve as either releaser pheromones, which initiate immediate behavioral responses, or primer pheromones, which are responsible for long-term physiological effects. The chemicals identified in these secretions and the involvement of bacteria in producing behaviorally active odorants have been reviewed (28). As a comparison with axillary odor, isovaleric acid in the deer and camel and androstenone in the pig and camel have been identified as products of mammalian skin glands.

The apocrine glands in humans are analogous to the apocrine/sebacious gland in animals and thus could be considered human scent glands. They are a secondary sexual characteristic, becoming active at puberty, and respond primarily to emotional stress. The location of the largest concentration of the glands

Table 3 Similarities in Presence of and Response to Androst-16-enes in Pigs and Humans

	Pig	Human
Presence		
Adispose tissue	3, 4[a]	3
Saliva	3, 4	3
Preputial gland	3	—
Apocrine gland/axilla	4	1–4
Urine	5	2, 4
Response		
Changes in olfactory acuity	1–4	4
Specific anosmia	—	1 and 4, 2 and 3
Mating position	1–4	—
Effect on judgment		4

[a]1, androstadienone; 2, androstadienol; 3, androstenone; 4, androstenol; 5, β-androstenol.

in the axilla provides an odor source in close proximity to the odor receptor (nose). The information content of axillary odors, including sexual and individual recognition, and the effects of the odors on social interaction have been the subjects of many studies over the past several years. These studies have been reviewed in detail (55,57–60). Although it appears that odors are qualitatively different in males and females, the ability to distinguish differences (select the male odor) correlates with intensity and unpleasantness rather than any unique odor characteristics. Other reports indicate that the odors (on T-shirts) of siblings or offspring can be distinguished from those of unrelated children. T-shirts scented with androstenol and cyclopentadecanolide, a musk odorant, were attributed to female wearers, while those scented with a synthetic analog of androstenone were judged as male (at higher concentrations) by female respondents (61).

Odor-behavioral studies have used face masks to present the odor of androstenol while asking the subject to perform a specified task. In an assessment-of-people test, there was a tendency of males to be assessed more favorably by women who were smelling the androstenol. Similarly, males and females rated women in photographs as more attractive in the presence of this odor. On exposure to androstenol throughout the menstrual cycle, women rated their mood as more submissive in midcycle. No other ratings, i.e., sexy/unsexy, were affected (62). However, in a study where androstenol, a musk perfume, and no odor were compared, no differences were found in physical attractiveness (63). In these studies only the latter used a control odor and none explored the use of objective measures of emotional response, i.e., heart rate, brain waves, or blood pressure. A report utilizing evoked response potentials showed specific changes in the cognitive evaluation of sexual stimuli (photographs) by androstenol. Male responses were depressed while female responses were enhanced in the presence of androstenol compared with alcohol (64). In a study using androstenone, females felt less sexy and males more passive to a target male in the presence of this odor compared to three other odor conditions (65). This study was extended to include androstenol, which for males increased the attractiveness of a target male. These studies involving personal ratings to various test stimuli in the presence of androstenol and androstenone have been critically reviewed (66).

More interesting is the possibility that axillary odor may have a primer pheromone effect on the menstrual cycle. In this case it is possible to have an objective measure of activity, i.e., the time of two or more individuals' menses drawing closer when compared to controls. The original observations of menstrual synchrony in women living in a college dormitory have been confirmed by several subsequent experiments (as well as anecdotal information). Although the manner in which this occurs is unknown, a human pheromone has been proposed as a likely possibility. This is supported by a study which suggested that axillary odor, delivered as ethanol extracts of axillary pads, placed on the upper lips of female subjects can cause their cycles to approach that of the donor (57,67). A subsequent double-bind study involved combined donor axillary extracts and a larger number of subjects demonstrated that the effect could be caused by axillary secretions (68). This study also reports an effect of male axillary odors on the length of the female cycle, with the male odors having a regularizing effect on aberant length cycles (<26 or >33 days). This is important since fertility is correlated with cycles of normal length (29.5 ± 3 days). A mechanism for menstrual synchrony has been demonstrated with rats (69). Odors from the follicular, luteal, and ovulatory phases of the cycle either shortened, did not change, or lengthened the estrus cycle, respectively.

The above studies of odor recognition and behavioral effects remain open to interpretation. Where objective measures were used a pheromonal effect was observed. Since the steroids involved in axillary odor are similar to steroid hormones and are active biologically in at least one mammalian species, a real possibility exists that they do convey some biological message. Experiments should be designed which avoid as much as possible subjective assessments and cultural associations (preference for musks) and take into account the proper context under which a sexual messenger would be expected to be active.

VIII. CONCLUSIONS

Axillary odors consist of at least the four androst-16-enes and isovaleric acid, which impart musky, urinous, and sweaty odors. Bacterial involvement in production of these odors has been clearly demonstrated and the presence of specific bacteria correlated in vivo and in vitro with the odor character. Methods are available for analyzing the odorous metabolites and thus monitoring the effectiveness of products for control of axillary odors, although close attention must be given to sampling techniques. Bacterial metabolism studies suggest that the androst-16-enes could be either formed or modified in the axilla. Questions still exist concerning the details of formation of the odorous steroids, the presence of other odorants, i.e., goat acids, and the amount and type of short-chain acids.

An increase in axillary odor above a baseline level in an individual can be considered a sign of emotional stress. Whether the odor has any positive influences on social interactions has not been clarified. In our culture, musk odors have positive sexual connotations, though it is questionable whether this arises naturally or is due to effective fragrance marketing. The effect of these odors on the female cycle is most intriguing and opens possibilities for natural odor control of a physiological process.

ACKNOWLEDGMENTS

The assistance of Scott Aldinger in the studies reported here is gratefully acknowledged. The studies on odor analysis and bacterial metabolism were done at the Monell Chemical Senses Center in collaboration with Dr. James Zechman and the Duhring Laboratories of the Department of Dermatology, University of Pennsylvania. I would also like to thank Ed Eigen and Dr. Claudia Froebe of Colgate-Palmolive and Dr. George Preti of Monell for helpful comments on the manuscript.

REFERENCES

1. Brody, B. The sexual significance of the axillae, *Psychiatry*, *38*: 278–289 (1975).
2. Ellis, H. *Studies in the Psychology of Sex*, Davis, Philadelphia (1905).
3. Amoore, J. Specific anosmia and the concept of primary odors, *Chem. Senses Flavor*, *2*: 267–281 (1977).
4. Comfort, A. The likelihood of human pheromones, *Pheromones* (M. C. Birch, ed.), North-Holland, Amsterdam, (1974), pp. 386–396.
5. McGinley, K., Labows, J., Zechman, J., Nordstrom, K., Webster, G., and Leyden, J. Analysis of cellular components, biochemical reactions and habitat of human cutaneous lipophilic diphtheroids, *J. Invest. Dermatol.*, *85*: 374–377 (1985).
6. Hurley, J., and Shelly, W. *The Human Apocrine Gland in Health and Disease*, Charles C. Thomas, Springfield, Illinois (1960).
7. Labows, J. N. Preti, G. Hoelzle, E., Leyden, J., and Kligman, A. Steroid analysis of human apocrine secretion, *Steroids*, *3*: 249–258 (1979).
8. Puhvel, S., Reisner, R., and Sakamoto, M. Analysis of lipid composition of isolated human sebaceous gland homogenates after incubation with cutaneous bacteria: Thin-layer chromatography, *J. Invest. Dermatol.*, *64*: 406–410 (1975).
9. Labows, J., McGinley, K., and Kligman, A. Perspectives on axillary odor, *J. Soc. Cosmet. Chem.*, *34*: 193–202 (1982).

10. Nasr, A. Histochemically demonstrable lipids in human eccrine and apocrine sweat, *J. R. Microsc. Soc., 86:* 427–432 (1967).
11. Hay, J., and Hodgins. M., Metabolism of androgens in vitro by human facial and axillary skin, *J. Endocrinol., 59:* 475–486 (1973).
12. Campbell, G., Burgdoff, W., and Everett, M. The immunohistochemical localization of lysozyme in human axillary apocrine glands, *J. Invest. Dermatol., 79:* 351–353 (1982).
13. Julesz, M. New advances in the field of androgenic steroidogenesis of human skin, *Acta Med. Acad. Sci. Hung., 25:* 273–285 (1968).
14. Brooksbank, B. Labelling of steroids in axillary sweat after administration of ^{3}H-Δ^{5}-pregnenolone and ^{14}C-progesterone to a healthy man, *Experentia, 26*: 1012–1014 (1970).
15. Gower, D. B. 16-Unsaturated C_{19} steroids: A review of their chemistry, biochemistry and possible physiological role, *J. Steroid Biochem., 3:* 45 (1972).
16. Gower, D. B., Hancock, M., and Bannister, L. Biochemical studies on the boar pheromones, *Biochemistry of taste and Olfaction* (R. H. Cagan and M. R. Kare, eds.), Academic Press, New York (1981), pp. 7–31.
17. Shehadeh, N., and Kligman, A. The bacteria responsible for apocrine odor, Part II, *J. Invest. Dermatol., 41:* 1–5 (1963).
18. Leyden, J., McGinley, K., Holzle, E., Labows, J., and Kligman, A. The microbiology of the human axilla and its relationship to axillary odor, *J. Invest. Dermatol., 77:* 413–416 (1981).
19. Jackman, R., and Noble, W. Normal axillary skin microflora in various populations, *Clin. Exp. Dermatol., 8:* 259–268 (1983).
20. Amoore, J., Pelosi, P., and Forrester, J. Specific anosmias to 5α-androst-16-en-3-one and α-androst-16-en-3-one and ω-pentadecalactone: The urinous and musky primary odors, *Chem. Senses Flavor, 2:* 401–425 (1977).
21. Zechman, J., and Labows, J. Axillary steroids: Formation and analysis, in preparation.
22. Nixon, A., Jackman, P. J. H., Mallet, A., and Gower, D. B. Steroid metabolism by human axillary bacteria, *Biochem. Soc. Trans., 12:* 1114–1115 (1984).

23. Nixon, A., Jackman, P. J., Mallet, A., and Gower, D. B. Testosterone metabolism by isolated human axillary *Corynebacterium* spp.: A gas-chromatographic mass-spectrometric study, *J. Steroid Biochem.*, *24:* 887–892 (1986).
24. Gower, D. B., Nixon, A., Jackman, P. J., and Mallett, A. Transformation of steroids by axillary coryneform bacteria, *Int, J. Cosmet, Sci.*, *8* 149–158 (1986).
25. Armstrong, A., and Kadis, B. Steroid dehydrations in porcine subcellular fractions, *Steroids*, *15:* 737–749 (1970).
26. Bird, S., and Gower, D. B. Axillary 5α-androst-16-en-3-one, cholesterol and squalene in men: preliminary evidence for 5α-androstenone being a product of bacterial action, *J. Steroid Biochem.*, *17:* 517–522 (1982).
27. Sastry, S., Buck, K., Janak, J., Dressler, M., and Preti, G. Volatiles emitted by humans, *Biochemical Applications of Mass Spectrometry* (G. Waller and O. Dermer, eds.), Wiley, New York (1980), pp. 1086–1133.
28. Albone, E. S. *Mammalian Semiochemistry*, Wiley, New York (1984).
29. Brooksbank, B., Brown, R., and Gustafusson, J. The detection of 5α-androst-16-en-3α-ol in human male axillary sweat, *Experentia*, *30:* 864–865 (1974).
30. Dravnieks, A. Evaluation of human body odors: Methods and interpretations, *J. Soc. Cosmet. Chem.*, *26:* 551–557 (1975).
31. Dravnieks, A., Krotoszynski, B., Lieb, W., and Jungermann, E. Influence of an antibacterial soap on various effluents from axillae, *J. Soc. Cosmet. Chem.*, *19:* 611–626 (1968).
32. Labows, J. N., Preti, G., Hoelze, E., Leyden, J., and Kligman, A. Axillary odors: Compounds of exogenous origin, *J. Chromatogr.*, *163:* 294–299 (1979).
33. Claus, R. Radioimmunological determination of 5α-androst-16-en-3-one in the adipose tissue of pigs, *C. R. Acad. Sci. Ser. D*, *278:* 299–302 (1974).
34. Thompson, R., Jr., and Pearson, A. Quantitative Determination of 5α-androst-16-en-3-one by gas cromatography—mass spectrometry and its relationship to sex odor intensity of pork, *J. Agric. Food Chem.*, *25:* 1241–1245 (1977).
35. Bird, S., and Gower, D. B. The validation and use of a radioimmunoassay for 5α-androst-16-en-3-one in human axillary collections, *J. Steroid Biochem.*, *14:* 213–219 (1981).

36. Claus, R., and Alsing, W. Occurrence of 5α-androst-16-en-3-one, a boar pheromone, in man and its relationship to testosterone, *J. Endocrinol.*, *68:* 483 (1976).
37. Preti, G., Cutler, W. B., Christensen, C. M., Lawley, H. J., Huggins, G. R., and Garcia, C. R. Human axillary extracts: Analysis of compounds from samples which influence menstrual timing, *J. Chem. Ecol.*, *13:* 717–731 (1987).
38. Kwan, T., Taylor, N., and Gower, D. The use of steroid profiling in the resolution of pregnenolone metabolites from porcine testicular preparations, *J. Chromatogr.*, *301:* 189–197 (1984).
39. Sasada, H., Sugiyama, T., Yamashita, K., and Masaki, J. Identification of specific odor components in mature male goat during the breeding season, *Nippon Chikusan Gakkai Ho*, *54:* 401–408 (1983).
40. Boelens, H., Haring, H., and deRijke, D. Threshold values of and human preferences for 4-ethyl octanoic and 3-methyl butanoic acid, *Perfum. Flav.*, *8:* 71–74 (1983).
41. Nicolaides, N. Skin lipids: Their biochemical uniqueness, *Science*, *186:* 9–26 (1974).
42. Perry, T., Hansen, S., Diamond, S., Bullis, B., Mok, C., and Melancon, S. Volatile fatty acids in normal human physiological fluids, *Clin. Chim. Acta*, *29:* 369–374 (1970).
43. Nitta, H., and Ikai, K. Studies on the body odor: separation of the lower fatty acids of the cutaneous excretion by paper chromatography, *Nagoya Med. J.*, *1:* 217–234 (1953); *Biol. Abs.*, *29:* 21022 (1955).
44. Osberghaus, R. Nonmicrobicidal deodorizing agents, *Cosmet. Toiletries*, *95:* 48–50 (1980).
45. Brooks, R., and Pearson, A. Steroid hormone pathways in the pig, with special emphasis on boar odor: A review, *J. Anim. Sci.*, *62:* 632–645 (1986).
46. Ohkubo, T., and Shigeharu, S. Mechanism of the action of β-glucuronidase inhibition upon apocrine sweat and sebaceous glands and its dermatological application, *Acta Determatol Kyoto*, *53:* 85–93 (1973).
47. Bird, S., and Gower, D. B. Estimation of the odorous steroid, 5α-androst-16-en-3-one, in human saliva, *Experentia*, *39:* 790–792 (1983); Brooksbank, B., Cunningham, A., and Wilson, D. The detection of androst-4,16-dien-3-one in peripheral plasma of adult men, *Steroids*, *13:* 29 (1969); Bicknell, D., and Gower, D. The development

and application of a radioimmunoassay for 5α-androst-16-en-3α-ol in plasma, *J. Steroid Biochem.*, *7:* 451–455 (1976).

48. Kingsbury, A., and Brooksbank, B. The metabolism in man of ^{3}H-5α-androst-16-en-3-ol and of ^{3}H-5α-androst-16-en-3-one, *Hormon. Res.*, *9:* 254–270 (1978).
49. Labows, J., and Wysocki, C. Individual differences in odor perception, *Perfurm. Flav.*, *9*: 21–26 (1984).
50. Wysocki, C., and Beauchamp, G. Ability to smell androstenone is genetically determined, *Proc. Natl. Acad. Sci. USA*, *81:* 4899–4902 (1984).
51. Wysocki, C. J., Beauchamp., G. K., Schmidt, H. J., and Dorries, K. M. Changes in olfactory sensitivity to androstenone with age and experience, *Chem. Senses* *12*: 637 (1987) (abstract).
52. Gower, D. B., Bird, S., Sharma, P., and House, F. Axillary 5-alpha-androst-16-en-3-one in men and women—relationships with olfactory acuity to odorous 16-androstenes, *Experentia*, *41*: 1134–1136 (1985).
53. Dodd, G., and van Toller, S. The biochemistry and psychology of perfumery, *Perfum. Flav.*, *8*: 1–12 (1983).
54. Ohloff, G., Maurer, B., Winter, B., and Giersch, W. Structural and configurational dependence of the sensory process in steroids, *Helv. Chim. Acta*, *66:* 192 (1983).
55. Doty, R. Olfactory communication in humans, *Chem. Sens.*, *6:* 351–376 (1981).
56. Amoore, J., Popplewell, J., and Whissell-Buechy, D. Sensitivity of women to musk odor: No menstrual variation, *J. Chem. Ecol.*, *1:* 291–297 (1975).
57. Russell, M. Human olfactory communication, *Chemical Signals in Vertebrates*, Vol. 3 (D. Muller-Schwarze and R. Silverstein, eds.), Plenum, New York (1983), pp. 259–273.
58. Schleidt, M., and Hold, B. Human axillary odor: Biological and culture variables, *Determination of Behavior by Chemical Stimuli* (J. Steiner and J. Ganchow, eds.), IRL Press, London (1982), pp. 91-104.
59. Labows, J. Social-sexual effects of pheromones, *The Psychology of Cosmetic Treatments for Dermatologists* (A. Kligman and J. Graham, eds.), Praeger, New York (1985), pp. 100–112.
60. Filsinger, E., and Fabes, R. Odor communication, pheromones, and human families, *J. Marriage Fam.*, *47:* 349–359 (1985).

61. Koster, E. P., Jellinek, J. S., Verhelst, N. D., Mojet, J., and Linschoten, M. R. Odorants related to human body odor, *J. Soc. Cosmet. Chem.*, *37:* 409–428 (1987).
62. Benton, D. The influence of androstenol—a putative human pheromone—on mood throughout the menstrual cycle, *Biol. Psychol.*, *15:* 249–256 (1982).
63. Black, S. and Biron, C. Androstenol as a human pheromone: No effect on perceived physical attractiveness, *Behav. Neural Biol.*, *34:* 326–330 (1982).
64. Johnston, V. S. The effects of androstenol on ERPS to sexual stimuli, *Psychophysiology*, *20:* 450 (1983).
65. Filsinger, E., Braun, J., Monte, W., and Linder, D. Human Responses to the pig sex pheromone, 5α-androst-16-en-3-one, *J. Comp. Psychol.*, *98:* 219-222 (1984).
66. Filsinger, E., Braun, J., and Monte, W. An examination of the effects of putative pheromones on human judgments, *Ethol. Sociobiol.*, *6:* 227–236 (1985).
67. Russell, M. Olfactory influences on the human menstrual cycle, *Pharmacol. Biochem Behav.*, *13:* 737–738 (1980).
68. Preti, G., Cutler, W. B., Garcia, C. R., Huggins, G. R., and Lawley, H. J. Human axillary secretions influence women's menstrual cycles: The role of donor extract of females, *Horm. Behav.*, *20:* 474–482 (1986); Cutler, W. B., Preti, G., Krieger, A., Huggins, G. R., Garcia, C. R., and Lawley, H. J. Human axillary secretions influence women's menstrual cycles: The role of donor extract from men, *Horm. Behav.*, *20:* 463–473 (1986).
69. McClintock, M. Synchronizing ovarian and birth cycles by female pheromones, *Chemical Signals in Vertebrates*, Vol. 3 (D. Muller-Schwarze and R. Silverstein, eds.), Plenum, New York (1983) pp. 159–178.
70. Toth, I., and Faredin, I. C19-steroid sulfate in human sweat, *Acta Med. Hung.*, *42:* 21–28 (1985).

10

Deodorant Ingredients

Earl Philip Seitz, Jr. and David I. Richardson* *The Dial Corporation, Scottsdale, Arizona*

I. INTRODUCTION

The search for new and more effective deodorant ingredients and formulations has been a very active research area of cosmetic chemistry for most of this century. There are literally thousands of references in the field from both patent and scientific literature. The research in this area has been in part inspired by a sizable retail market for these products. In 1986, for example, the sales of deodorant soaps, antiperspirants, and deodorants alone made up approximately 14% of the \$8.6 billion U.S. health and beauty aids market (1,2).

In addition to deodorant soaps, antiperspirants, and underarm deodorants, this product category includes deodorant preparations for the whole body (e.g., deo-colognes), hair, breath, feminine hygiene, and feet. Deodorant laundry products could also be on this list in light of the potential for treatment of clothes to play a role in malodor reduction (3,4).

Cosmetic deodorants are preparations which modify, reduce and/or remove, or prevent the development of body odors, or

---

**Current affiliation:* Vipont Pharmaceuticals, Inc., Fort Collins, Colorado do all of these (adapted from a definition by S. Plenchner (5)). The chapter is organized on the basis of this definition and by type of deodorant ingredient. Deodorant fragrances are discussed first because they generally act by modifying odor. Odor-removing or -reducing ingredients, such as sodium bicarbonate, are discussed next, and odor-preventing materials, which are in large part antimicrobial agents, are discussed last. (The antimicrobial agents are the most important deodorant ingredients from a commercial standpoint.)

Where possible, theories about mechanisms of action of the deodorant materials are given. Generally, if more than one mechanism of action is involved, classification is under the major type of deodorant ingredient (e.g., the discussion of deodorant fragrances covers odor-modifying, odor-absorbing, and antimicrobial fragrances). The focus of the chapter is on two of three major deodorant products, deodorant soaps and underarm deodorants. Some of the deodorant properties of antiperspirants will be discussed, but for the most part the properties of antiperspirants are left to other chapters of this book.

II. ODOR MODIFICATION AND DEODORANT FRAGRANCES

Perhaps the most popular method of achieving deodorancy (until modern times) was the use of fragrances to mask body odor (6). For a man or woman in the Middle Ages, perfumes represented a much better alternative than bathing for the control of body odor, since neither bathing nor using soap to wash off perspiration odor was done regularly during that time. The reasons for this lack of personal hygiene include the lack of bathing facilities, the expense of soap, and religious and civil taboos in some areas of Europe and Colonial America (7–9). While not the only means of odor control today, deo-fragrances are likely to be found in all classes of deodorant products, either as the primary active ingredient or in conjunction with another type of deodorant material.

A significant portion of the odor control literature has been written from the perspective of controlling odors in the environment. Nevertheless, the theories presented in this body of literature can provide insight for understanding deodorants for personal care. Summarized below are several mechanisms of action which have been put forth to explain the effects that deodorant fragrances can have on malodors.

A. Odor Modification

Odor modification involves changing a malodor to a more pleasant character (*masking*) and/or reducing the odor intensity to a more acceptable level (*counteraction*) (10,11).

1. *Odor Masking and Reodorization*

Odor masking may involve a simple overpowering of the malodor with a pleasant fragrance (10) or a more sophisticated approach utilizing a deodorant fragrance designed such that it includes body malodors as components. The malodorous components are left out of the fragrance and the combined effect of the deodorant fragrance and body malodors is pleasant. According to Sturm (12), "deo-fragrances incorporate the sweaty smell as the fatty-sweaty component of the perfume composition." Anonis (13) commented that with perfuming deodorants and antiperspirants in general the fragrance of the product must complement body odor but not reinforce it.

The intensity of the odor need not decrease for odor-masking effects to work. Binary odor mixtures have been the most well studied, and a terminology has been developed to express the perceived magnitude of the odors of these mixtures. For odor masking, a mixture may show a stronger intensity (*hyperaddition*), the same intensity (*complete addition*), or a lower intensity (*hypoaddition*) than the sum of the components. The most common observation is hypoaddition. Regardless of the effects on intensity, the odor-masking effect works primarily by modifying the perceived character of malodor.

Certain terpenes (e.g., α-ionone, α-methylionone, citral, geranyl formate, and geranyl acetate) have been found to enhance the odor-masking effects of other compounds in deodorant formulations for the mouth. These materials have been called "reodorants." The inventors of this concept claim that the use of odor-masking fragrances such as flower oils alone may give rise to "peculiar and, in many cases, unpleasant sensations." The use of reodorants in concentrations of 10 to 2000 parts per million is purported to circumvent this problem. The two factors considered important for a compound to work as an effective reodorant are its propensity to "volatilise competitively with whatever malodour may be present" and its high odor-masking capability at low concentrations (14,15).

2. *Odor Counteraction* (11,16–18)

Odor counteraction is a phenomenon in which two or more odors are mixed and give a combined odor intensity less than that of either of the individual components. Odor counteraction has been further subclassified as compromise or compensation (10,11). For illustration, consider a mixture of two odors, A and B. Using an arbitrary perceived odor intensity scale, assume A is 7 on the scale and B is 3. If the perceived intensity of the mixture were less than 7 (odor A alone) but greater than 3 (odor B alone), the mixture would be said to exhibit *odor compromise*. If the perceived odor intensity were less than both of the individual components (i.e., less than 3), the mixture would be said to exhibit *odor compensation*. Odor compromise is by far more common (11).

B. Odor Densensitization (15,16)

Some materials may temporarily deactivate nasal sensory receptors and thus act as a deodorant. An example of this effect may be found in a recent patent by Torii and Egma (19). They prepared isolates from the Theaceae plant (presumed to contain flavanols) and found deodorant activity. They described the deodorant effect of the active components as follows:

> Their deodorant effectiveness may be due to a complex mechanism consisting of clathrating, addition and neutralization reactions with the active ingredient of malodorous sources in addition to biological reactions in human beings, such as inhibition of olfactory receptors.

From this description it is evident that this deodorant material may have effects which could also be classified as chemical odor removal and odor absorption/adsorption, discussed later in the chapter.

C. Odor Absorption (by Deo-Fragrances)

An odor absorption effect has been reported for some deodorant fragrances. As an example, one of the criteria for the deodorant components claimed by Hooper et al. in several patents (4, 20) is that the component should reduce the partial vapor pressure of morpholine by at least 10% more than that required by Raoult's law. While the basis of deodorant action for materials exhibiting this effect is not explicitly discussed in the patents, it seems likely that the effect is based on malodor absorption. Odor absorption and adsorption will be discussed in greater detail later in the chapter.

D. Biochemical Effects (by Deo-Fragrances)

In addition to the absorption effect mentioned above, Hooper et al. (4,20) required that their deodorant components show a significant capacity to inhibit lipoxidase. As with the previous topic, the theoretical connection between this property and the deodorant action of the material is not covered in detail. The following discussion is offered to explain the purported effect. A lipoxidase (or lipoxygenase) can catalyze the hydroperoxidation of polyunsaturated fatty acids and esters containing a *cis,cis*-1,4-pentadiene moiety (21,22). Linoleic acid and esters of linoleic acid, which contain this pentadienyl group, are present in the epidermis (23,24). The hydroperoxides which could be generated by the action of lipoxidase on epidermal linoleates are known to decompose or undergo further oxidation to short- and medium-chain aldehydes, ketones, and acids, most of which have strong, unpleasant odors (25). Lipoxygenase activity associated with the metabolism of arachidonic acid is known to exist in human epidermis (26), but whether or not it (or some other lipoxygenase) is involved in odor generation on the skin is not clear. Osberghaus (27) also mentions the potential for oxidation of unsaturated compounds on the skin to be involved in odor development.

E. Antimicrobial Fragrances

Sturm (12) described "deosafe" fragrances which not only reduce the perception of odor with odor masking but also stop odor development through antimicrobial action. Morris et al. (28) concluded on the basis of screening 521 fragrance chemicals that even the best antimicrobial fragrance of their study was 100 to 1000 times less effective than common soap antimicrobials. They further concluded that "creation of a practical antimicrobial soap fragrance does not appear to be possible." Sturm (12), however, maintains that the use of a deosafe fragrance in deodorant products is workable because the lower antimicrobial efficacy of the fragrance materials is compensated for by the odor-masking effect of the other active components of the deosafe fragrance.

In any case, the most successful applications of antimicrobial fragrances would probably be with products that are intended to be left on the skin (29).

III. ODOR REDUCTION AND/OR REMOVAL

A. Physical or Mechanical Odor Removal

In their pioneering work in the study of perspiration and odor, Killian and Panzarella (30) found that simple daily bathing with soap and water reduced the "capacities of perspiration to develop offensive perspiratory odors." Blank et al. (31) also recognized the role of washing to remove odor:

> There can be little doubt that good skin hygiene can not only reduce the total number of bacteria on the cutaneous surface, but can also remove odorous substances, no matter what their origin. Therefore, the first step in odor control is good skin hygiene.

B. Chemical Odor Removal

One of the earliest-used deodorant ingredients is sodium bicarbonate, as is evident from a 1883 patent by Slocomb and Day (32) and a later patent by Smith (33). In 1946 Lamb (34) strongly recommended a bicarbonate-based deodorant for general use and for treatment of cases of extreme bromidrosis "so repulsive that neither the doctor nor the nurse could be other than cognizant of the fact during the whole examination." Lamb verified the effectiveness of his formulation with a clinical study. He also suggested a mechanism of action which is probably correct (35), i.e., that at least part of the deodorizing action of sodium bicarbonate is to chemically neutralize odoriferous short-chain fatty acids in the axilla.

Sodium and potassium bicarbonates are still in use as deodorant ingredients and are the basis of several recent patents. According to Marschner (36–38), formulation of bicarbonates into a stick deodorant form is difficult because of "limited solubility," "compatibility with the other ingredients," problems with dimensional stability, and "esthetic appearance and feel on the skin." Marschner's patents outline the formulation of deodorant sticks with a polyhydric alcohol, a fatty acid soap, and up to 70% sodium bicarbonate. He also developed a suspension formula containing up to 20% bicarbonate for use in "deodorant products such as roll-ons, pumps and, on substrates such as deodorant pads" (39). The suspension was based on alcohol and hydroxyethyl cellulose. An example of an aerosol product based on sodium bicarbonate is provided in a patent by Stanave (40). The formulation involved the use of "micro-pulverized" bicarbonate and an organopolysiloxane.

There are other deodorant ingredients which also may exhibit the chemical odor removal effect. As will be mentioned later in the chapter, it has been suggested that part of the deodorant action of antiperspirant salts is due to a chemical effect on odorous materials produced in the axilla (41). Zinc glycinate has been patented (42) as a material which "chemically neutralizes odiferous compounds" as well as inhibiting bacterial growth. The deodorant action of zinc glycinate was thought to be similar to that of sodium bicarbonate, namely an acid/base reaction. The purported advantage is that the product, $Zn(OH)_2$, is a milder base with less skin irritation potential. Other examples of deodorant ingredients which probably work by chemical neutralization of odor are zinc carbonate and magnesium or lanthanum oxides, hydroxides, or carbonates, or mixtures thereof (43,44). These materials may have odor adsorptive properties as well.

A number of articles have been written which describe the efficacy and mechanism of action of Grillocin®, a combination of zinc ricinoleate and synergistic additives. This material neither inhibits the growth of bacteria or fungi nor prevents perspiration (45). The clinical deodorant efficacy of alcohol solutions of Grillocin has been demonstrated (35,46). Sartori et al. (47) state that "The deodorizing process is quite clearly of a chemical nature." Lowicki et al. (45) postulated the formation of a clathrate adduct to explain the mechanism of action of this material. If the clathrate mechanism is operative, Grillocin should be classified as an odor absorber, as suggested by Osberghaus (27) and as required by the nature of clathration. [Clathration does not involve strong bonding between host and guest, nor does it depend on chemical properties (48).]

The final example of this type of deodorant ingredient is a material of "natural origin" which is an extract of "brewed wine or residual fermentation materials produced in the manufacture of brewed wine" (49). While not clearly explained in the patent, this material seems to have a chemical or enzymatic effect in removing bad odors. A variety of applications is suggested, including cosmetic deodorants.

C. Odor Absorbants or Adsorbants

The use of deodorant fragrances which can absorb odors has been discussed above. In addition, there are a number of other deodorant substances which are thought to act by odor absorption or adsorption. (An odor absorbant is a material which captures and retains odor molecules in its interior, while an odor adsorbant retains odor molecules on its surface.) This is to be contrasted with chemical odor removal, discussed above.

In 1953 Thurmon (50) patented the use of cation ion-exchange resins as deodorant ingredients. The next year Ikai (51) published results of in vitro tests in which he demonstrated the deodorizing effects of anionic and cationic ion-exchange resins (and mixtures thereof) on lower fatty acids, ammonia, odor of feces, and odor of axillary secretions. The author conducted in vivo tests, with patients suffering from osmidrosis, using ion-exchange powders and various formulations. Generally, the deodorant effect was good and relatively long-lasting in the case of some of the formulations, but the powders alone washed off with perspiration. He also identified the deodorization as an adsorptive effect. Winters (52) summarized the properties of various ion-exchange resins and discussed their suitability for formulation of deodorant products. Although he said that the discovery of the deodorant properties of the resins "opened a new avenue of investigation for cosmetic chemists," little commercial use has been made of these findings (5). Schrader (53) and Osberghaus (54) have obtained relatively recent patents using ion exchangers in deodorant formulations.

Some examples of odor-absorbing deodorant ingredients from the recent patent literature include aluminum and potassium double sulfate (55), 2-naphtholic acid dibutylamide (56), and isononanoyl-2-methylpiperidide (57). A final example in this category is contained in a recent patent by Jacquet et al. (58). The active component of their invention is a zinc or magnesium salt of a polycarboxylic acid which the authors purport to "have an odour-absorbing action by trapping the small molecules, which are responsible for the bad odours, in the crystal lattice formed by the said metal salts of an organic polyacid."

IV. PREVENTION OF ODOR DEVELOPMENT

A. Bacteria and Odor

The connection between axillary bacterial action and body odor was firmly established in the late 1940s and early 1950s (30,41). This basic finding has been continually refined since then, and, as discussed in the preceding chapter and elsewhere, the following is known. There is good evidence that the bacteria responsible for axillary odor are Gram-positive micrococci and lipophilic diphtheroids. The main substrate for the odor-causing bacterial action is apocrine sweat. (It is not known yet whether the secretion from the recently discovered "apoeccrine" sweat gland (59,60) is involved in odor production.) The micrococci are thought to produce a "sweaty, acid" odor and the lipophilic diphtheroids a "pungent, acrid" odor (61–65). It is also thought that the bacterial level must be kept below $10^4/cm^2$ for good odor control (64,66).

B. Commercial Antimicrobial Agents—Historical Perspective

1. *Early Products*

The use of antimicrobial agents which are effective against these axillary bacteria has offered formulators a way to make effective deodorant products without the use of strong deofragrances. Even before the works of Killian and Panzarella (30) and Shelley et al. (41), deodorant formulators thought that bacteria played a role in body odor. This is evident from the following passage taken from a cosmetic text of the time. In 1934 Chilson (67) placed this requirement on deodorant formulations:

> The normal body exudes from one to three quarts of perspiration a day. The water content of this perspiration evaporates immediately, depositing waste matter on the surface of the skin which rapidly becomes rancid through the catalyzing action of bacteria. Consequently a deodorant must kill the bacteria on the skin in order to prevent the formation of decomposition products.

Harry (68) and de Navarre (69) also discussed the relationship between bacteria and body odor in their early works on cosmetics in 1940 and 1941, respectively.

Two of the earliest deodorant products were Mum® (ca. 1890), which contained zinc oxide as the active ingredient, and an antiperspirant (1902) which was a simple water solution of 25% aluminum chloride (70). Some of the other early ingredients used for deodorants were boric acid, benzoic acid, chloroamine-T, chlorothymol, formaldehyde, hexamine, oxyquinoline sulfate, sodium perborate, zinc salicylate, zinc sulfocarbolate, zinc sulfide, zinc peroxide. These materials were purported to have deodorizing and/or antimicrobial properties (67,68).

Some of these early deodorant/antiperspirant products undoubtedly had antimicrobial properties. It took some time however, before formulations were developed which were both effective and pleasant for the consumer to use (71). McDonough (72) described skin irritation and fabric damage problems with some of the early products. He outlined a remedy for the unfortunate woman who had just applied a 15% solution of aluminum chloride to her underarm after using a depilatory. He warned of damage to fabrics from contact with antiperspirant/deodorant products of his day. He also described an "excellent" deodorant lotion which contained ferric chloride to oxidize body odors but which had the disadvantage of staining fabrics with a rust color.

Perhaps one of the best-known early deodorant products was Lifebuoy® soap, which was introduced in 1895 (73). This product contained cresylic acid, an antiseptic with a strong phenolic odor, which was probably more effective as an odor-masking agent than as an antimicrobial ingredient (74). This strong odor was a problem with early consumer acceptance of deodorant soaps and opened opportunities for the development of germicides which were more cosmetically appealing as well as more effective (75,76).

2. *Development of Hexachlorophene*

The first of these effective, low-odor antimicrobial agents was hexachlorophene, 2,2'-methylenebis[3,4,6-trichlorophenol], discovered in 1941 (75,77,78). The development of the first nationally-marketed bar soap to use this material is a rather interesting story of serendipitous discovery (79).

In 1943, W. S. Gump made a business tour of the United States as a representative of the Givaudon-Delawanna company. The purpose of his tour was to persuade soap manufacturers to try the newly invented hexachlorophene in bar soaps. He met with disappointing results because most soapmakers of the time felt that consumers had a very negative conception of deodorant soaps, i.e., that they were "medicinal smelling" and "irritating to the skin" (79). Gump left a sample of the new material with Armour and Company chemists in Chicago, but it sat on the shelf for 3 years before it was tried.* In 1946 it was put at the bottom of a list of several chemicals to be tried for development of a new deodorant soap for the postwar market.

R. E. Casely was the chemist on the project and had spent months trying the chemicals at the top of the list with no success. He then got an idea that there might be a link between skin bacteria and body odor. He also recalled from Gump's visit 3 years earlier that hexachlorophene, a potent antimicrobial, retained its activity in soap. Casely confirmed his hunch by testing a formulation on himself and saw dramatic results. (He used small cotton pads to absorb odor in his armpits, and then evaluated the odor after storing the pads in airtight containers.) The new deodorant soap, Dial®, was launched in 1948 and started a new category of bar soaps which remains one of the strongest in today's market.

By comparison to the level of effort expended, the number of successful antimicrobial ingredients developed for consumer products since 1950 is small. According to Jungermann (75), the main reason for this is the following number of criteria a bar soap bacteriostat must pass to be considered for use in the market:

> 1) broad spectrum antibacterial activity in the presence of soap; 2) skin substantivity; 3) effective deodorancy; 4) efficacy in skin degerming... 5) chemical stability in soap; 6) compatibility with color and odor of finished products; 7) nonreactivity with other components in the soap... 8) mildness and safety for general use of the finished product; 9) satisfactory economics.

Despite these rigorous criteria, several new antimicrobial agents for soap were introduced in the 1950s and 1960s. A brief history of these (adapted from Refs. 74, 75, and 80 is presented below.

---

*There is disagreement in the literature about when the sample of hexachlorophene was left at Armour and Company and how long it remained on the shelf before being tried. In Ref. 76 it was stated that the sample was received in "late 1946 or 1947" and "sat on the shelf for at least six months" before being tried.

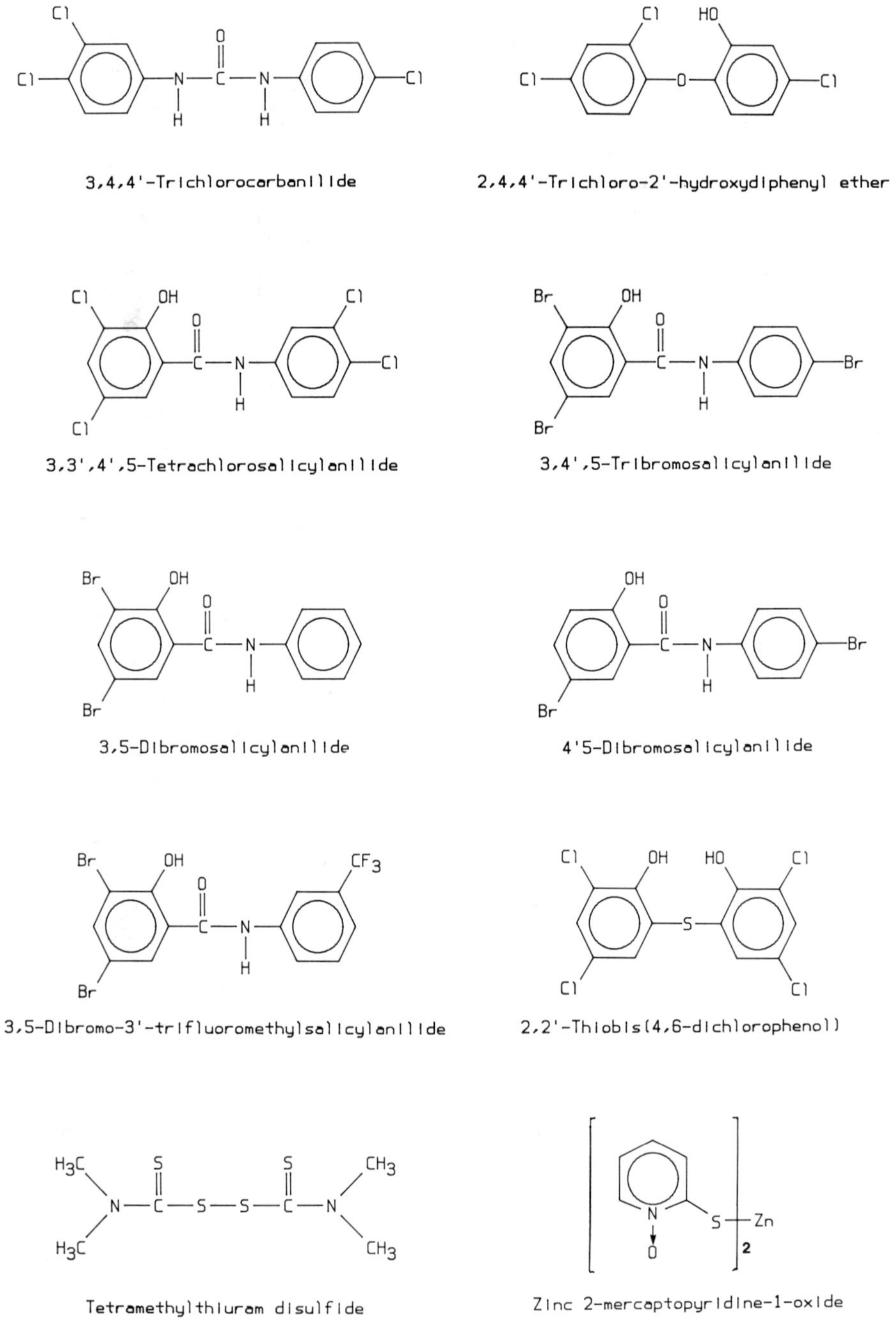

Figure 1 Antimicrobial agents developed after hexachlorophene.

3. *Deodorant Materials Developed after Hexachlorophene*

Many antimicrobial ingredients were developed after the introduction of hexachlorophene. Some of the most important of these compounds are 3,4,4'-trichlorocarbanilide (triclocarban or TCC®), 2,4,4'-trichloro-2'-hydroxy diphenyl ether (triclosan), 3,3',4'5-tetrachlorosalicylanilide (TCSA), 3,4'5-tribromosalicylanilide (tribromsalan or TBS), 3,5- and 4'5-dibromosalicylanilide, 3,5-dibromo-3'-trifluoromethyl salicylanilide, 2,2'-thiobis(4,6-dichlorophenol) (bithionol), tetramethylthiuram disulfide (thiram), and zinc 2-mercaptopyridine-1-oxide. The molecular structures for these compounds are given in Figure 1.

Bithinol, which has a molecular structure similar to that of hexachlorophene, was introduced as a new bacteriostat for soap in late 1952 by Shumard et al. (81). Thiram, a disulfide, was introduced for the same purpose in 1954 (82). Both of these compounds also found some use in deodorants (83). Several of the quaternary ammonium compounds which are antibacterial were also found useful for deodorants in the 1950s (84). About the same time some deodorants used antibiotics such as neomycin. These ingredients did not gain wide acceptance, however, because of "cost, sensitization, and the possibility of the development of resistant strains of organisms" (83,85).

4. *Safety Issues with Halogenated Salicylanilides and Hexachlorophene*

Tetrachlorosalicylanilide was used in bar soaps marketed in England and the United States in 1960 and 1961. The use of this compound gave rise to the recognition of a problem of photosensitivity for TCSA, related salicylanilides, and bithionol. These agents were subsequently removed from the market. The use of Thiram was stopped in the United States because of problems with the odor and with contact allergy (74).

The use of hexachlorophene was banned in the United States in 1972 by the Food and Drug Administration (FDA). The following excerpts describe the circumstances surrounding the FDA's action:

> Although there was no evidence that toilet soaps containing reasonable amounts of hexachlorophene (up to 1%) had any harmful effects on adults or babies, FDA proposed limiting its use in 1972 (86,87), This was followed by the banning of hexachlorophene several months later when a French

manufacturer of baby powder inadvertently added 6% of hexachlorophene to his product, resulting in more than 30 deaths (88).

Although the extensive use of hexachlorophene over the years has not shown toxic symptoms in humans (89,90), neurotoxicity has been demonstrated in rats with large doses (91,92), also in monkeys (93) and in premature infants (94). In view of these findings, the FDA (95) banned the over-the-counter sale of soaps, cosmetics, and drugs containing an amount exceeding 0.1%. All products with a higher percentage were put on a prescription basis (96).

For a review of toxicological studies on hexachlorophene see Kaul and Jewett (97).

The effects of these events on the industry have been to 1. dramatically reduce the number of antimicrobial agents available to formulators of deodorant products, 2. increase the rigor with which new and currently existing antimicrobial agents are screened for safety, and 3. increase the cost of developing new bacteriostats.

C. Safety Testing of Currently Used Antimicrobial Ingredients

Three antimicrobial agents which have survived close safety scrutiny by industry and FDA (98,99) and are most commonly used in deodorant products today are 3,4,4'-trichlorocarbanilide (triclocarban), 2,4,4'-trichloro-2'-hydroxydiphenyl ether (triclosan), and antiperspirant ingredients such as aluminum zirconium tetrachlorohydrex glycine.

In addition to the many tests which have been run on these materials to verify their lack of toxicity, an important element of their suitability for market is the assessment of their clinical irritation/sensitization potential. This is an especially difficult hurdle in view of the perception, which still persists with many consumers, that deodorant products are harsh. While it is still a good idea for persons new to the use of antiperspirant products to test the product on themselves before regular use, modern underarm deodorants and antiperspirants are considerably milder than the aqueous solutions of aluminum chloride used a few decades ago. In the case of deodorant soaps, the small amount of germicide in the soap has little to do with the mildness of the product. Generally it is the soap base which is the primary determinant of this property. The fragrance of a deodorant product may have a strong influence on the perception of its irritation potential, as was learned with the harsh-smelling deodorant soaps in the first part of this century.

D. Antimicrobial Efficacy of Modern Deodorant Ingredients

1. *Triclocarban*

Triclocarban is effective against Gram-positive organisms including odor-causing micrococci and diphtheroids (100,101). In the United States it has mainly been marketed for use in bar soaps, and typically is used at concentrations of 0.5 to 1.5% by weight (102).

Beaver et al. (103) outlined the relationship of structure to activity for a wide variety of substituted carbanilides including triclocarban. They discovered that the antimicrobial properties of these materials were "remarkably specific in that activity was greatly enhanced or lost completely with slight changes in chemical structure." They further suggested that a "lock-and-key" mechanism may be operative for the bacteriostatic activity of these compounds. Some other of the authors' interesting observations include the following.

1. No appreciable activity was found until both rings were chlorinated and activity was at a maximum with chlorine on the 3 and 4 positions of one ring and the 3 or 4 position of the other ring.
2. Activity was reduced to a minimum by *ortho* substitution regardless of the electronic nature of the group.
3. Using 3,4,4'-trichlorocarbanilide as a model, substitution of the 4' chloro group with a variety of substituents markedly reduced activity. Use of a 3' chloro group (or other substituents) instead of the 4' chloro group also reduced activity.
4. Replacing the 4'-chlorophenyl group with other aryl and and alkyl groups greatly reduced activity.
5. Thio analogs were less active than the carbanilides.

While no explanation was offered to account for these observations, the authors' findings offer strong support for the notion that structure specificity plays a role in the mechanism of action.

According to Hamilton (104), the antibacterial action of triclocarban is similar to that of several other membrane-active compounds, i.e., it depends on a reversible, "relatively nonspecific physiochemical" adsorption to the cell membrane. The author further speculates that the action of these compounds involves "changes in the membrane structure, resulting from the breakage of hydrogen bonds, etc., rather than the loss or inactivation of specific chemical or enzymatic groupings." In the case of triclocarban, bacteriostasis results from the adsorption of about 5×10^5 molecules of germicide per bacterium. While most of the work in this paper was done using tetrachlorosalicylanilide (TCS) against *Staphylococcus aureus*, the author indicates that the mechanisms of action are probably similar for trichlorocarban and the other compounds he studied. His results showed that bacteriostasis results from disruption of energy-dependent transport of materials across the cell membrane and not energy-independent entry of nutrients or cell leakage alone. With respect to cell leakage, Hamilton states that "If leakage plays a part in the bacteriostatic action of TCS, it cannot simply be the loss of cellular material which results in the inhibition of growth, but the continuing inability of the cell to make good the loss in the presence of the germicide."

It was mentioned above that the adsorption of antimicrobials like triclocarban involved a nonspecific mechanism. Once adsorbed, however, these compounds can exhibit specific effects on the plasma membrane of the cell (105). These specific effects include uncoupling of proton translocation, increases in both H^+ and Cl^- permeability (106), and inhibition of the cell's ability to accumulate isoleucine by active transport (107).

Resistance to triclocarban, such as that seen by *Escherichia coli* (Gram-negative bacteria), results from a property of the cell wall which inhibits "the penetration through the cell wall to the combining sites on the membrane" (103). In later work, Hamilton (108) showed that *E. coli* cells take up about as much triclocarban as a sensitive species, *S. aureus* (Gram-positive). The cell wall of the Gram-negative organism absorbs triclocarban but must somehow prevent its penetration to the sensitive cell membrane.

2. *Triclosan*

Triclosan is effective against Gram-positive and Gram-negative bacteria and has found application in deodorants, bar soaps, and other deodorant products. The use levels of triclosan in bar soaps are usually lower than those of triclocarban —the recommended level being between 0.2–0.5% (109).

Triclosan levels for underarm deodorants are somewhat lower because these products are left on the skin (range of 0.03–0.3%).

Regös and Hitz (110) studied the mechanism of action of triclosan and suggested that, as with triclocarban, the primary site of action is the cell membrane. The authors describe the antimicrobial action more specifically as follows:

> At low bacteriostatic concentrations, this action on the membrane interferes with the uptake of amino acids, uracil or other nutrients from the medium. At higher bactericidal concentrations the membrane lesions lead to the leakage of cellular content and the death of the cell.

Meincke et al. (111) speculated that the uptake of triclosan by both cell walls and whole cells is due to the "hydrophobic and lipophilic" nature of the antibacterial agent and its greater solubility in the cell components than in the surrounding water-rich environment. These authors further suggested that resistance of such microorganisms as *Pseudomonas aeruginosa* may be due to a higher lipid content in the cell wall. They mentioned that "a higher lipid content of the cell wall would tend to adsorb more irgasan [triclosan], thus hindering the access of the drug to its proposed site of action, the cytoplasmic membrane." An earlier speculation about the resistance of *S.marcescens* (*Serratia marcescens*) *and P.aeruginosa* was that it may be due to an enzyme, irgasanase, which "breaks down irgasan [triclosan] before it reaches its site of action" (112).

3. *Antiperspirant Salts*

Antiperspirant aluminum salts are included in this category because, in addition to their high activity in stopping perspiration wetness, they have good antimicrobial activity. These salts are much less antimicrobially active than triclocarban and triclosan when compared on an equal weight basis (101). However because they are used at such a high concentration in antiperspirants (up to 27%) and are left on the skin, they are active against axillary bacteria and work well as deodorants. The reduction of moisture provided by antiperspirant products also makes the axilla a much less favorable environment for bacterial growth (66). Shelley et al. (41) suggested that part of the deodorant action of aluminum salts may be due to a chemical change which would render "the normal odiferous products of bacterial decomposition inoffensive." The activity has also been attributed to the property of these aluminum salts to denature essential bacterial proteins (113). Other antimicrobial ingredients (e.g., triclosan) may be put into an antiperspirant formula to give additional control of bacteria (114,115). This may be especially important for an aerosol formulation (116).

There have been numerous methods developed to sample bacteria from the skin and many examples of applications of these methods for determination of the antimicrobial efficacy of deodorant products. A relatively new method using the Thran spray sampler has been found particularly well suited for determining the reduction in vivo of axillary bacteria by deodorant products (117,118). Using this method, Theiler et al. (117) found an average 24-hour reduction of axillary bacteria of 0.35 log (55%) with a soap bar containing 1.5% triclocarban and 1.82 log (98.5%) with a stick antiperspirant containing 20% aluminum zirconium tetrachlorohydrex glycine.

E. Substantivity of Antimicrobial Agents

For antimicrobial ingredients intended for use in "wash-off" deodorant products such as bar soaps, there is an additional requirement beyond the necessary activity against odor-causing bacteria. Since much of the product is removed during rinsing, some of the antimicrobial must remain after the rinse, i.e., be substantive to the skin. This is a special challenge of formulation chemists since the product must have acceptable cleaning properties as well. Taber et al. (119) aptly stated this challenge as "It is a peculiarity of washing with antibacterial soaps or detergents that the antibacterial agents are deposited onto skin from vehicles whose function it is to remove substances from skin surfaces."

Based on studies with a soap bar containing 1.5% TCC applied to human skin, the amount of triclocarban deposited is about 0.33 $\mu g/cm^2$ (118). The wash protocol used for this experiment was approximately representative of a typical use situation for bar soaps, i.e., 30 seconds application, 90 seconds delay, then a thorough water rinse. A value for triclosan, estimated from a similar procedure using an animal model and a liquid soap containing 0.5% triclosan, is about 2 $\mu g/cm^2$ (121). (The data above are presented only as sample deposition values and should not be used for comparison of triclocarban and triclosan, since deposition can be very dependent on formulation.)

A simple calculation, using an estimate of 10^6 bacteria/cm^2 in the axilla (117), suggests that there are at least 10^8 molecules of triclocarban or triclosan deposited on the skin for each bacterium present.

Comparing the above estimate with the number of molecules per bacterium of triclocarban (5×10^5) needed for bacteriostasis (see above), it might seem that there is more than adequate antimicrobial agent to control odor-causing bacteria. While soaps containing antimicrobial deodorant ingredients are effective in controlling odor, there is some room for improvement in reducing the number of odor-causing bacteria in the axilla. [As mentioned earlier, Theiler et al. found a 0.35 log (55%) reduction of axillary bacteria with a soap containing 1.5% triclocarban (117).] A number of ideas have been proposed in the literature to explain the phenomenon of germicide deposition and to increase the deposition and/or activity of germicides in bar soaps and other formulations.

In a study published in 1948, Fahlbert et al. (122) demonstrated several things about the nature of the deposition of hexachlorophene on the skin from liquid and bar soaps. Their observations were that 1. hexachlorophene could be detected on the skin up to 48 hours after treatment with a liquid soap, 2. their liquid soap formulation deposited more germicide than the bar soap, and 3. the residual effect of hexachlorophene was not due to its solubility in skin lipids. Goldemberg (123) discussed a number of factors which influence substantivity of materials to keratin, including 1. molecular weight, 2. ionic nature (symmetry and sign of charge), 3. surface activity, 4. ability to form hydrogen bonds, 5. reactivity of ionic materials with keratin, and 6. binding of nonpolar, neutral materials. (This discussion used principles of dye substantivity to wool keratin as a model for substantivity to skin.) Another possible factor mentioned as having influence on the deposition of germicides on the skin was pore size. Linfield et al. (3) suggested that the larger pore areas of the underarm may lead to accumulation of antibacterial chemicals "as a function of soap retention." Parran (124) theorized that the deposition of antimicrobial particles from detergent bases might be a function of the psi and zeta potentials of the particles and skin. Another important observation that he made was that "particle deposition is strongly affected by the type of detergent base employed."

While skin substantivity is certainly a necessary element of the performance of an antimicrobial deodorant ingredient, it may not be the main determinant of efficacy. In an evaluation of skin substantivity versus antibacterial activity, McNamara and Steinbach (125) found no direct correlation. Based on preliminary data, they also saw no direct relationship between chemical structure and skin substantivity. Their study included triclocarban and a number of other antimicrobial agents of varying structures. Manowitz and Johnston (126), in a study of the deposition of hexachlorophene from bar soaps, reached the following conclusions:

> It was found that the quantity of hexachlorophene applied to the skin was a major factor in controlling the amount retained. Increasing quantities were deposited by the following methods:
> (a) Raising the concentration of hexachlorophene in the soap
> (b) Increasing the number of washes
> (c) Increasing the amount of soap applied during a single wash.

From their results with bathing studies they suggested that deposition of hexachlorophene depended on "wet pickup of the skin"; i.e., it was not limited by the amount of material in the bath water. Further, they thought that the deposition might be due to the "physical entrapment" of hexachlorophene on the skin and that "it is probably retained both as individual particles and solubilized in the soap left on the skin after washing."

In 1973 Elkhouly and Woodroffe (127) studied the effects of a surfactant, polyethylene glycol 400 monolaurate (PEG 400 ML), on the deposition of TCC® from solutions and suspensions with and without added salt. They made the following observations about the performance of this germicide with a solubilizing surfactant:

1. With excess triclocarban and any level of PEG 400 ML, "TCC will be distributed between the nonionic micelles and aqueous phase which also contains the excess insoluble TCC as finely suspended particles." There was more triclocarban deposited on the skin as the concentration of PEG 400 ML increased, but the relationship was only random. Experiments with a fixed amount of triclocarban solubilized in varying amounts of PEG 400 ML indicated that "TCC solubilized in the nonionic micelles was not retained by the skin and was easily rinsed off."

2. From studies in which all the triclocarban was solubilized the authors concluded that "solubilized TCC is not substantive to skin and that solubilization provides no improvement in the uptake of TCC unless the nonionic micelles are almost saturated with it."
3. This work also showed that 0.5% added salt also increased deposition.

In addition, these authors studied the antibacterial effects of PEG 400 ML and salt on triclocarban in their system. They observed that concentrations of PEG 400 ML below the critical micelle concentration did not affect the minimum inhibitory concentration of triclocarban, but just above this point some of the germicide was dissolved in micelles and resulted in poor performance. They further stated that the optimal range of activity was a ratio of 1:5 to 1:10 triclocarban to PEG 400 ML. They also found that 0.5% salt improved the antimicrobial activity of the triclocarban-PEG 400 ML mixtures.

At about the same period as the work of Elkhouly and Woodruffe, Parran patented the use of water-soluble cationic polymers as deposition-enhancing agents. Examples of the polymers he used include polyethylenimine and the reaction product of polyethyleneimine and either ethylene or propylene oxide (128), quarternary ammonium-substituted cellulose derivatives (129), and a polymer formed from tetraethylene pentamine and epichlorohydrin (130). The range of applications was wide for both detergents used in the base formulation and the material to be deposited on the skin. Examples included the use of triclocarban in a bar soap. While the mechanism was not fully understood, the author believed that "the polymer coats or attaches itself in some way on the involved particles imparting a net positive charge thereto which increases the affinity of the particles for the generally negatively charged washed surfaces" (128).

A recent patent (131) is concerned with the application of cationic polymers for the enhancement of antimicrobial activity of underarm deodorants and/or antiperspirants. The polymer used was a poly-dimethyldialkyl ammonium chloride-acrylamide copolymer or a dimethyldialkyl ammonium chloride polymer and the antimicrobial agents used included triclocarban and triclosan. The enhanced activity was thought to be due to "improved substantivity, retention on skin surface, or improved diffusion of the deodorant's active ingredient."

In a patent contemporary with those of Parran (described above), Cheng et al. (132) described the discovery of a variety of solvents which acted as potentiators to enhance the activity of several antimicrobial agents. The solvents were both hydrophilic and hydrophobic and included alkylphosphoramides, alkylureas, mixtures of C_8 to C_{22} fatty acids, polyethylene glycols, anionic-cationic complexes, and amines. It was thought that for hydrophilic solvents the potentiator affected "the manner of flocculation and subsequent deposition of the anti-bacterial during washing." For hydrophobic solvents the solvent was thought to act as a vehicle to enhance the deposition of the material. A more recent example of the use of a similar effect is the work of Hoppe et al. (133), who utilized an alkyl phenol polyglycol ether and small amounts of wool wax alcohols. They both claimed both reduction of "sandiness" from the germicide and better deposition onto the skin.

Grand (134,135) has patented the use of an "aminopolyureylene" resin to enhance the deposition of a number of water-soluble and water-insoluble active materials on a variety of desired surfaces. The suggested applications included bar soap products and triclocarban as one of the active materials. Seilinger (136) employed mica to improve the skin adherence of active ingredients in deodorants and antiperspirants. The final example of deposition enhancement, though quite interesting, was designed for hard surface disinfection. Brenner and Laufer (137) silanized a number of standard antimicrobial agents (including triclosan) and found not only excellent antimicrobial activity but also tenacioius substantivity in some cases. For example, silanized triclosan, after being applied to petri dishes, had "strong antimicrobial properties even after repeated hot alkali washing."

F. Synergistic Mixtures of Antimicrobial Agents

About 1960, Casely and Noel (3,138) discovered that certain pairs of "chemically unlike antibacterial agents" gave synergistic performance in bar soaps and with other surfactant formulations. Linfield et al. (3), in studies with synergistic mixtures of triclocarban and hexachlorophene, found that a 50:50 mixture was the optimum ratio for that pair. Over the years since then there have been a number of patents, such as the following, for synergistic mixtures of triclocarban and/or triclosan with or without other ingredients:

1. 1-Hydroxy-2-pyridinethione and at least two other antimicrobials selected from a group of salicylanilides and carbanilides (including triclocarban) (139)
2. Nine to 40 parts triclocarban to 1 part triclosan (140)
3. Two to 100 parts of an typical antimicrobial agent (e.g., hexachlorophene, bithionol, triclocarban) plus 1 part of an organomercury salt (141)
4. Hexachlorophene, triclocarban, and 1-(2-ethylhexyl)-3-[3,5-bis(trifluoromethyl)phenyl] urea (142)
5. Triclocarban or 4,4'-dichloro-3-trifluoromethyl carbanilide plus 2,2'-methylene-bis[3,4,6-trichlorophenol]-di-(*N*-methylcarbamate) (143)
6. Approximately equal portions of triclocarban, triclosan, and 4,4'-dichloro-3-trifluoromethyl carbanilide (144)
7. Triclosan or similar compounds plus triclocarban or similar compounds in varioius ratios (145)
8. Triclosan, triclocarban, or 4,4'-dichloro-3-trifluoromethyl carbanilide plus citronellyl senecioate in various ratios (146)

G. Effects of Surfactants on Antimicrobial Activity

It was observed as early as 1911 that surfactants can have effects on the activity of antimicrobial agents (147). Between that time and 1950 a number of attempts were made to characterize these effects, but the results appeared to be contradictory (148). In the 1950s Bean and Berry (148,149) and Berry and Briggs (150) pointed out that the earlier, seemingly contradictory results could be accounted for by the micellar behavior of soaps and other surfactants. Their studies included a sparingly soluble germicide, benzylchlorophenol (5-chloro-2-hydroxydiphenylmethane), in relatively dilute solutions of potassium laurate. In one series of experiments they observed the following:

1. An enhancement of antimicrobial activity below the critical micelle concentration (CMC), which they accounted for by decreasing surface tension and increasing solution saturation
2. A decrease in activity just above the CMC, which they believed was due to the nascent soap micelles being "undersaturated" in benzylchlorophenol*

---

*It had been shown previously that the antimicrobial activity of solutions of phenol in soap was related to the concentration of phenol in the aqueous phase and not its concentration in the micelles (151).

3. A final rapid increase and leveling off of activity well above the CMC, which they suggested was due to the size and relatively saturated state of the micelles

In 1960 Linfield et al. (3) studied the effects of nonionic surfactants on the antimicrobial activity of triclocarban in bar soaps. They found a potentiating effect on triclocarban for low levels of several nonionics, but an inhibitory effect at higher levels. Generally they found monoesters of polyglycols ineffective but polyglycol ethers of tridecyl alcohol and nonylphenol particularly effective. They observed little potentiation for hexachlorophene or bithinol.

Molnar (152) conducted a broad study of the effects of surfactants on activity of several antimicrobial agents including triclocarban. The surfactants he used were anionics (including a commercial fatty acid soap and a neutral bar made principally from acyl isethionate and stearic acid), amphoterics, nonionics, nonionics with some cationic character, and cationics. Because of the large variety of surfactants, antimicrobial agents, and bacteria studied, it is not convenient to summarize the author's results. As the author himself stated, however, "The work demonstrates that in any technical appraisal of the antimicrobial potency of a compound, the influence of the surfactant vehicle is of paramount importance."

H. Chemistry of Triclocarban and Triclosan

1. *Triclocarban—Preparation/Formulation*

According to Beaver and Stoffel (153), triclocarban may be prepared by adding a solution of 4-chlorophenyl isocyanate in an aprotic solvent to a 3,4-dichloroaniline in the same solvent. A variety of aprotic solvents may be used including ethers and liquid alkanes. The reaction is exothermic and will proceed with good yields (88–100%) at room temperature. In a later patent, Herber (154) describes a somewhat different procedure which involves the use of *o*-dichlorobenzene as a solvent, simultaneous addition of reactants to the reaction vessel, and reaction temperatures of 100 to 170°C.

The product is sold as a micronized powder for incorporation into bar soaps. The micronization is done to prevent the formation of a gritty-feeling soap bar and also to aid in dispersion. A typical approach to formulation for bar soaps is to suspend the micronized triclocarban in a watery slurry, coat soap chips with the slurry, and then disperse the triclocarban and other ingredients throughout the soap mass by milling and/or plodding.

Although triclocarban is marketed mainly for use in bar soaps, there have been several patents which describe formulation techniques for incorporation of the ingredient into clear liquid products such as shampoos and liquid detergents. One of the main problems with using triclocarban in these products is that it is practically insoluble in water. Russell (155) suggested the use of solubilizing agents such as polyethoxylated lanolin alcohol, polyethoxylated fatty acid salts of sorbitan, and polyethoxylated phenols. Another approach, invented by Lavril (156), involved the use of an amphoteric surfactant such as dodecylbetaine for the same purpose. The advantage claimed was that the betaine interfered less with foaming than some of the previously used solubilizers had. Another scheme for solubilizing triclocarban in a liquid detergent, described by Wilson (157), involved various combainations of the following: alkaline mixtures of water and alkane diol, polyethers, fatty acid alkanolamides, alkanolamines, nonionic surfactants, and alkaline solutions of anionic surfactants.

2. *Triclosan—Preparation/Formulation*

Model and Bindler (158) describe several ways in which triclosan and related compounds may be prepared. Two of their example preparations (shown in Figs. 2 and 3) involve well-known aromatic nucleophilic substitution chemistry, diazotization, and subsequent standard conversions of nitro groups to the desired substituents.

In two relatively recent patents, Lund (159) and Lund and Brown (160) discuss processes to improve the yield and purity of triclosan produced by the path outlined in Fig. 2. According to these authors, before their improved method the hydrolysis was run in 75% sulfuric acid at 155–160°C. The crude reaction product contains triclosan plus 24–28% 2,4,8-trichlorodibenzofuran and traces of several other compounds. The purification of the crude reaction product was conducted as follows: 1. separation of the organic layer from the acid at 120°C., 2. treatment with caustic to form a salt of triclosan, 3. separation and extraction of organic impurities with toluene, and 4. acification, separation of the organic layer, distiallation, and recrystallization to form the commercially pure product. The yield of this step was about 53–56% from 2,4,4'-trichloro-2-aminodiphenyl ether.

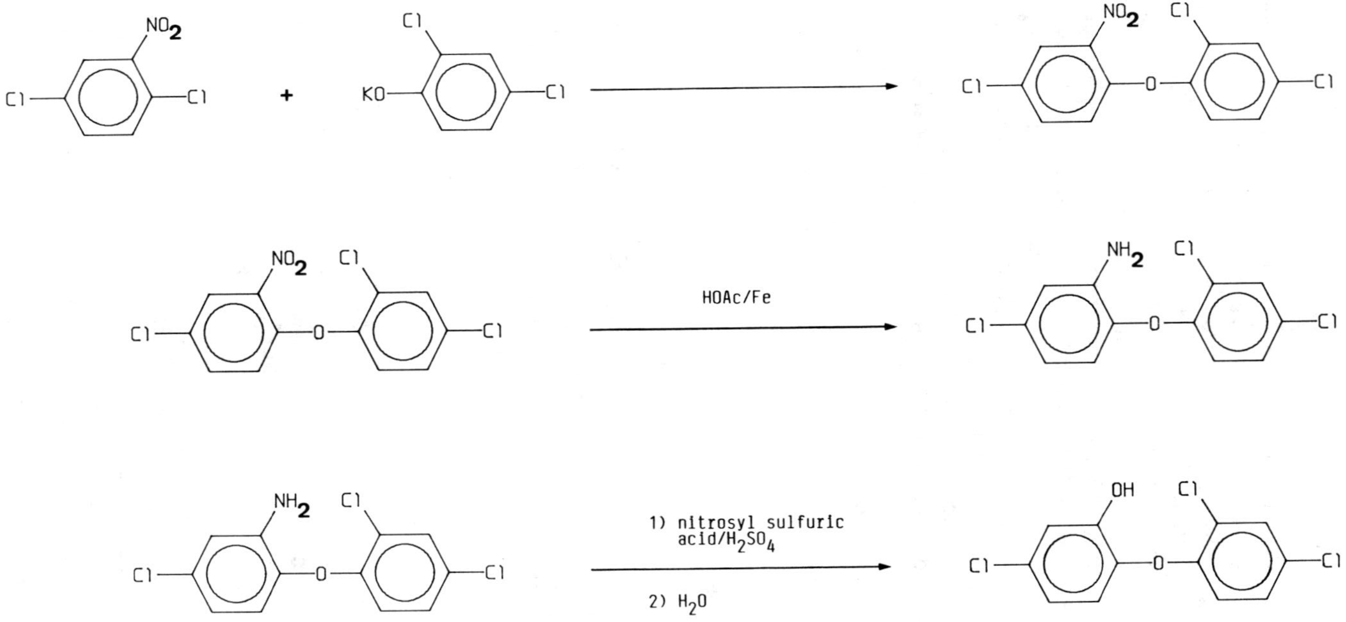

Figure 2 Preparation of triclosan from 1,4-dichloro-2-nitrobenzene and 2,4-dichlorophenol.

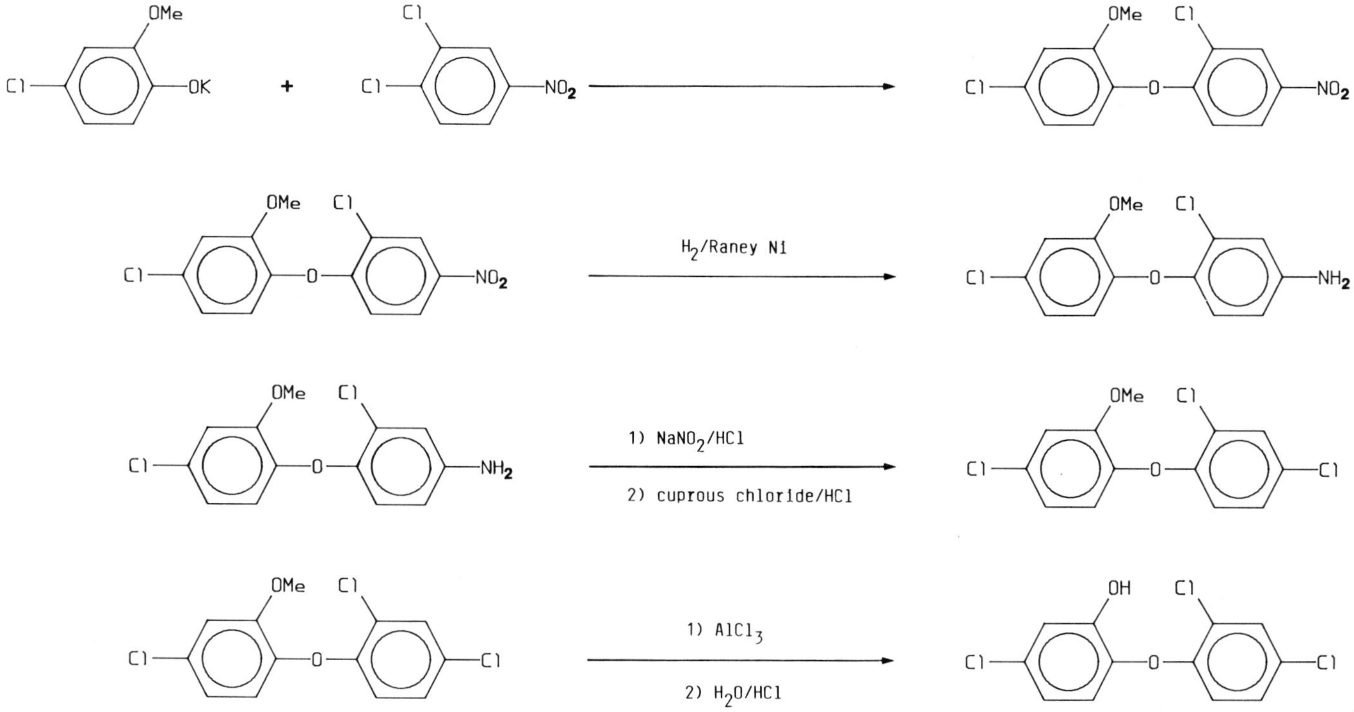

Figure 3 Preparation of triclosan from 4-chloro-2-methoxyphenol and 1,2-dichloro-4-nitrobenzene.

The improvements developed by Lund and Brown increased the yield by approximately 5% to 58–63% by raising the hydrolysis temperature to 175–180°C and modifying the purification scheme. The new purification involved a procedure similar to that outlined above except that tetrachloroethylene was the preferred extraction solvent and the distillation/recrystallization steps were replaced with the use of a molecular still. Schreiber and Marky (161) suggested the addition of a wet-grinding step to obtain a dust-free product.

Triclosan is marketed in powder form and is used in a variety of deodorant products. Although neither triclocarban nor triclosan is soluble in water, triclosan is soluble in a wider range of organic solvents and thus offers more latitude for incorporation into products. For deodorant bar soaps, triclosan may be dissolved in the perfume for subsequent addition to soap chips as well as the slurry method described above for triclocarban.

As described in Ref. 109, soap bars which contain triclosan "do not discolor significantly under normal exposure to tungsten or fluorescent light, but may darken on exposure to direct sunlight." Several solutions to this problem are offered in the literature, and an example is a patent by Watanabe and Arisawa (162). The inventors claim that the use 0.05 to 5% of at least one organic phosphate ester improves the sunlight stability of soap bars containing triclosan. One of the esters used was monosodium mono(polyoxyethylene(3))lauryl phosphate.

G. Other/New Deodorant Ingredients Based on Prevention of Odor Development

1. *Antimicrobial Agents*

Surfactants

Blank and Coolidge (163) presented an early (1950) discussion of the effectiveness of cationics for application to skin degerming. They mentioned that quaternary ammonium compounds had been known as germicides since 1922 and had gained interest in the medical profession as antiseptics after 1935. The authors presented evidence that this type of compound was poor as a skin animicrobial agent because it enhanced the retention of bacteria on the skin. The effect they proposed was that these cationics altered the surface charge of the skin from negative to positive and, since "most bacteria and dirt particles are negatively charged," gave the opposite of the desired effect.

The thoughts of Blank and Coolidge about the lack of suitability of quaternary ammonium compounds for skin-degerming formulations may be more appropriate for wash-off products. Quaternary ammonium compounds have been used effectively in leave-on deodorant products for some time (5), particularly benzethonium chloride and methylbenzenethonium chloride (116). Examples of recent work on this type of antimicrobial agent are given below.

Tolgyesi and Guth (164) showed utilization of a bifunctional quaternary compound, $RN^{+}Me_2(CH_2CH_2S)nCH_2CH_2N^{+}Me_2R\ 2X^{-}$ ($R = C_6$ to C_{16} alkyl and $n = 1-3$), and claimed good antimicrobial activity plus resistance to loss of the material from perspiration. Another discovery which could be included in this category is a 1984 patent by Carrillo (165) involving the use of polyhexamethylene biguanide hydrochloride as a deodorant material. The material was found to perform well in clincial deodorancy tests in standard formulations against similar products containing benzethonium chloride or triclosan. A somewhat similar material for deodorant applications (a urea-guanidine-formaldehyde condensation product) is described in an earlier patent (166).

Hewitt (167) has patented a novel combination of a polyquaternary ammonium compound with the phenolate anion of a standard antimicrobial agent like triclosan. The combination, when applied to the skin, is said to "provide for a slow steady release of the antimicrobial agent in relatively low, but, nevertheless, antimicrobially inhibiting quantities" (167).

In addition to the quaternary ammonium antibacterials described above, deodorant ingredients utilizing other types of surfactant amterials have been discovered. Michaels (168–171) has a series of patents which describe the deodorant properties of various mixtures of an alkyl-*N*,*N*-dimethylamine oxide, an alkyl-*N*,*N*-dihodroxyethylamine, or an acylamido *t*-amine oxide plus a betaine and a protonating agent. When used as a body wash, some of the example formulations were purported to control body odor for as long as 72 hours.

The last example of this category comes from 1983–1984 patents of Lowicki and Desai. They found "gentle bacterial and fungicidal action" from a surfactant which is a polyhydric alcohol partial ester of the adduct of maleic anhydride and undecylenic acid (172) and high antimicrobial efficacy with the zinc salt of a hydrolyzed tricarboxylic acid from ene-adducts of maleic anhydride to undecylenic acid (173).

Fatty Acid Derivatives

It is interesting to note that the skin has a remarkable "self-sterilizing" power which is responsible in part for the generally good resistance of most individuals to pathogenic bacteria on the skin. Some skin lipid components, especially unsaturated free fatty acids, have been mentioned as factors contributing to this resistance (174). Other researchers have found good antimicrobial activity with fatty acids and compounds derived therefrom, including acids not found on the skin.

In 1941 Pohle and Stuart (175) published work indicating that rosin soap could be used to enhance the germicidal effects of commercial fatty acid soaps. [Rosin is primarily made up of abietic and isomeric acids (176).] Notwithstanding the lack of antimicrobial activity found with zinc ricinoleates (45) (see above), Novak and co-workers (177–180) found antibacterial activity for a number of ricinoleic, oleic, and other fatty acids and some of their derivatives.

Kabara (181) has conducted a number of studies on the antimicrobial activity of fatty acids and their derivatives and has generalized the activity as follows:

> Saturated fatty acids have their highest activity when the chain length is twelve carbons (C_{12}) long; monounsaturated fatty acids reach their peak with palmitoleic acid ($C_{16:1}$); the most active polyunsaturated fatty acid is linoleic. *Trans* isomers are not active against microorganisms.

He has also patented the use of lauroyl monoesters of polyols (e.g., glycerin) (182) and synergistic mixtures thereof (183) for food and cosmetic antimicrobial agents.

The final example is from a 1985 publication by Pandey et al. (184). The authors found in vivo substantivity and antimicrobial activity with sodium 2-lauroyloxy propionate. The studies were done with soap solutions.

Natural Products

In addition to the multitude of synthetic bacteriostats described in the literature, there are a number of antimicrobial deodorant ingredients based on natural product extracts. One of the most well investigated of these comes from a plant group widely found in nature called lichens. [Lichens are symbiotic associations of fungi and algae (185).] Extracts from lichens have been used for years as a treatment for superficial wounds and infections. One component of the extracts, usnic acid, a substituted dibenzofuran, was found to have antimicrobial properties especially against Gram-positive bacteria. This property made it suitable for investigation as a deodorant material. To make it useful for cosmetic applications the usnic acid had to be complexed with triethanolamine (TEA) to make it soluble enough to give bacteriostatic concentrations in a product. Because of problems with pH stability, however, it is not suited for use in bar soaps (186). Other modifications using this type of material include 1. adding 0.05% hexachlorophene to a formulation with the TEA/usnic acid complex (187), 2. derivitization (188), 3. use of hydroxy amino compounds (189), and 4. use of other lichen acids such as evernic, vulpinic, and psoromic as well as usnic (190).

Other examples of deodorant-active materials in the natural product category are

1. An extract from either a wood-infesting fungus or mushroom species, which contains agaricic acid (a tricarboxylic acid) (191,192)
2. An extract of hops (193)
3. A lipopolyamino acid (194)

Miscellaneous Antimicrobials

There is a continuing effort to develop new antimicrobial agents for use in cosmetic deodorant products, and the approaches used in the recent literature range from the use of relatively simple ingredients to modification of antimicrobial agents used in the past to relatively exotic new synthetic agents. Several of these are listed below.

1. Liquid deodorant containing 47 to 87% propylene glycol as the active ingredient (195)
2. Deodorant cream with 6% hydrogen peroxide (196)
3. Deodorant with hexamethylenetetramine and zinc oxide (197)
4. Alkenylidene bisphenols (structurally similar to hexachlorophene) (198)
5. Alkyl salicylanilides (199) and new halosalicylanilides (200)
6. Prenylamine in deodorant compositions (201)
7. Bicyclic oxazolidines (202)

8. 2-Oxo-1,2,3-oxathiazolidines (203)
9. 5-(3,3-Dimethyl-2-norbornylidene)-3-penten-2-one (204)
10. O-(Nitroalkyl) thiocarbamates (205)
11. 1-Aza-3,7-dioxabicyclo[3.3.0]octanes (206)
12. 1-(3-Chloroallyl)-3,5,7-triaza-1-azoniaadamantane chloride (207)

Antioxidants—Enhancement of Effects of Antimicrobial Agents and Other Deodorant Effects

Antioxidants (especially vitamin E) have been claimed to enhance the antimicrobial activity of several deodorant ingredients including

1. A formulation containing triclosan and zinc oxide (208)
2. Esters of mono- and dibasic aliphatic hydroxycarboxylic acids (209)
3. Esters of citric acid and/or acetylcitric acid (210)
4. A combination of the esters in 2 and 3 above with antiperspirant Al salts (211)

Goldschmiedt (212) has described another use of antioxidants in deodorants. He summarized the activity of Mennen E®, a deodorant introduced in the United States in 1972, as follows:

> Odor is only formed when resident bacteria act in conjunction with oxygen and cause lipid components of perspiration to become rancid. Vitamin E, or tocopherol, in a personal deodorant acts as a preservative rather than as a germicide. It functions as an antioxidant to prevent the formation of odor.

Encapsulated Antimicrobial Agents

Some inventors have changed the physical form of conventional antimicrobial agents to achieve a special effect. Fang (213) encapsulated water-soluble cationic germicides in a carboxylic acid polymer matrix for a deodorant product which would give controlled release of the germicide. Adler et al. (214) developed a deodorant powder for sensitive skin by encapsulating triclosan in whey particles.

2. *Miscellaneous Odor Prevention Agents*

Enzyme Inhibitors

Osberghaus (27) proposed a novel hypothesis for controlling body odor without antibacterial agents. The scheme involves the use of esters of hydroxycarboxylic acids as deodorizing agents. When applied to the axilla these materials are purported to be attacked by nonspecific esterases and hydrolyzed to free carboxylic acids. The author further states that the skin pH is lowered with the acids, and the skin esterases lose activity and "thus checkmate themselves." Osberghaus believes this cycle inhibits the liberation of odorous free fatty acids from the axilla. Reese and Osberghaus (215,216) patented some applications of this idea. Geks and Schmidt-Kastner (217) patented another deodorant based on enzyme inhibition using a biological protease inhibitor. The authors claim that decomposition of proteinaceous material in the axilla contributes to body odor.

Bacterial Antibodies

The final example of this chapter is a very clever application of biotechnology to the development of new deodorant products. Beck (218) patented the use of bacterial antibodies for deodorants. The author raised the antibodies by injecting cows with killed cells of common skin bacteria, isolated the immunoglobulin G (IgG), and prepared a deodorant formulation from talc and 1.5% of the antibody. The formulation showed antimicrobial efficacy in a clinical test. Beck suggested that the mechanism of action could be bactericidal action or prevention of the bacteria from adhering to the skin.

V. SUMMARY/CONCLUSION

This chapter has covered the three major mechanisms of action for deodorant ingredients: odor modification, reduction and/or removal, and prevention. These mechanisms were discussed in relation to many types of deodorant ingredients. Of all of these materials, antimicrobial agents (which prevent the formation of odor) are the most commercially important. The need for development of new deodorant ingredients should continue in the foreseeable future because of 1. the current attitudes linking personal hygiene with social acceptability, 2. the size of the cosmetic product market, and 3. the ever-present prospect of change in government regulations for deodorant products. It is difficult to predict where the next successful idea will come from in the multitude of deodorant product concepts. Perhaps we should look to a past legend that newly married couples in the South Sea Islands took potions, dictated by ritual, to systemically produce a pleasant body fragrance on their wedding night.

ACKNOWLEDGMENTS

The author gratefully acknowledges the expert assistance and special efforts of the following people at the Dial Technical Center in gathering literature and patent references for this chapter: Shirley Blazer, Ray Drozda, Tom Jackson, Linda Monroe, and Lorraine Nesvig. Further, I wish to thank the following people for their efforts in providing editorial and/or technical assistance: Ray Drozda, Dr. John Heinze, Dr. Aaron Herrick, Dr. Richard Murahata, Dr. John Murphy, Dr. Helen North-Root, Dr. D. I. Richardson, and Catherine Schmit.

REFERENCES

1. "Personal Care Products Market in the U.S.," Report 1569, Frost & Sullivan, Inc., New York (July 1986).
2. *Soap, Cosmet. Chem. Spec., 62:* 32 (December 1986).
3. Linfield, W. M., Casely, R. E., and Noel, D. R. Studies in the development of antibacterial surfactants. II. Performance of germicidal and deodorant soaps, *J. Am. Oil Chem. Soc., 37:* 251–154 (1960).
4. Hooper, D. C., Johnson, G. A., and Peter, D. U.S. Patent 4,304,679 (1981).
5. Plechner, S. Antiperspirants and deodorants, *Cosmetics: Science and Technology*, 2nd ed. (M. S. Balsam and E. Sagarin, eds.), Wiley-Interscience, New York (1972), pp. 373–416.
6. Fiedler, H. P. *Der Schweiss*, Editio Cantor KG, Aulendorf i. Württ (1968), pp. 413–414.
7. Rosenbury, T. *Life on Man*, Viking, New York (1969), p. xiv.
8. Frazier, G. Bath. (History), *The World Book Encyclopedia*, Vol. 2, World Book, Inc., Chicago (1985), p. 119.

9. Osteroth, D. Soap through the ages, *Dragoco Report, 4:* 87–97 (1981).
10. Buckenmayer, R. H. Odor control, *Kirk-Othmer Encyclopedia of Chemical Technology,* Vol. 16, 34d ed. (M. Grayson and D. Eckroth, eds.), Wiley-Interscience, New York (1981), p. 297.
11. Cain, W. S., and Drexier, M. Scope and evaluation of odor counteraction and masking, *Ann. N.Y. Acad. Sci., 237:* 427-439 (1974).
12. Sturm, W. Deosafe fragrances: Fragrances with deodorizing properties, *Cosmet. Toiletries, 94:* 35–48 (February 1979).
13. Anonis, D. P. Perfumes in deodorants/antiperspirants, *Drug Cosmet. Ind., 119:* 46–49, 126–127 (September 1976).
14. British Patent 1,311,060 (1972), assigned to Warner-Lambert Co.
15. Read, J. Odor control, *The Sense of Smell* (R. H. Wright, ed.), CRC Press, Boca Raton, Florida (1982), pp. 157–159.
16. Amoore, J. E. Odor theory and odor classification, *Fragrance Chemistry. The Science of the Sense of Smell* (E. T. Theimer, ed.), Academic Press, New York (1982), pp. 28–76.
17. Cain, W. S. Odor intensity: Mixtures and masking, *Chem. Senses Flavor, 1:* 339–352 (1975).
18. Köster, E. P. Intensity in mixtures of odorous substances, *Olfaction and Taste III* (C. Pfaffmann, ed.), Rockefeller Univ. Press, New York (1969), pp. 142–149.
19. Torii, K., and Egma, C. U.S. Patent 4,501,730 (1985).
20. Hooper, D. C., Johnson, G. A., and Peter, D. U.S. Patent 4,322,308 (1982); British Patent 2,012,302 A (1982); British Patent 2,016,507 A (1982); U.S. Patent 4,289,641 (1981); U.S. Patent 4,288,341 (1981); U.S. Patent 4,469,848 (1984).
21. Hayaishi, O., Nozaki, M., and Abbot, M. T. Oxygenases: Dioxygenases, *The Enzymes,* 3rd ed., Vol. 12, Part B (P. D. Boyer, ed.), Academic Press, New York (1975), p. 150.
22. Jackson, H. W. Oil favor quality assessment, *Bailey's Industrial Oil and Fat Products,* Vol. 3 (T. H. Applewhite, ed.), Wiley-Interscience, New York (1985), p. 245.
23. Yardley, H. J. Epidermal lipids, *Biochemistry and Physiology of the Skin,* Vol. 1 (L. A. Goldsmith, ed.), Oxford University Press, New York (1983), p. 373.

24. Downing, D. T., Stewart, M. E., Wertz, P. W., Coltin, VI, Abraham, W., and Strauss, J. S. Skin lipids: An Update, *J. Invest. Dermatol.*, *88:* 2s–6s (1987).
25. Sonntag, N. O. V. Reactions of fats and fatty acids, *Bailey's Industrial Oil and Fat Products*, Vol. 1, 4th ed. (D. Swern, ed.), Wiley-Interscience, New York (1979), p. 140.
26. Hammarström, Lindgren, J. A., Marcelo, C., Duell, E. A., Anderson, T. F., and Voorhees, J. J. Arachidonic acid transformations in normal and psoriatic skin, *J. Invest. Dermatol.*, *73:* 180–183 (1979).
27. Osberghaus, R. Nonmicrobicidal deodorizing agents, *Cosmet. Toiletries*, *95:* 48–50 (July 1980).
28. Morris, J. A., Khettry, A., and Seitz, E. W. Antimicrobial activity of aroma chemicals and essential oils, *J. Am. Oil Chem. Soc.*, *56:* 595–603 (1979).
29. Schmit, C. L. unpublished work (1987).
30. Killian, J. A., and Panzarella, F. P. Comparative studies of samples of perspiration collected from clean and unclean skin of human subjects, *Proc. Sci. Sect. Toilet Goods Assoc.* *7:* 3–11 (May 1947).
31. Blank, I. H., Moreland, M., and Dawes, R. K. The antibacterial activity of aluminum salts, *Proc. Sci. Sec. Toilet Goods Assoc.* *27:* 24–28 (May 1957).
32. Slocomb, A. B., and Day, J. W. U.S. Patent 279,195 (1883).
33. Smith, A. G. U.S. Patent 1,558,405 (1924).
34. Lamb, J. H. Sodium bicarbonate: An excellent deodorant, *J. Invest. Dermatol.*, *7:* 131–133 (1946).
35. Baxter, P. M., and Reed, J. V. The evaluation of underarm deodorants, *Intl. J. Cosmet. Sci.*, *5:* 85–96 (1983).
36. Marschner, F. W. U.S. Patent 4,382,079 (1983).
37. Marschner, F. W. U.S. Patent 4,440,741 (1984).
38. Marschner, F. W. U.S. Patent 4,440,742 (1984).
39. Marschner, F. W. U.S. Patent 4,534,962 (1985).
40. Stanave, F. J. Canadian Patent 1,088,428 (1980).
41. Shelley, W. B., Hurley, H. J., and Nichols, A. C. Axillary odor, *Arch. Dermatol. Syphilol.*, *68:* 430–446 (1953).
42. Marschner, F. W. U.S. Patent 4,565, 693 (1986).
43. Bews, B., Critchley, P., Durrant, J. A., Stebles, M. R. D., and Tipping, L. R. H. European Patent 24,175 (1981)

44. Bews, B., Critchley, P., Durrant, J. A., Stebles, M. R. D., and Tipping, L. R. H. European Patent 24,176 (1981).
45. Lowicki, N., Sidillo, M., and Neunhoeffer, O. Ein neues Prinzip der Desodorierung, *Fette Seifen Anstrichm.*, *75:* 647–651 (1973).
46. Meyer-Rohn, J. Experimentelle und klinische Untersuchungen zur desodorierenden Wirkung von GRILLOCIN, *Parfüm. Kosmet.*, *57:* 239–246 (1976).
47. Sartori, P., Lowicki, N., and Sidillo, M. Action of metal ricinoleates as deodorizing agents, *Cosmet. Toiletries*, *92:* 45–48 (May 1977).
48. Makin, E. C. Clathration, *Kirk-Othmer Encyclopedia of Chemical Technology*, Vol. 6, 3rd ed., (M. Grayson and D. Eckroth, eds.), Wiley-Interscience, New York (1981), p. 179.
49. Moroe, M. U.S. Patent 4,634,588 (1987).
50. Thurmon, F. M. U.S. Patent 2,653,902 (1953).
51. Ikai, K. Deodorizing experiments with ion exchange resins, *J. Invest. Dermatol.*, *23:* 411–422 (1954).
52. Winters, J. C. Potential utility of ion-exchange resins in antiperspirant-deodorant formulations, *J. Soc. Cosmet. Chem.*, *7:* 256–267 (1956).
53. Schrader, K. German Patent 2,731,520 (1979); *Chem. Abstr.*, *90:* 174515v (1979).
54. Osberghaus, R. German Patent 2,721,297 (1978); *Chem. Abstr.*, *90:* 76415t (1979).
55. Douant, J. French Patent 2,551,658 (1985); *Chem. Abstr.*, *103:* 59113n (1985).
56. Moeller, H., and Osberghaus, R. German Patent 3,009,544 (1981); *Chem. Abstr.*, *95:* 209387n (1981).
57. Moeller, H., and Osberghaus, R. German Patent 3,009,545 (1981); *Chem. Abstr.*, *96:* 11522e (1982).
58. Jacquet, B., Lang, G., and Malaval, A. U.S. Patent 4,425,321 (1984).
59. Sato, K., Leidal, R., and Sato, F. Morphology and development of an apoeccrine sweat gland in human axillae, *Am. J. Physiol.*, *252:* R166–R180 (1987).
60. Sato, K., and Sato, F. Sweat secretion by human axillary apoeccrine sweat gland in vitro, *Am. J. Physiol.*, *252:* R181–R1187 (1987).
61. Labows, J. N. Chapter 9, this volume.

62. Jackman, P. J. H. Body odor—the role of skin bacteria, *Semin. Dermatol.*, *1:* 143–148 (1982).
63. Labows, J. N., McGinley, K. J., and Kligman, A. M. Perspectives on axillary odor, *J. Soc. Cosmet. Chem.*, *34:* 193–202 (1982).
64. Leyden, J. J., McGinley, K. J., Hölzle, E., Labows, J. N., and Kligman, A. M. The microbiology of the human axilla and its relationship to axillary odor, *J. Invest. Dermatol.*, *77:* 413–416 (1981).
65. Shehadeh, N., and Kligman, A. M. The bacteria responsible for axillary odor. II, *J. Invest. Dermatol.*, *41:* 3 (1963).
66. Hölzle, E., and Neubert, U. Antimicrobial effects of an antiperspirant containing aqueous aluminum chloride hexahydrate, *Arch. Dermatol. Res.*, *272:* 321–329 (1982).
67. Chilson, F. Deodorants, *Modern Cosmetics,* Drug and Cosmetic Industry, New York (1934), pp. 167–177.
68. Harry, R. G. Deodorants, *Modern Cosmeticology. The Principles and Practice of Modern Cosmetics,* Chemical Publishing Co., Brooklyn, New York (1940), pp. 123–131.
69. de Navarre, M. G. Deodorant creams. Patent digest, *The Chemistry and Manufacture of Cosmetics,* Van Nostrand, New York (1941), pp. 259–267.
70. Jass, H. E. Rationale of formulations of deodorants and antiperspirants, *Principles of Cosmetics for Dermatologists* (P. Frost and S. Horwitz, eds.), Mosby, St. Louis (1982), pp. 98–104.
71. Laden, K. Chapter 1, this volume.
72. McDonough, E. G. Sweat glands and preparations, *Truth about Cosmetics,* Stratford Press, New York (1937), pp. 248–256.
73. Lever Brothers, Consumer Services, private communication.
74. Marzulli, F. N., and Bruch, M. "Antimicrobial Soaps: Benefits versus Risks," presented at seminar on Skin Microbiology: Relevance to Clinical Infection, San Francisco (August 29–31, 1979).
75. Jungerman, E. Soap bacteriostats, *J. Am. Oil Chem. Soc.*, *45:* 345–350 (1968).
76. History of the cosmetics and toiletries industry in words and pictures, *Cosmet. Toiletries,* *97:* 33–100 (April 1982).
77. Kunz, E. C., and Gump, W. S. Argentine Patent 52,542 (1941).
78. Gump, W. S. U.S. Patent 2,250,480 (1941).

79. Chase, N. How Armour produced, *Advertising Agency and Advertising and Selling, 190:* 58–59 (March 1950).
80. Herman, P. S., and Sams, W. M., Preface and Salicylanalides and photosensitivity, *Soap Photodermatitis: Photosensitivity to Halogenated Salicylanilides,* Thomas, Springfield, Illinois (1972), pp. vii–viii, 17–61.
81. Shumard, R. S., Beaver, D. J., and Hunter, M. C. New bacteriostat for soap, *Soap Sanit. Chem., 29:* 34–37,90 (1953); (paper presented at 39th annual meeting of the Chemical Manufacturers Association, New York (December 8, 1952).
82. Vinson, L. J. New antiseptic for soap, *Soap Sanit. Chem., 30:* 44–47, 103 (April 1954).
83. Kennon, L. Some aspects of toiletries technology, *Am. Perfum. Cosmet., 80:* 37–55 (November 1965).
84. Plechner, S. Antiperspirants and deodorants, *Cosmetics Science and Technology* (E. Sagarin, ed.), Interscience, New York (1957), pp. 717–739.
85. Shehadeh, N. H., and Kligman, A. M. The effect of topical antibacterial agents on the bacterial flora of the axilla, *J. Invest. Dermatol., 40:* 61–71 (1963).
86. *Fed. Regist., 37:* 219 (1972); *Fed. Regist., 37:* 20160 (1972).
87. Lockhart, J. D. How toxic is hexachlorophene, *Pediatrics, 50:* 229 (1972).
88. Jungermann, E. Soap, *Bailey's Industrial Oil and Fat Products,* Vol. 1, 4th ed. (D. Swern, ed.), Wiley-Interscience, New York (1979), p. 571.
89. Gump, W. S. Toxicological properties of hexachlorophene, *J. Soc. Cosmet. Chem., 20:* 173 (1969).
90. Vaterlaus, B. P., and Hostynek, J. J. Tolerability of hexachlorophene, *J. Soc. Cosmet. Chem., 24:* 291 (1973).
91. Kimbrough, R. D., and Gaines, T. B. Hexachlorophene effects on the rat brain-study of high doses by light microscopy and electron microscopy, *Arch. Environ. Health, 23:* 114 (1971).
92. Gaines, T. B., Kimbrough, R. D., and Linder, R. E., The oral and dermal toxicity of hexachlorophene in rats, *Toxicol. Appl. Pharmacol., 25:* 332 (1973).
93. Santolucito, J. A. The electro encephalogram and visual evoked potential of the squirrel monkey fed hexachlorophen, *Toxicol. Appl. Pharmacol., 22:* 276 (1972).
94. Powell, H. M. *J. Indiana State Med. Assoc., 38:* 303 (1945).

95. *Fed. Regist.*, *37:* 160, 219 (January 7, 1972).
96. Gump, W. Disinfectants and antiseptics, *Kirk-Othmer Encyclopedia of Chemical Technology*, Vol. 7, 3rd ed. (M. Grayson and D. Eckroth, eds.), Wiley-Interscience, New York (1981), p. 814.
97. Kaul, A. F., and Jewett, J. F. Agents and techniques for disinfection of the skin, *Surg. Gynecol. Obstet.*, *152:* 677–685 (1981).
98. Food and Drug Administration, Docket No. 75N-0183.
99. Food and Drug Administration, Antiperspirant drug products for over-the-counter human use; tentative final monograph (proposed rule), *Fed. Regist.*, *47:* 36492–36505 (1982).
100. Monsanto Technical Bulletin O/SS-1 (March 1970) "A Review of the Use of TCC Soap Bacteriostat in Bar Soaps," Detergent Materials Special Report, Monsanto Industrial Chemicals, St. Louis, Missouri (1972).
101. Heinze, J. E., and Schmit, C. L., unpublished results.
102. "TCC Bacteriostat (3,4,4'-Trichlorocarbanilide) Product Data Sheet," Monsanto Chemical Company, St. Louis, Missouri (1983).
103. Beaver, D. J., Roman, D. P., and Stoffel, P. J. The preparation and bacteriostatic activity of substituted ureas, *J. Am. Chem. Soc.*, *79:* 1236–1245 (1957).
104. Hamilton, W. A. The mechanism of the bacteriostatic action of tetrachlorosalicylanilide: A membrane-active antibacterial compound, *J. Gen. Microbiol.*, *50:* 441–458 (1968).
105. Hamilton, W. A. Membrane-active antibacterial compounds, *Biochem. J.*, *118:* 46P–47P (1970).
106. Hamilton, W. A., and Jeacocke, R. E. The ion-specific increases in membrane permeability with a group of membrane-active antibacterial agents, *Biochem. J.*, *127:* 56P–57P (1972).
107. Niven, D. F., and Hamilton, W. A. The mechanism of energy coupling in the active transport of amino acids by *Staphylococcus aureus*, *Biochem. J.*, *127:* 58P (1972).
108. Hamilton, W. A. The mode of action of membrane-active antibacterials, *FEBS Symp.*, *20:* 71–79 (1970).
109. "Irgasan DP300 Broad Spectrum Bacteriostat," Ciba-Giegy Corporation, Greensboro, North Carolina (1983).
110. Regös, J., and Hitz, H. R. Investigations on the mode of action of triclosan, a broad spectrum antimicrobial agent, *Zentralbl. Baktinol. Hyg. I Abt. Orig. A*, *226:* 390–401 (1974).

111. Meincke, B. E., Kranz, R. G., and Lynch, D. L. Effect of irgasan on bacterial growth and its adsorption into the cell wall, *Microbios, 28*: 133–147 (1980).
112. Kranz, R. G., Lynch, D. L., Darveau, R. P., and Russel, M. Absorption and concentration of irgasan by two resistant micro-organisms, *Microbios Lett., 5*: 109–114 (1978).
113. Blank, I. H., and Dawes, R. K. Antibacterial activity of weak solutions of aluminum salts, *AMA Arch. Dermatol., 81:* 565–569 (1960).
114. Ciba Geigy Corporation, unpublished results (1986).
115. Rubino, A. M., Gilman, W. S., and Jones, J. L., U.S. Patent 3,860,705 (1975).
116. Raymond, R. L., Deodorants and antiperspirants, *The Science and Technology of Aerosol Packaging* (J. J. Sciarra and L. Stoller, eds.), Wiley-Interscience, New York (1974), p. 369.
117. Theiler, R. F., Schmit, C. L., and Roheim, J. R. Application of a new microbiological technique to the study of antiperspirant and deodorant soap efficacy, *J. Soc. Cosmet. Chem., 34:* 351–359 (1983).
118. Fearnley, C., and Cox, A. R., A new microbiological approach to the assessment of underarm deodorants, *Intl. J. Cosmet. Sci., 5:* 97–109 (1983).
119. Taber, D., Lazanas, J. C., Fancher, O. E., and Calandra, J. C., The accumulation and persistence of antibacterial agents in human skin, *J. Soc. Cosmet. Chem., 22*: 369–377 (1971).
120. North-Root, H., Demetrulias, J., Wester, R., Maibach, H., and Corbin, N. Deposition of 3,4,4'-trichlorocarbanilide on human skin, *Toxicol. Lett., 22:* 235–239 (1984).
121. J. Demetrulius, unpublished results.
122. Fahlberg, W. J., Swan, J. C., and Seastone, C. V. Studies on the retention of hexachlorophene (G-11) in human skin, *J. Bacteriol., 56:* 323–328 (1948).
123. Goldemberg, R. L. Keratin substantivity, *Drug. Cosmet. Ind., 85:* 618–619, 694–696 (1959).
124. Parran, J. J., Jr., Deposition on the skin of particles of antimicrobial agents from detergent bases, *J. Invest. Dermatol., 45:* 86–88 (1965).
125. McNamara, T. F., and Steinbach, M. In vitro evaluation of the tissue substantivity of selected antibacterial agents, *J. Am. Oil Chem. Soc., 44:* 478–480 (1966).
126. Manowitz, M., and Johnston, V. D. Deposition of hexachlorophene on the skin, *J. Soc. Cosmet. Chem., 18:* 527–536 (1967).

127. Elkhouly, A. E., and Woodroffe, R. C. S., Effects of polyethylene glycol 400 monolaurate and sodium chloride mixtures on the antibacterial activity of solubilized trichlorocarbanilide, *J. Appl. Bacteriol.*, *36:* 387–395 (1973).
128. Parran, J. J., Jr., U.S. Patent 3,489,686 (1970).
129. Parran, J. J., Jr., U.S. Patent 3,580,853 (1971).
130. Parran, J. J., Jr., U.S. Patent 3,723,325 (1973).
131. Klein, W. L., and Sykes, A. R., European Patent 200,548 (1986).
132. Cheng, W. M., Davies, J. F., and Pethica, B. A., U.S. Patent 3,835,057 (1974).
133. Hoppe, U., Wittern, K.-P., Sauermann, G., Stelling, O., and Rockl, M., U.S. Patent 4,547,307 (1985).
134. Grand, P. S., U.S. Patent 3,726,815 (1973).
135. Grand, P. S., U.S. Patent 3,875,071 (1975).
136. Seilinger, A., U.S. Patent 3,923,971 (1975).
137. Brenner, M. W., and Laufer, L., U.S. Patent 3,940,430 (1976).
138. Casely, R. E., and Noel, D. R., German Patent 1,017,335 (1957); Casely, R. E., and Noel, D. R., British Patent 792,538 (1958).
139. Judge, L. F., and Kooyman, D. J., U.S. Patent 3,281,366 (1966).
140. Jungermann, E., and Taber, D., U.S. Patent 3,445,398 (1969).
141. Barr, F. S., and Collins, G. F., U.S. Patent 3,475,540 (1969).
142. Jungermann, E., Taber, D., and Raphaelian, L. A., U.S. Patent 3,652,767 (1972).
143. Taber, D., and Raphaelian, L. A., U.S. Patent 3,705,104 (1972).
144. Cheng, W. M., Davies, J. F., and Stuttard, L. W., U.S. Patent 3,723,326 (1973).
145. Apostolatos, G. N., Bohrer, J. C., and Inamorato, J. T., U.S. Patent 4,118,332 (1978).
146. Lewis, R. G., U.S. Patent 4,305,930 (1981).
147. Hamilton, H. C., Soaps from different glycerides—their germicidal and insecticidal values alone and associated with active agents, *J. Ind. Eng. Chem.*, *3:* 582 (1911).
148. Bean, H. S., and Berry, H. The bactericidal activity of phenols in aqueous solutions of soap. Part I. The solubility of a water-insoluble phenol in aqueous solutions of soap, *J. Pharm. Pharmacol.*, *2:* 484–490 (1950).

149. Bean, H. S., and Berry, H., The bactericidal activity of phenols in aqueous solutions of soap. Part II. The bacterial activity of benzylchlorophenol in aqueous solutions of potassium laurate, *J. Pharm. Pharmacol.*, *3:* 639–655 (1951).
150. Berry, H., and Briggs, A., The influence of soaps on the bactericidal activity of a sparingly water-soluble phenol, *J. Pharm. Pharmacol.*, *8:* 1143–1154 (1956).
151. Alexander, A. E., and Tomlinson, A. J. H., *Surface Chemistry*, Butterworth's Scientific Publications, London (1949) and Ref. 150.
152. Molnar, N. M., Influence of various surfactants on the antimicrobial activity of bromsalans and other ring-halogenated substances, *J. Am. Oil Chem. Soc.*, *45:* 729–737 (1968).
153. Beaver, D. J., and Stoffel, P. J., U.S. Patent 2,818,390 (1957).
154. Herber, J. F., U.S. Patent 3,320,313 (1967).
155. Russell, K. L., U.S. Patent 3,431,207 (1969).
156. Lavril, M., U.S. Patent 3,660,296 (1972).
157. Wilson, J. H., U.S. Patent 3,594,322 (1971).
158. Model, E., and Bindler, J., U.S. Patent 3,506,720 (1970).
159. Lund, R. B., U.S. Patent 4,486,610 (1984).
160. Lund, R. B., and Brown, G. W., U.S. Patent 4,467,117 (1984).
161. Schreiber, W., and Märky, M., U.S. Patent 4,490,562 (1984).
162. Watanabe, H., and Arisawa, M., U.S. Patent 4,235,733 (1980).
163. Blank, I. H., and Coolidge, M. H., Degerming the cutaneous surface. I. Quaternary ammonium compounds, *J. Invest. Dermatol.*, *15:* 249–256 (1950).
164. Tolgyesi, E., and Guth, J. J., Canadian Patent 1,157,777 (1983); *Chem. Abstr.*, *100:* 197663f (1984).
165. Carrillo, A. L., U.S. Patent 4,478,821 (1984).
166. Schmitz, R. A., and Anderson, A. F., U.S. Patent 3,016,327 (1962).
167. Hewitt, G. T., U.S. Patent 4,010,252 (1977).
168. Michaels, E. B., U.S. Patent 4,062,976 (1977).
169. Michaels, E. B., U.S. Patent 4,075,350 (1978).
170. Michaels, E. B., U.S. Patent 4,145,436 (1979).
171. Michaels, E. B., U.S. Patent 4,183,952 (1980).
172. Lowicki, N., and Desai, N. B., U.S. Patent 4,376,789 (1983).

173. Lowicki, N., and Desai, N. B., U.S. Patent 4,454,153 (1984).
174. Ricketts, C. R., Squire, J. R., and Topley, E., *Clin. Sci.*, *10:* 89–111 (1951).
175. Pohle, W. D., and Stuart, L. S., *Oil Soap*, *18:* 2–7 (1941).
176. Windholz, M., Budavari, S., Blumetti, R. F., and Otterbein, E. S., (eds.), *The Merck Index*, Merck & Co., Rahway, New Jersey (1983), p. 1191.
177. Novak, A. F., Clark, G. C., and Dupuy, H. P., *J. Am. Oil Chem. Soc.*, *38:* 321–324 (1961).
178. Novak, A. F., Fisher, M. J., Fore, S. P., and Dupuy, H. P. *J. Am. Oil Chem. Soc.*, *41:* 503–505 (1964).
179. Novak, A. F., Solar, J. M., Mod, R. R., Magner, F. C., and Skau, E. L. *Appl. Microbiol.*, *18:* 1050–1056 (1969).
180. Sumrell, G., Mod, R. R., Magne, F. C., and Novak, A. F., *J. Am. Oil Chem. Soc.*, *55:* 395–397 (1978).
181. Kabara, J. J., Antimicrobial agents derived from fatty acids, *J. Am. Oil Chem. Soc.*, *61:* 397–403 (1984).
182. Kabara, J. J., U.S. Patent 4,002,775 (1977).
183. Kabara, J. J., U.S. Patent 4,067,997 (1978).
184. Pandey, N. K., Natraj, C. V., Kalle, G. P., and Nambudiry, M. E. N., Antibacterial properties of soap containing some fatty acid esters, *Intl. J. Cosmet. Sci.*, *7:* 9–14 (1985).
185. *Encyclopaedia Britannica*, Vol. 10, Encyclopaedia Britannica, Inc. Chicago (1974), p. 882.
186. Bergerhausen, H., Deodorizing action of a complex of usnic acid, *Cosmet. Toiletries*, *91:* 25–26 (February 1976).
187. German Patent 2,432,484 (1976), assigned to Orissa Drebing G.m.b.H.; *Chem. Abstr.*, *84:* 169556f (1976).
188. German Patent 2,351,927 (1975), assigned to Orissa Drebing G.m.b.H.; *Chem. Abstr.*, *83:* 84728t (1975).
189. Marks, A., German Patent 2,354,517 (1975); *Chem. Abstr.*, *83:* 152203v (1975).
190. German Patent 2,351,864 (1975), assigned to Orissa Drebing G.m.b.H.; *Chem. Abstr.*, *83:* 103167c (1975).
191. British Patent 1,477,882 (1977), assigned to Oreal S.A.; *Chem. Abstr.*, *87:* 172723a (1977).
192. Pasero, R., and Koulbanis, C., Canadian Patent 1,040,101 (1978); *Chem. Abstr.*, *90:* 92266a (1979).
193. Owades, J. L., German Patent 2,749,274 (1978); *Chem. Abstr.*, *89:* 11981j (1978).

194. Lauzanne-Morelle, E., and Morelle, J., French Patent 2,122,284 (1972); *Chem. Abstr., 78:* 101883f (1973).
195. Rothwell, P. J., Forshaw, S. J., Smith, S. R., and Norman, C., British Patent 2,113,090 (1983).
196. Zaki, S. A., Preparation of deodorant cream containing hydrogen peroxide, *Bull. Fac. Pharm. Cairo Univ., 12:* 47–51 (1973).
197. Hlavin, Z., and Hlavin, L., Israeli Patent IL 49,107 (1979); *Chem. Abstr., 92:* 203418q (1980).
198. Conradi, R. A., Vander Wyk, J. C., and Bowlus, S. B., *J. Med. Chem., 22:* 1000–1002 (1979).
199. Coburn, R. A., Evans, R. T., Genco, R. J., and Batista, A., U.S. Patent 4,358,443 (1982).
200. Ozawa, I., Ito, T., Hamada, Y., and Takeuchi, I., U.S. Patent 4,310,682 (1982).
201. Szatloczky, E., Boros, K., Csakvari, G., Gyarmati, I., and Zalay, L., World Intellectual Property Organization, International Application published under the Patent Cooperation Treaty (PCT), International Publication Number: WO 83/03755 (1983).
202. Schnegelberger, H., and Bellinger, H., U.S. Patent 3,824,309 (1974).
203. Stracke, H. U., and Koppensteiner, G., German Patent 2,456,874 (1976); *Chem. Abstr., 85:* 112656e (1976).
204. German Patent 2,653,186 (1978), assigned to Dragoco Gerberding und Co. G.m.b.H.; *Chem. Abstr., 89:* 220760a (1978).
205. Andres, H., and Koppensteiner, G., German Patent 2,304,581 (1974); *Chem. Abstr., 81:* 135773x (1974).
206. Schnegelberger, H., and Bellinger, H., German Patent 2,218,348 (1973); *Chem. Abstr., 80:* 52285t (1974).
207. Belgian Patent 781,603 (1972), assigned to N. V. Blendax–Belgium S. A.; *Chem. Abstr., 78:* 88531t (1973).
208. McGee, A. G., and Williams, D. F., German Patent 2,358,121 (1974); *Chem. Abstr., 81:* 68362x (1974).
209. Osberghaus, R., German Patent 2,826,759 (1979); *Chem. Abstr., 92:* 169059m (1980).
210. Osberghaus, R., German Patent 2,826,758 (1979); *Chem. Abstr., 92:* 169058k (1980).
211. Osberghaus, R., German Patent 2,826,757 (1979); *Chem. Abstr., 92:* 169057j (1980).
212. Goldschmiedt, H., Vitamin E in cosmetics, *Soap Cosmet. Chem. Spec., 48:* 40,42 (1972).
213. Fang, F. S., U.S. Patent 4,415,551 (1983).

214. Adler, E., Blachut, M., Golz, K., Neumann, R., Raddatz, F., and Walter, I., East German Patent 229,304 (1985); *Chem. Abstr.*, *105:* 120496z (1986).
215. Reese, G., and Osberghaus, R., U.S. Patent 4,010,253 (1977).
216. Reese, G., and Osberghaus, R., U.S. Patent 4,005,189 (1977).
217. Geks, F.-J., and Schmidt-Kastner, G., U.S. Patent 3,950,509 (1976).
218. Beck, L. R., European Patent 127,712 (1984).

11

Clinical Testing of Deodorants

John E. Wild *Hill Top Research, Inc., Cincinnati, Ohio*

I. INTRODUCTION

The use of clinical trials to determine efficacy of products has been a long-standing practice. In the case of drugs, clinical trials are required as a condition prior to approval of a new drug for marketing.

Clinical efficacy testing has increased substantially since the passage of the amendments to the Food, Drug and Cosmetics Act in 1962. At that time, the Food and Drug Administration (FDA) was given authority to review, prior to marketing, the safety and efficacy of all drugs not generally recognized as safe and effective. In addition, the FDA was given authority to review efficacy claims for new drugs which had been introduced into the market between 1938 and 1962. Because of their potential for greatest risk, prescription drugs were given the first priority for review. After 10 years of extensive review, the emphasis shifted to over-the-counter drugs in 1972. That review is still under way and involves some 48 rule making sections which are proposing and/or finalizing FDA regulations. The need for clinical research to substantiate safety and efficacy of new drugs has spawned a series of other regulatory requirements which are either finalized or in the proposal stage.

In 1981 the FDA issued regulations (1) which provide for the protection of human subjects used in clinical investigations that are conducted as a requirement for prior submission to the FDA or in support of applications for permission to conduct research or to market regulated products. These regulations were used to clarify the FDA regulations on informed consent and provide protection of the rights and welfare of human subjects participating in research that falls within the jurisdiction of the FDA. In the proposal stages are regulations on the obligations of sponsors and monitors (2) and also the obligations of clinical investigators of regulated articles (3).

The current regulatory atmosphere for drugs and over-the-counter drugs has set the operational standards for most clinical research being carried out today, especially in the United States. For instance, a deodorancy claim is in itself a cosmetic claim; however, when it is in combination with an antiperspirant (4) or antimicrobial (5) there may be a need to carry out a clinical project under the guidelines of the FDA regulations for drugs. Knowledge of the regulatory status of a product being tested is a necessary step in preparing for a clinical research project.

Manufacturers of all kinds of products are asked to substantiate claims they wish to make about their products. Federal, state, and local governments may require claim support data prior to approval or purchase of certain products. In addition to government agencies, other groups such as advertising agencies, television networks, and the Better Business Bureau make requests for claim support information. In recent years suppliers to the cosmetic manufacturing industry have been requested to supply claim support data as part of their marketing background for their compounds.

New job categories have been created because of the need to understand, interpret, and monitor clinical efficacy testing. It is not uncommon for manufacturers to have one or more employees working directly on the clinical safety and efficacy of their products. In addition, most manufacturers today have a research and development staff to aid in supporting the safety and efficacy of their products.

As clinical efficacy testing has expanded, the contract laboratory/clinical investigator has emerged as a useful tool for the manufacturer to employ as a consultant to design and conduct clinical investigations.

It is the objective of this chapter to present various clinical methods available to an investigative staff to support deodorant efficacy claims for products used as axillary deodorants or antiperspirant/deodorants. Many of the techniques and principles used in axillary deodorancy evaluations could also apply to other types of personal care and household products making deodorancy claims; however, specific protocols for establishing claim support data for those products will not be addressed in this chapter.

II. BACKGROUND

A deodorant is described as an agent that destroys odors, especially disagreeable odors (6). Many odors are found on the human body, but one which is particularly offensive is axillary malodor. Axillary malodor is primarily the result of microbiological degradation of apocrine sweat, as shown by the work of Shelley et al. (7).

Several approaches are available to the formulator for controlling axillary malodor. Inhibition of apocrine sweat through the use of active antiperspirant materials can be used to reduce the volume of substrate available for metabolism by the axillary microorganisms. Direct control can be attained by the use of antimicrobials which inhibit the resident flora of the axilla. Absorbent materials and/or fragrances can be used to mask odor.

Procedures for the in vitro determination of antimicrobial activity are useful for determining effects of compounds for potential use in deodorant products. However, these procedures do not seem to be accurate predictors of the actual potential for malodor control.

The use of gas chromatography in deodorant evaluations is limited by the complex nature and the variation of composition of axillary malodor between individuals, as reported by Baxter and Reed (8).

Clinical evaluation for deodorant effectiveness is the current method of choice for this purpose. Although the nature of the clinical study is subject to variation, proper statistical analysis of the data and control of as many of the variables as possible should lead to an informative result (9). Clinical methodology can best be divided, according to subject-odor judge interaction, into two groups:

1. Direct axillary odor evaluation
2. Indirect axillary odor evaluation

Selection of the direct or indirect method of evaluating the axilla is dependent on the judges' training and the ease with which the judges can perform their duties most consistently. The investigator should be able to determine which method is most suitable for a group of judges based on experience and their attitudes toward the task at hand.

III. DIRECT AXILLARY ODOR EVALUATION METHOD

The governing document for the entire study plan should be a detailed protocol. The protocol should contain the objective of the study, the materials to be tested, the criteria for including and for excluding subjects, the procedures to be used, the parameters to be observed or measured, and the methods to be used for analyzing the results.

The study objective may vary depending on whether the trial is being done to screen prototype samples, to support a deodorancy claim, or to comparatively rank several deodorant products. The variation will consist primarily in number of subjects and intervals of odor assessment. The overall study procedure will be basically the same.

Trained odor judges evaluate the axillary odor level of the subjects by directly sniffing their axillae. The method consists of evaluating the deodorant materials following use by selected subjects.

Odor intensity of the subjects is determined following a control period of approximately 2 weeks duration, during which subjects have abstained from the use of any deodorant materials such as antimicrobial soaps, deodorants, or antiperspirants. Subjects are selected for their ability to develop axillary malodor during this control period.

Following selection onto the study, the subjects are assigned test material to be used at prescribed intervals according to test instructions. At preset intervals following treatment, the subjects' axillae are evaluated by the trained judges using a predetermined scale.

Usually three or four odor judges who have been trained and are proficient in the scale to be used are employed for the evaluation. Appropriate statistical analysis of the judges' scores which meets the design of the overall experiment is carried out.

A more detailed description of the actual events of the clinical testing will follow.

A. Objective

The objective of the study should be clearly defined so that the proper experimental design of the study can be determined. If the study objective is to determine just the presence of deodorant activity, a design including active versus placebo or no treatment in just a few subjects may be sufficient. However, if the objective is claim support for superior activity of new or improved formulations or for performance over competitive products, the design should include direct comparisons with those other products or materials and have sufficient numbers of subjects and observations to statistically support a claim. Financial and legal implications of inadequate support data for a product change or competitive challenge could be disastrous.

B. Test Materials

Deodorant products include a wide variety of product forms, each requiring some specificity in controlling the application to the subject during the study. In general, products should be applied in a similar manner. They should be without identifying characteristics as much as practical. Product containers should be labeled with subject identification as well as axillae to be treated. The appropriate dosage should be established for each product type. Preferably, usage should be based on consumer use rates. Disposable washcloths and towels for application, washing, or drying of the axillary vault are most suitable. Supervision and control of product application is recommended to ensure proper use.

1. *Aerosol and Pump Deodorants, Antiperspirant and Fragranced Sprays*

Spray-on materials are most uniformly applied by a single supervisor directly to the test axilla for a specified time and at a recommended distance, i.e., 2 seconds from a distance of 6 inches from the axilla. Proper ventilation of the treatment area is recommended for the protection of the subject and the application technicians. Delivery rate should be approximately 1 gram of product per axilla.

2. *Deodorant Soap Bars and Deodorant Liquids and Gels*

Supervision of axillary washing with soap products is best carried out by the supervisor instructing the panelists on time to lather the soap, time of application to the axilla, thorough removal of the soapy lather from the axilla, and drying. Using disposable washcloths and towels is most helpful in preventing cross-contamination of the axilla. Times for washing and lathering usually range from 10 to 30 seconds.

3. *Deodorant and Antiperspirant Creams, Roll-ons, and Lotions*

Supervision and direct application by the supervisor of metered amounts of these forms of products are recommended. Spread of the test material in the axilla is best accomplished by either the roll-on applicator or a glass rod or other device.

4. *Deodorant and Antiperspirant Sticks*

Stick products are most uniformly applied by an application technician. Weights of the applied product are taken immediately in order to control the volume used. Usual application rates are 0.3 to 0.5 gram per treatment.

5. *Deodorant Powders*

Powders are best applied by distributing a weighed amount of sample as cautiously as possible into the axilla using a square of fabric or similar device.

Prior to application of deodorant products other than soap, the axilla of the subjects should be washed with a nondeodorant soap under the supervision of the treatment technician.

C. Subject Selection

The population of subjects should be determined by the experimental design. If specific claims are being made toward a given population, that population should be represented in the selection. In general, deodorancy claims are most meaningful to the population who will develop odor when active products are not being used. Therefore the subject population should be one that will develop moderate to strong malodor at the control baseline/ screening odor evaluation. Subjects with low or extremely high odor levels may be excluded. Those with extremely high levels may have other complications or have generally poor hygiene and will not make good cooperative subjects.

The subjects should be 18 years of age or older and should exhibit interest in the project. This can usually be determined by their willingness to adhere to the rules of the test, to be on time for their scheduled test sessions, and to cooperate with the investigator's staff and the other subjects.

A distribution of the average odor scores for a pooled population of 1000 subjects following 10–14 days of control is shown in Fig. 1.

For the subjects to participate in deodorancy studies, they must comply with the following rules:

1. Abstain from the use of all deodorants, antiperspirants, and perfumed or medicated products on the underarm areas for the entire conditioning and test periods.
2. Use the control soap bars for all bathing and washing during both periods and abstain from all underarm washing during the test period except for supervised washing at the test location.
3. Abstain from wearing (on body or clothing) perfumes, powders, or any other scented products including laundry rinses, hair sprays, and after-shaves during the entire test period.
4. Abstain from smoking for at least 1 hour prior to each visit to the test location during the test period and abstain from the use of alcoholic beverages and/or chewing gum, breath mints, mouth rinses, coffee, tea, etc. during the test period.
5. Female subjects should shave their underarms 24 hours just prior to the test period and abstain from any underarm shaving during the test period.
6. Abstain from swimming for the final week of the conditioning period and the entire test period. Also minimize other physical exertion such as tennis and jogging during the test periods.
7. Agree to withdraw from the study at the start of the test period if odor scores obtained during the control period indicate that the axillary odor intensity is not within the criteria for the study.
8. Abstain from eating highly spiced foods such as onions and garlic for 24 hours prior to odor evaluations.

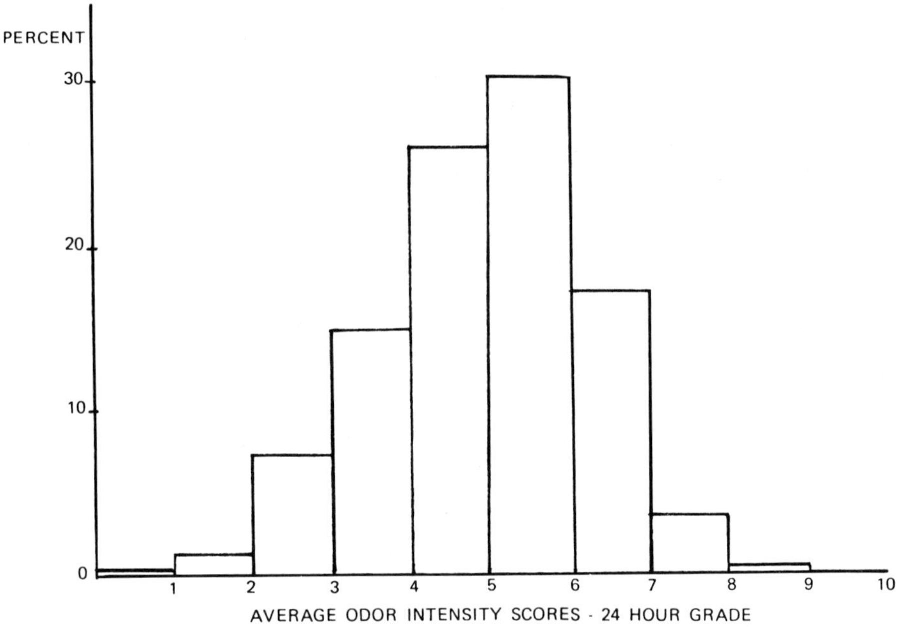

Figure 1 Distribution of average axillary odor grades following 10–14 days on nonantibiotic soap and no antiperspirants or deodorants—1000 male and female subjects.

In addition, some investigators require a subject to sign an informed consent statement which itemizes what is required of the subject and also describes the risks and benefits involved in the subject's participation in the study. The required elements of informed consent are described in the *Federal Register*, July 27, 1981 (1).

D. Odor Evaluation Method

Axillary odor evaluations should be made at specified times during the study which correspond to the anticipated claim for deodorant efficacy. The baseline assessment should be conducted at the maximum specific hour interval to be used in the study so that the investigator can see the subjects' potential for highest odor development.

Odor evaluations should be made by employing a specified scoring system. This system is based on scoring the magnitude of the odor and encompasses no body odor to strong disagreeable odor. Various forms of scaling will be discussed later in the chapter.

Odor assessments should be carried out by three or four experienced evaluators. This will permit replicate evaluation and enhance the power of the assay. Evaluations of the panelists' axillary odors are made as the panelists file by the judges in a random order. The men and women are processed in separate groups. The men remove all clothing above their waists prior to approaching the judges. Women panelists remove all clothing above their waists except for brassieres. Prior to the panelists' approach to the judges, both arms are held tightly against their sides. At the evaluation the panelists elevate alternately the right and left arm to an almost vertical position. As the arm is elevated, the odor judge sniffs the axillary surface by placing the nose very close to the surface. Each axilla is sniffed twice by each judge, who then records a single score for each armpit. The order of sniffing always remains consistent. The panelists hold their arms against their sides except during the actual sniffing. When this process is completed by one judge, the panelists proceed to the next judge, where the process is repeated. Each evaluation is entered on a record form which does not show subjects' sample assignments or any previous odor scores.

Evaluations are made at approximately 1-minute intervals between panelists. The order of evaluation by the odor judges is changed periodically during the test.

To enhance the accuracy of the deodorant treatment and to protect the axilla from outside contaminants, most investigators supply the subjects with freshly laundered T-shirts following the treatment just prior to each odor evaluation. T-shirts are laundered using an unfragranced detergent.

As an addition or alternative to the judges' direct evaluation method, the subjects can perform a direct evaluation of their own axillae and either rank intensity of odor or select the armpit with the least malodor.

E. Test Schedule

The test schedule consists of a conditioning phase, a treatment phase and a regression phase.

1. *Conditioning Phase*

The conditioning phase is the period when the prospective subjects abstain from use of all deodorant products and use only a nondeodorant conditioning bar for all bathing. Usual washout periods range from 10 to 14 days for deodorant products. Washout periods for antiperspirants are usually 17 days. The conditioning phase is concluded when the baseline/control odor evaluation is conducted and the subjects with preestablished odor levels are selected to enter into the treatment phase. Experience has shown that approximately 20% of the prospective subject candidates are eliminated at the control odor evaluation.

2. *Treatment Phase*

The treatment phase usually begins almost immediately after the subject has been selected to continue on the study. Treatments are usually supervised at the laboratory or are conducted unsupervised at home, according to directions. The duration of treatment may range from one application of product to as many as one or two applications a day for 2 weeks, depending on the test protocol.

Odor assessments are generally at specified hourly increments following one, two, or more treatments during the study. The investigator may wish to select one or more odor assessment intervals. The intervals are often determined by label claims to be made. The most used odor assessment intervals are 5, 8, 12, and 24 hours following a given treatment.

3. *Regression Phase*

The duration of this portion of the study is determined at the option of the sponsor. It is the portion of the study when treatment has been concluded but the study conditions continue until the final odor assessment. That interval, based on currently effective products, is usually no longer than 48 hours.

IV. INDIRECT AXILLARY ODOR EVALUATION METHOD

Indirect methods for evaluation of deodorant effects are based on the transfer of axillary malodor to a material or an implement which has been placed in direct contact with the axilla.

The indirect methods employ the basic principles and techniques used in the direct method for subject selection, odor judge qualification, treatments, and testing schedules. The primary difference is the odor evaluation method. The odor evaluation method described by Henry et al. (10), which employs the transfer of odor to the end of a 20 × 150 mm borosilicate glass tube, has been used by these investigators for several years and provides them with meaningful data. In addition, they have found that this alternative to direct odor assessment has several advantages:

The subject and judge do not come into proximity and therefor there is little or no embarrassment, bias, or revulsion.

The judges can perform their task in a more comfortable sitting position, concentrating more effectively on the odor evaluation.

The capped tubes can be stored for a short time under refrigeration for later evaluation at a more convenient time for the judge.

A. Objective

The objective of the study should be stated. It may simply be a comparison of the effectiveness of the test materials.

B. Test Materials

The test materials, mode, and rate of application should be described.

C. Subject Selection

Subjects who have been prescreened and are known to develop moderate malodor when not using deodorant products should be used. Subjects are instructed as follows:

Abstain from the use of all deodorants, antiperspirants, and perfumed or medicated products for 7 days before start of the test and throughout the test period.

Abstain from smoking, chewing gum, or use of alcohol for at least 15 minutes prior to each visit to the test location.

Use Ivory soap for all bathing and washing during the test, including the pretest period. Abstain from washing the axillae except at supervised test sessions during the test period.

Female subjects are to shave their axillae 24 hours just prior to the test period and not to shave again until after completion of the test.

Male subjects are required to wear a clean shirt after each product application each day. Female subjects are required to wear a clean blouse.

D. Odor Evaluation Method

Odor evaluations are performed by trained judges, who use a scale of 0 (no malodor) to 10 (strong, disagreeable odor). When fragranced materials are involved the judge is instructed to disregard the fragrance. As part of their training, judges are screened on their ability to rank levels of malodor in the presence of fragrances. Judges are not to smoke or eat within 15 minutes of an odor evaluation. They are not to use fragrances themselves and should be present for each evaluation.

The specifics of the indirect test tube method are as follows:

1. One 20 × 150 mm borosilicate glass test tube is inserted into the axillary vault, with the rounded bottom end touching the skin. The subject holds the tube in place by holding the arms tight against the body.
2. At the direction of a supervisor, the subject vigorously rubs the bottom of a labeled tube in the axilla and hands the tube to the supervisor, who will cover the rounded end with a plastic tube closure (KIMKAP®, Kimble, Division of Owens-Illinois) and place the tube in a rack.
3. The opposite axilla is sampled in the same fashion.
4. Following both samplings, the tubes are then passed to two odor judges, who independently evaluate the odor present on the rounded tube ends. If there are more than two judges a second set of tube samples should be obtained.
5. The judges indicate to the supervisor that there is sufficient sample (odor) obtained from the tube, after which time the subject is dismissed until the next sample interval. If judges are not satisfied with the tube sample they may request that an additional sample be taken.

6. There is approximately a 1-minute interval between subjects.
7. Judges' scores are entered on record sheets which do not show treatment assignments or previous scores.

E. Test Schedule

The following test schedule is described in the literature (10):

Test day 1. Subject selection is made. Subjects are selected who achieve a control score of at least 4 (0–10 scale) in both axillae (by no less than half the judges). This baseline score represents the achievement of at least a moderate degree of malodor from which improvement can be determined.

Following the baseline score the axillae are washed with tepid water and Ivory soap, using a gauze pad or washcloth to remove excess soap, and dried with a paper towel. The designated product is applied to the axillae and the subjects are instructed to return for odor evaluations at specific intervals (normally 3, 6, and 24 hours). Subjects are reminded not to wash axillae.

Test day 2. A 24-hour posttreatment evaluation is followed by an axillary wash with Ivory soap and treatment.

Test day 3. Subjects report for the final 24-hour odor evaluation.

F. Odor Evaluation Method—Cloth Method

The basic methodology is similar to that described in the previous methods. However, in this transfer-to-cloth method, T-shirts and/or cloth swatches which have been sewn into T-shirts are used to pick up the axillary odor. The odor evaluations are performed by trained judges, who score the malodor on the T-shirt or piece of fabric on a scale of 0 (no malodor) to 10 (very strong, disagreeable odor).

The outstanding details of the method are as follows:

1. Following axillary treatment, the subjects are provided with a freshly laundered (nonfragranced detergent) T-shirt to wear for the interval from treatment until odor evaluation.

2. At the time of evaluation the subject reports to the laboratory and the T-shirts are removed and turned inside out so that the surface touching the axillary skin is exposed. If axillary swatches have been sewn into the axillary area of the T-shirt, they are now removed by a technician and placed in appropriately identified odor-free containers.
3. Following removal, the armpit areas of the shirt are sniffed and scored for malodor by the judges.

V. ODOR JUDGE SELECTION AND TRAINING

The area of sensory evaluation has grown rapidly in the past several years. Sensory evaluations are made continually by the consumer when making judgments concerning products. Many product attributes are related to sensory responses. Deodorancy is one such attribute.

The use of trained judges as a measuring tool is analogous to the use of scientific instruments to measure certain parameters (11). In the case of the sensory judges' evaluation, there is the added capacity of qualification of the parameter into the total human experience.

Just as scientific instrumentation must be calibrated to give standard measurements, the sensory judge must also be calibrated to the sensory measurement to be performed. The sensory judge must be familiarized with the test procedures and be trained by actual experience with the attribute(s) to be identified and measured.

A. Judge Recruitment

Judges who are willing to participate and be trained for this function should be recruited. The task of smelling armpits lacks aesthetic quality. However, experience has shown that judges with proper training and motivation can perform this task with no hesitation or impairment of their ability to assess axillary odor levels.

Judges should be in general good health with no conditions which might impair their ability to make reliable judgments. Certain temporary conditions such as colds, flu, seasonal allergy, or fatigue may temporarily disqualify a judge from participation in a test. Other factors may also influence a judge's potential, such as smoking, age, and gender.

A personal interview with the potential judge in which all of the previously discussed information is taken into account will be beneficial in recruitment of reliable judges.

B. Screening Tests

In general, screening tests are performed by the candidate judges to determine their normal olfactory acuity, interest in odor evaluation technique, ability to discriminate and reproduce results, and conscientiousness and motivation.

The judge should demonstrate a propensity for both qualitative and quantitative olfactory discrimination. This ability can be tested by a series of olfactory discrimination tests (12) that are available or can be prepared using standard odors.

Judges should not be anosmic to compounds which are known to be components of axillary odor. Many investigators test their judges for anosmias to compounds such as androstenone, androstenal, and isovaleric acid.

Ranking of several concentrations of isovaleric acid is the most frequently used test of olfactory ability. The concentrations used are shown in Sec. VI, A on category scaling.

C. Specific Training

Attribute rating methods measure intensities of specific characteristics defined in the test objective. In the case of deodorancy the attribute is axillary malodor.

After judges have demonstrated their ability with standard materials, on-the-job training with experienced judges is recommended to further enhance their ability.

Judges should be instructed prior to each study about the sensory technique to be sure they understand the method and scoring records to be kept.

When sniffing, judges should avoid long, deep inhalation. They should sniff enough times to reach a decision (usually twice) but not so often as to become confused. The mouth should be closed during sniffing.

Judges should be made aware that adaptation can make an odor seem weaker on subsequent sniffs. Adaptation can also make it difficult to detect weak odors after smelling a stronger odor. This effect is temporary and should only last a few seconds (<30 seconds). Therefore, allowing short intervals between sniffs and taking short breaks between groups of subjects being judged should allow sufficient time for return to initial sensitivity.

The score data from these training tests are subjected to statistical analysis. The method utilizes either of two scaling methods; category scaling or magnitude estimation (ratio scaling).

After several successful comparative performances in which the judge trainee is not significantly different from the previously qualified judges, the judge trainee is considered qualified.

D. Monitoring of Judge Performance

The odor judges' performance should be evaluated frequently. In many cases the data analysis of the project will identify the judges' individual performance. If a judge seems to be straying from the remainder of the panel of judges, he or she should be consulted by the project leader and/or investigator to determine whether there are any underlying causes for an inadequate or erratic performance. Experience has shown that loss of interest, ill health, and personal problems can lead to a decline in performance. Assistance to restore the judge's previous skill should be given when possible.

In addition, periodic evaluation of the judges' olfactory ability should be measured employing the same technique used in the screening process. Olfactory ability is known to be impaired with age (12). Prepared tests can measure this impairment to determine whether the judge has lost smell identification ability. These tests also supply interpretation criteria for the test results.

Keeping a panel of trained judges in tune is a worthwhile endeavor.

VI. SCALING TECHNIQUES

Attribute scaling can employ either category of ratio scaling techniques.

A. Category Scaling

Category scaling methods are among the oldest and most frequently used techniques. Category techniques were used in the developmental stages of deodorant evaluation (13,14).

The category scale may be structured or unstructured. Structured scales are anchored by descriptive terms or standards for each interval of the scale. Unstructured category scales are anchored at each end of the scale.

The most frequently used category scales employed in deodorancy testing are the 0 to 10 and 0 to 5 scales. The availability of judges trained in a specific scale will be the determining factor in the selection of the scale by the investigator.

The 0 to 10 scale has been used as both a structured and an unstructured scale. The unstructured scale is described by two and sometimes three intervals: 0, no odor; 5, moderate odor; 10, strong, disagreeable odor. The structured scale has been represented in the following way. Some descriptive terms may vary slightly depending on the investigator.

Score	Malodor intensity scale
0	None, no odor
1	Threshold odor
2	Very slight odor
3	Slight odor
4	Slight to moderate odor
5	Moderate odor
6	Slightly strong odor
7	Moderately strong odor
8	Strong odor
9	Very strong odor
10	Extremely strong odor

The 0 to 5 scale has been standardized by using reference dilutions of isovaleric acid which correspond to each of the scores (15).

Score	Odor level	Concentration of aqueous solution of isovaleric acid (ml/liter)
0	No odor	0
1	Slight	0.013
2	Definite	0.053
3	Moderate	0.22
4	Strong	0.87
5	Very strong	3.57

B. Ratio Scaling

Ratio scaling or magnitude estimation uses number amounts to assign ratios or proportions to indicate intensity of two correponding measurements.

Judges are tested for their ability to make correct ratio judgments with known materials such as various sizes of geometric figures. They then demonstrate their ability to make ratio judgments of various concentrations of odors.

In the actual odor evaluation technique the judge is asked to rate the subject's right axilla with a number and then rate the left axilla either higher or lower at some ratio relative to the right axilla.

VII. EXPERIMENTAL DESIGN AND DATA ANALYSIS

Three typical test designs are used in deodorant tests. The panel size for deodorant evaluation is 30 to 50 subjects per treatment. However, the investigator may choose a panel size which will be most suitable for the power of the statistical analysis.

A. Single-Pair Design

This design is used when two treatments are compared. One treatment is assigned to the right or left axilla and the other treatment is assigned to the contralateral axilla. Each treatment should appear on the right or left axilla an equal number of times.

B. Each-versus-Control Design

This design is used when three or more treatments are to be compared and one specific treatment can be assigned as the control treatment. In this design a group of subjects will be assigned a test treatment pair (i.e., A versus B, A versus C, etc.). Again, within each treatment group there will be an equal number of right and left axillary assignments for each test treatment.

C. Round-Robin Design

This design is used when three or more treatments are to be compared and no treatment will be identified as a control. With this design all of the possible treatment pairs are assigned to a balanced right and left distribution over the entire subject population.

D. Data Analysis

Standard statistical techniques can be applied to odor score data when assumptions regarding the population from which the data were obtained are known.

If the data meet the assumptions for parametric analysis, then Student's *t*-test or analysis of variance is usually performed. If data do not meet these criteria, nonparametric analysis such as the distribution-free signed rank test or Kruskal-Wallis one-way analysis of variance is carried out.

The deodorant value (15) can be calculated if the criteria for establishing the value are met. The number representing the deodorant value is the result of subtracting the average score for the test treatment from the average score for the control treatment.

VIII. CONCLUDING REMARKS

This chapter has presented the current state of the art of clinical methods available to the investigator in support of efficacy of deodorants and/or deodorant/antiperspirants. These methods can be used to assist the investigator in the design and performance of clinical trials for deodorancy.

Additional guides will be forthcoming as more work is accomplished in the area. The American Society for Testing and Materials is in the process of preparing a Standard Practice for the Sensory Evaluation of Axillary Deodorancy (16).

REFERENCES

1. Department of Health and Human Services, Food and Drug Administration. Protection of human subjects/informed consent/standards for institutional review boards for clinical investigations; and clinical investigations which may be reviewed through expedited review procedure, *Fed. Regist.*, Part IX (July 27, 1981).
2. Department of Health and Human Services, Food and Drug Administration. Proposed establishment of regulations on obligations of sponsors and monitors, *Fed. Regist.*, Part IV (September 27, 1977).
3. Department of Health and Human Services, Food and Drug Administration. Proposed establishment of regulations on obligations of clinical investigators fo regulated articles, *Fed. Regist.*, Part V (August 8, 1978).
4. Department of Health and Human Services, Food and Drug Administration. Tentative final monograph for antiperspirant drug products for over-the-counter human use; *Fed. Regist.*, Part II (August 20, 1982).
5. Department of Health and Human Services, Food and Drug Administration. OTC topical antimicrobial products, over-the-counter drugs generally recognized and safe, effective and not misbranded; *Fed. Regist.*, Part II (January 6, 1979).
6. *Illustrated Stedman's Medical Dictionary*, 24th ed., Williams & Wilkins, Baltimore, Maryland (1982), p. 377.
7. Shelley, W. B., Hurley, H. J., and Nichols, A. C. Axillary odor: Experimental study of the role of bacteria, apocrine sweat, and deodorants, *Arch. Dermatol. Syphilol.*, *68*: 430–466 (October 1953).
8. Baxter, P. M., and Reed, J. U. The evaluation of underarm deodorants, *Int. J. Cosmet, Sci.*, *5*: 85–95 (1983).
9. Carabello, F. B. Guidelines for the clinical study of antiperspirant and deodorant efficacy, *Cosmet. Toiletries*, *95*: 33–40 (July 1980).

10. Henry, S. M., Jacobs, G., and Cotty, V. F. An alternate to direct panelist-judge interaction in evaluating underarm deodorants, *J. Soc. Cosmet. Chem., 35:* 283–295 (September/October 1984).
11. American Society for Testing and Materials. "Guidelines for the Selection and Training of Sensory Panel Members," ASTM Publ. 758, ASTM, New York (1981).
12. Doty, R. L., Shaman, P., Applebaum, S. L., Gilbertson, R., Siksorski, L., and Rosenberg, L. Smell identification ability: Changes with age, *Science, 226:* 1441–1442 (1984).
13. Gee, A. A., and Seidenberg, I. I. Evaluation of germicidal soaps, *Soap Cosmet. Chem. Spec., 30:* 42–48 (1954).
14. Whitehouse, H. S., and Carter, R. Evaluation of deodorant toilet bars, *Proc. Sci. Sect. Toilet Goods Assoc., 48:* 31–37 (December 1967).
15. Hooper, D. C., Johnson, G. A., and Peter, D., Detergent product containing deodorant compositions, U.S. Patent 4,304,679 (December 8, 1981).
16. ASTM-E-18.04.19 Committee. "Task Group Report: Standard Practice for the Sensory Evaluation of Axillary Deodoracy," in preparation.

Index